Useful Astronomical Data

W9-ATF-023

(1 AU = mean distance from earth to sun = 1.50×10^{11} m)

Object	Mass	Radius	Mean Orbital Radius	Orbital Period	Orbital Eccentricity
Sun	1.99×10^{30} kg	696,000 km	—	—	—
Moon	7.36×10^{22} kg	1,740 km	384,000 km	27.3 d	0.055
Mercury	$0.0558 M_E$	2,439 km	0.387 AU	0.241 y	0.206
Venus	$0.815 M_E$	6,060 km	0.723 AU	0.615 y	0.007
Earth	5.98×10^{24} kg $\equiv M_E$	6,380 km	1.000 AU	1.000 y	0.017
Mars	$0.107 M_E$	3,370 km	1.524 AU	1.88 y	0.093
Jupiter	$318 M_E$	69,900 km	5.203 AU	11.9 y	0.048
Saturn	$95.1 M_E$	58,500 km	9.539 AU	29.5 y	0.056
Uranus	$14.5 M_E$	23,300 km	19.182 AU	84.0 y	0.047
Neptune	$17.2 M_E$	22,100 km	30.058 AU	165 y	0.009
Pluto/Charon	$0.0025 M_E$	3500/1800 km	39.785 AU	248 y	0.254
Ceres (asteroid)	1.2×10^{21} kg	500 km	2.768 AU	4.61 y	0.077
Halley's comet	1.2×10^{14} kg	≈ 7 km	17.94 AU	76.0 y	0.967

Based mostly on data in D. Halliday and R. Resnick, *Fundamentals of Physics*, 3/e, New York: Wiley, p. A6.

Six Ideas That Shaped Physics

Unit N: The Laws of Physics Are Universal

Physics

Second Edition

Thomas A. Moore

Boston   Burr Ridge, IL   Dubuque, IA   Madison, WI   New York   San Francisco   St. Louis
Bangkok   Bogotá   Caracas   Kuala Lumpur   Lisbon   London   Madrid   Mexico City
Milan   Montreal   New Delhi   Santiago   Seoul   Singapore   Sydney   Taipei   Toronto

McGraw-Hill Higher Education

A Division of The **McGraw-Hill** Companies

SIX IDEAS THAT SHAPED PHYSICS, UNIT N: THE LAWS OF PHYSICS ARE UNIVERSAL
SECOND EDITION

Published by McGraw-Hill, a business unit of The McGraw-Hill Companies, Inc., 1221 Avenue of the Americas, New York, NY 10020. Copyright © 2003, 1998 by The McGraw-Hill Companies, Inc. All rights reserved. No part of this publication may be reproduced or distributed in any form or by any means, or stored in a database or retrieval system, without the prior written consent of The McGraw-Hill Companies, Inc., including, but not limited to, in any network or other electronic storage or transmission, or broadcast for distance learning.

Some ancillaries, including electronic and print components, may not be available to customers outside the United States.

This book is printed on recycled, acid-free paper containing 10% postconsumer waste.

1 2 3 4 5 6 7 8 9 0 QPD/QPD 0 9 8 7 6 5 4 3 2

ISBN   0–07–239712–8

Publisher: *Kent A. Peterson*
Sponsoring editor: *Daryl Bruflodt*
Developmental editor: *Spencer J. Cotkin, Ph.D.*
Marketing manager: *Debra B. Hash*
Senior project manager: *Susan J. Brusch*
Lead production supervisor: *Sandy Ludovissy*
Media project manager: *Sandra M. Schnee*
Lead media technology producer: *Judi David*
Designer: *David W. Hash*
Cover/interior designer: *Rokusek Design*
Cover image: *© PhotoDisc*
Senior photo research coordinator: *Lori Hancock*
Photo research: *Chris Hammond/PhotoFind LLC*
Supplement producer: *Brenda A. Ernzen*
Compositor: *Interactive Composition Corporation*
Typeface: *10/12 Palatino*
Printer: *Quebecor World Dubuque, IA*

The credits section for this book begins on page 257 and is considered an extension of the copyright page.

Library of Congress Cataloging-in-Publication Data

Moore, Thomas A. (Thomas Andrew)
    Six ideas that shaped physics. Unit N. The laws of physics are universal / Thomas A. Moore. — 2nd ed.
       p.   cm.
    Includes bibliographical references and index.
    ISBN 0–07–239712–8 (acid-free paper)
    1. Mechanics. I. Title: Laws of physics are universal. II. Title

QC125.2 .M66   2003   2002070885
531.7—dc21   CIP

www.mhhe.com

Dedication

For Brittany,

whose intuitive understanding of newtonian mechanics is part of what makes her awesome.

Table of Contents for

Six Ideas That Shaped Physics

Contents: Unit N
The Laws of Physics Are Universal

About the Author

Thomas A. Moore graduated from Carleton College (magna cum laude with Distinction in Physics) in 1976. He won a Danforth Fellowship that year that supported his graduate education at Yale University, where he earned a Ph.D. in 1981. He taught at Carleton College and Luther College before taking his current position at Pomona College in 1987, where he won a Wig Award for Distinguished Teaching in 1991. He served as an active member of the steering committee for the national Introductory University Physics Project (IUPP) from 1987 through 1995. This textbook grew out of a model curriculum that he developed for that project in 1989, which was one of only four selected for further development and testing by IUPP.

He has published a number of articles about astrophysical sources of gravitational waves, detection of gravitational waves, and new approaches to teaching physics, as well as a book on special relativity entitled "A Traveler's Guide to Spacetime" (McGraw-Hill, 1995). He has also served as a reviewer and an associate editor for American Journal of Physics. He currently lives in Claremont, California, with his wife Joyce and two college-aged daughters. When he is not teaching, doing research in relativistic astrophysics, or writing, he enjoys reading, hiking, scuba diving, teaching adult church-school classes on the Hebrew Bible, calling contradances, and playing traditional Irish fiddle music.

Preface

This volume is one of six that together comprise the text materials for *Six Ideas That Shaped Physics*, a fundamentally new approach to the two- or three-semester calculus-based introductory physics course. *Six Ideas That Shaped Physics* was created in response to a call for innovative curricula offered by the Introductory University Physics Project (IUPP), which subsequently supported its early development. In its present form, the course represents the culmination of more than a decade of development, testing, and evaluation at a number of colleges and universities nationwide.

Opening comments about *Six Ideas That Shaped Physics*

This course is based on the premise that innovative approaches to the presentation of topics and to classroom activities can help students learn more effectively. I have completely rethought from the ground up the presentation of every topic, taking advantage of research into physics education wherever possible, and have done nothing just because "that is the way it has always been done." Recognizing that physics education research has consistently emphasized the importance of active learning, I have also provided tools supporting multiple opportunities for active learning both inside and outside the classroom. This text also strongly emphasizes the process of building and critiquing physical models and using them in realistic settings. Finally, I have sought to emphasize contemporary physics and view even classical topics from a thoroughly contemporary perspective.

I have not sought to "dumb down" the course to make it more accessible. Rather, my goal has been to help students become *smarter*. I intentionally set higher-than-usual standards for sophistication in physical thinking, and I then used a range of innovative approaches and classroom structures to help even average students reach this standard. I don't believe that the mathematical level required by these books is significantly different from that in most university physics texts, but I do ask students to step beyond rote thinking patterns to develop flexible, powerful conceptual reasoning and model-building skills. My experience and that of other users are that normal students in a wide range of institutional settings can, with appropriate support and practice, meet these standards.

The six volumes that comprise the complete *Six Ideas* course are

The six volumes of the *Six Ideas* text

Unit C (**Conservation laws**):	Conservation Laws Constrain Interactions
Unit N (**Newtonian mechanics**):	The Laws of Physics Are Universal
Unit R (**Relativity**):	The Laws of Physics Are Frame-Independent
Unit E (**Electricity and magnetism**):	Electric and Magnetic Fields Are Unified
Unit Q (**Quantum physics**):	Particles Behave Like Waves
Unit T (**Thermal physics**):	Some Processes Are Irreversible

I have listed these units in the order that I recommend they be taught, although other orderings are possible. At Pomona, we teach the first three units during the first semester and the last three during the second semester of a year-long course, but one can easily teach the six units in three quarters or even over three semesters if one wants a slower pace. The chapters of all these texts have been designed to correspond to what one might realistically discuss in a single 50-minute class session at the *highest possible pace*. A reasonable course syllabus will therefore set an average pace of not more than one chapter per 50 minutes of class time.

For more information than I can include in this short preface about the goals of the *Six Ideas* course, its organizational structure (and the rationale behind that structure), the evidence for its success, and guidance on how to cut and/or rearrange material, as well as many other resources for both teachers and students, please visit the *Six Ideas* website (see the next section).

Important Resources

Instructions about how to use this text

I have summarized important information about how to read and use this text in an *Introduction for Students* immediately preceding the first chapter. Please look this over, particularly if you have not seen other volumes of this text.

The *Six Ideas* website

The *Six Ideas* website contains a wealth of up-to-date information about the course that I think both instructors and students will find very useful. The URL is

www.physics.pomona.edu/sixideas/

Essential computer programs

One of the most important resources available at this site is a number of computer applets that illustrate important concepts and aid in difficult calculations. In critical places, this unit draws on one of these programs, and past experience indicates that students learn the ideas much more effectively when these programs are used both in the classroom and for homework. These applets are freeware and are available for both the Mac (Classic) and Windows operating systems.

Some Notes Specifically About Unit N

The purpose and place of this unit in the course

This particular unit is primarily focused on Newton's second law and its application to both terrestrial and celestial physics. Its goal is to help students appreciate the power and breadth of the newtonian perspective as well as the historical importance of Newton's work.

This unit is structured on the premise that students have already studied unit C, and indeed it draws on ideas from almost all the chapters of that unit. It in turn is needed as basic background for all the other units in the course.

The unit's spiral structure

This unit is designed to teach newtonian mechanics by using a "spiral learning" approach. The first four chapters provide a mostly qualitative introduction to the concepts and techniques of newtonian mechanics, while the remaining chapters explore quantitative applications of these ideas in depth. Instructors can help students get the most out of this approach by helping them see the connections between the earlier and later spirals through the given material.

Motion maps, trajectory diagrams, and the Newton program

An unusual feature of the first part of this text is the exploration of motion using *motion maps* and *trajectory diagrams*. Both of these tools are designed to deepen students' intuitive understanding of motion, and trajectory diagrams in particular are a powerful tool for qualitatively predicting an object's motion in advance of using mathematics. If students spend enough time practicing the use of both of these tools, their understanding of newtonian mechanics will become much deeper and more flexible.

I have also discussed the trajectory diagram in such depth because it provides an excellent conceptual basis for computer programs that calculate trajectories. The Newton computer program, which is available for free download from the *Six Ideas* website, does exactly this. For this edition, I have built this program into the presentation and have provided a number of new problems that use it. This program makes it possible to explore a number of realistic applications of Newton's second law that would otherwise be inaccessible, and greatly simplifies the presentation of planetary motion in chapter N13. I strongly recommend to instructors that you use this program consistently starting in chapter N4, so that your students get used to what the program does and how it operates before getting to chapter N13.

Unit N, like unit C, is a mostly indivisible whole. Chapter N5 (which looks at torque and statics problems) could probably be omitted if cuts are absolutely necessary: it is not essential for anything else in the course. Chapter N13 also covers material that is not needed in the rest of the course, but I would recommend against omitting this chapter: dropping chapter N13 means that students would not see the fulfillment of the unit's "great idea." In short, if cuts need to be made, start with chapter N5 and/or N13, but all the other chapters have important roles to play.

Please see the instructor's manual for more detailed comments about this unit and suggestions about how to teach it effectively.

Possible cuts

Appreciation

A project of this magnitude cannot be accomplished alone. I would first like to thank the others who served on the IUPP development team for this project: Edwin Taylor, Dan Schroeder, Randy Knight, John Mallinckrodt, Alma Zook, Bob Hilborn and Don Holcomb. I'd like to thank John Rigden and other members of the IUPP steering committee for their support of the project in its early stages, which came ultimately from an NSF grant and the special efforts of Duncan McBride. Users of the texts—especially Bill Titus, Richard Noer, Woods Halley, Paul Ellis, Doreen Weinberger, Nalini Easwar, Brian Watson, Jon Eggert, Catherine Mader, Paul De Young, Alma Zook, David Tanenbaum, Alfred Kwok, Dan Schroeder, and Dave Dobson—have offered invaluable feedback and encouragement. I'd also like to thank Alan Macdonald, Roseanne Di Stefano, Ruth Chabay, Bruce Sherwood, and Tony French for ideas, support, and useful suggestions. Thanks also to Robs Muir for helping with several of the indexes. My editors Jim Smith, Denise Schanck, Jack Shira, Karen Allanson, Lloyd Black, J.P. Lenney, and Daryl Bruflodt as well as Spencer Cotkin, Donata Dettbarn, David Dietz, Larry Goldberg, Sheila Frank, Jonathan Alpert, Zanae Roderigo, Mary Haas, Janice Hancock, Lisa Gottschalk, Debra Hash, David Hash, Patti Scott, Chris Hammond, Brittney Corrigan-McElroy, Rick Hecker, and Susan Brusch have all worked very hard to make this text happen, and I deeply appreciate their efforts.

I'd like to thank all the reviewers, including Edwin Carlson, David Dobson, Irene Nunes, Miles Dressler, O. Romulo Ochoa, Qichang Su, Brian Watson, and Laurent Hodges, for taking the time to do a careful reading of various units and offering valuable suggestions. Thanks to Connie Wilson, Hilda Dinolfo, and Connie Inman, and special student assistants Michael Wanke, Paul Feng, Mara Harrell, Jennifer Lauer, Tony Galuhn, and Eric Pan, and all the Physics 51 mentors for supporting (in various ways) the development and

Thanks!

teaching of this course at Pomona College. Thanks also to my Physics 51 students, and especially Win Yin, Peter Leth, Eddie Abarca, Boyer Naito, Arvin Tseng, Rebecca Washenfelder, Mary Donovan, Austin Ferris, Laura Siegfried, and Miriam Krause, who have offered many suggestions and have together found many hundreds of typos and other errors. Eric Daub and Daniel Villalon were indispensable in helping me put this edition together. Finally, very special thanks to my wife, Joyce, and to my daughters, Brittany and Allison, who contributed with their support and patience during this long and demanding project. Heartfelt thanks to all!

Thomas A. Moore
Claremont, California

Introduction for Students

Introduction

Welcome to *Six Ideas That Shaped Physics*! This text has been designed using insights from recent research into physics learning to help you learn physics as effectively as possible. It thus has many features that may be different from science texts you have probably encountered. This section discusses these features and how to use them effectively.

Why Is This Text Different?

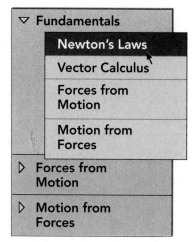

Research consistently shows that people learn physics most effectively if they participate in *activities* that help them *practice* applying physical reasoning in realistic situations. This is because physics is not a collection of facts to absorb, but rather is a set of *thinking skills* requiring practice to master. You cannot learn such skills by going to factual lectures any more than you can learn to play the piano by going to concerts!

This text is designed, therefore, to support *active learning* both inside and outside the classroom by providing (1) resources for various kinds of learning activities, (2) features that encourage active reading, and (3) features that make it easier for the text (as opposed to lectures) to serve as the primary source of information, so that more class time is available for active learning.

The Text as Primary Source

To serve the last goal, I have adopted a conversational style that I hope will be easy to read, and I tried to be concise without being so terse that you need a lecture to fill in the gaps. There are also many text features designed to help you keep track of the big picture. The unit's **central idea** is summarized on the front cover where you can see it daily. Each chapter is designed to correspond to one 50-minute class session, so that each session is a logically complete unit. The two-page **chapter overview** at the beginning of each chapter provides a compact summary of that chapter's contents to consider before you are submerged by the details (it also provides a useful summary when you review for exams). An accompanying **chapter-location diagram** uses a computer-menu metaphor to display how the current chapter fits into the unit (see the example at the upper right). Major unit subdivisions appear as gray boxes, with the current subdivision highlighted in color. Chapters in the current subdivision appear in a submenu with the current chapter highlighted in black and indicated by an arrow.

All technical terms are highlighted using a **bold** type when they first appear, and a **glossary** at the end of the text summarizes their definitions. Please also note the tables of useful information, including definitions of common symbols, that appear inside the front cover.

A physics *formula* is both a mathematical equation and a *context* that gives the equation meaning. Every important formula in this text appears in a **formula box.** Each contains the equation, a **Purpose** (describing the formula's

Features that help the text serve as the primary source of information meaning and utility), a definition of the **Symbols** used in the equation, a description of any **Limitations** on the formula's applicability, and possibly some other useful **Notes**. Treat everything in such a box as an *indivisible unit* to be remembered and used together.

Active Reading

What it means to be an *active reader*

Like passively listening to a lecture, passively scanning a text does not really help you learn. *Active* reading is a crucial study skill for effectively learning from this text (and other types of technical literature as well). An active reader stops frequently to pose internal questions such as these: *Does this make sense? Is this consistent with my experience? Am I following the logic here? Do I see how I might use this idea in realistic situations?* This text provides two important tools to make this easier.

Tools to help you become an active reader

Use the **wide margins** to (1) record *questions* that occur to you as you read (so that you can remember to get them answered), (2) record *answers* when you receive them, (3) flag important passages, (4) fill in missing mathematics steps, and (5) record insights. Doing these things helps keep you actively engaged as you read, and your marginal comments are also generally helpful as you review. Note that I have provided some marginal notes that summarize the points of crucial paragraphs and help you find things quickly.

The single most important thing you can do

The **in-text exercises** help you develop the habits of (1) filling in missing mathematics steps and (2) posing questions that help you *practice* using the chapter's ideas. Also, although this text has many examples of worked problems similar to homework or exam problems, *some* of these appear in the form of in-text exercises (as you are more likely to *learn* from an example if you work on it a bit yourself instead of just scanning someone else's solution). Answers to *all* exercises appear at the end of each chapter so you can get immediate feedback on how you are doing. Doing at least some of the exercises as you read is probably the *single most important thing you can do* to become an active reader.

Active reading does take effort. *Scanning* the 5200 words of a typical chapter might take 45 minutes, but active reading could take several times as long. I personally tend to "blow a fuse" in my head after about 20 minutes of active reading, so I take short breaks to do something else to keep alert. Pausing to fill in missing math also helps me to stay focused longer.

Class Activities and Homework

End-of-chapter problems support active learning

The problems appearing at the end of each chapter are organized into categories that reflect somewhat different active-learning purposes. **Two-minute problems** are short, concept-oriented, multiple-choice problems that are primarily meant to be used *in* class as a way of practicing the ideas and/or exposing conceptual problems for further discussion. (The letters on the back cover make it possible to display responses to your instructor.) The other types of problems are primarily meant for use as homework *outside* class. **Basic** problems are simple drill-type problems that help you practice in straightforward applications of a single formula or technique. **Synthetic** problems are more challenging and realistic questions that require you to bring together multiple formulas and/or techniques (maybe from different chapters) and to think carefully about physical principles. These problems define the level of sophistication that you should strive to achieve. **Rich-context** problems are yet more challenging problems that are often written in a narrative format and ask you to answer a practical, real-life question rather than explicitly asking for a numerical result. Like situations you will encounter in real life, many provide too little information and/or too much information, requiring you to make estimates and/or discard irrelevant data (this is true of some *synthetic* problems as well). Rich-context problems are generally too difficult for most students to solve alone; they are designed for *group* problem-solving sessions. **Advanced** problems are very sophisticated problems that provide supplemental discussion of subtle or advanced issues related to the material discussed in the chapter. These problems are for instructors and truly exceptional students.

Read the Text *Before* Class!

You will be able to participate in the kinds of activities that promote real learning *only* if you come to each class having already read and thought about the assigned chapter. This is likely to be *much* more important in a class using this text than in science courses you may have taken before! Class time can also (*if* you are prepared) provide a great opportunity to get your *particular* questions about the material answered.

Class time works best if you are prepared

N1

Newton's Laws

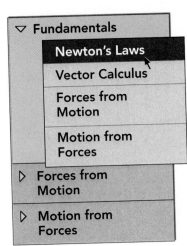

Chapter Overview

Introduction

In unit C, we focused on the behavior of systems of objects. In this unit, we will focus on how interactions affect the motion of an *individual* object. This unit has three major subdivisions (see the diagram to the left). This chapter opens the "Fundamentals" subdivision by discussing Newton's three laws of motion and how they are connected to our previous work in unit C. Chapter N2 introduces some concepts of vector calculus that we need to solve Newton's second law, which provides the key understanding of how forces affect motion. Chapters N3 and N4 provide an overview of how we can use Newton's second law and observations of an object's motion to determine useful things about the forces acting on it, or use the second law and a complete knowledge of the forces acting on the object to predict its motion.

The two subsequent subdivisions explore each of these applications of Newton's second law in much greater detail. Chapter N13 discusses Newton's greatest achievement: his explanation of planetary motion using a theoretical model that applies equally well to *both* terrestrial *and* celestial motion. This provided an unprecedented and historically important demonstration that *the laws of physics are universal*.

Section N1.1: The Newtonian Synthesis

This section reviews the history of physics before Newton to underline the astonishing nature of this achievement. Aristotle, whose work provided the general background for Western thinking about motion until the 1500s, thought that when undisturbed, objects moved in ways consistent with their basic *nature*. In his view, it was natural for a heavy object to be at rest on the earth's surface (or to seek that state of being) and natural for celestial objects to move endlessly in perfect circles. Since terrestrial and celestial objects have different natures in this view, it makes sense that they obey fundamentally different rules of motion. Ptolemy's 2nd-century geocentric astronomical model and even Copernicus's 16th-century heliocentric model assumed this basic distinction. But Newton, building on theoretical work by Galileo and Descartes and observational work by Kepler, was able to show that a theoretical model based on three simple laws of motion and a law of gravitation could explain *both* terrestrial motion and celestial motion in a simpler and more accurate way than any model previously proposed. This **newtonian synthesis** of celestial and terrestrial physics had an enormous impact on Western thought and led to the birth of physics as a science.

Section N1.2: Newton's First Law

Newton's first law states that

> In the absence of external interactions, an object's (or system's) center of mass moves at a constant velocity.

Currently, physicists consider this law to be a consequence of conservation of momentum (and/or a law that distinguishes inertial from noninertial reference frames: see section C4.5), but Newton had important rhetorical reasons for stating this idea as a fundamental law: it makes very clear that in his model the *natural* state of motion of objects (both celestial and terrestrial) is motion in a straight line at a constant velocity.

Section N1.3: Newton's Third Law

Newton's third law states that

> When two objects interact, the force the interaction exerts on one object is equal in magnitude and opposite in direction to the force that it exerts on the other object.

The section shows how this law directly follows from conservation of momentum and the definition of force. The section also shows how this law can lead to counter-intuitive results.

Section N1.4: Newton's Second Law

Conservation of momentum and the definition of force also directly imply **Newton's second law:**

$$\vec{F}_{net,ext} = M\vec{a}_{CM} \qquad\qquad (N1.7)$$

Purpose: This equation links the **acceleration** $\vec{a}_{CM}$ of an object's center of mass (CM) to its total mass M and the vector sum $\vec{F}_{net,ext}$ of external forces acting on it.

Symbols: $\vec{a}_{CM}$ is defined to be the *rate* at which the velocity $\vec{v}_{CM}$ of the object's center of mass changes: $\vec{a}_{CM} \equiv d\vec{v}_{CM}/dt$.

Limitations: The object's mass must be fixed. The object must also have a speed that is small compared to the speed of light (as we will see in unit R).

Note: Read this equation verbally as follows: "The net external force on an object *causes* its mass to accelerate."

We will spend the rest of the unit exploring applications of this law.

Section N1.5: Classification of Forces

This section further develops the force classification scheme first discussed in chapter C1, to help us talk more descriptively about the external forces acting on an object. Figure N1.1 pictorially describes the classification scheme. Throughout this text, we will denote force vectors by using an $\vec{F}$ with a descriptive subscript chosen from this classification scheme.

Section N1.6: Free-Body Diagrams

A **free-body diagram** is a tool for visually representing the external forces acting on an object. The diagram displays the object in question (abstracted from its surroundings), the external forces acting on it (depicted as labeled arrows), perhaps some coordinate axes, and (to avoid confusion) *nothing else.* The force arrows are drawn so that they either begin or end at roughly the point on the object where the force acts. This section provides detailed rules for constructing such diagrams and shows some examples. As we will see throughout the unit, such a visual representation makes it much easier to understand what goes into the left side of Newton's second law.

N1.1   The Newtonian Synthesis

In this unit, our focus shifts to the behavior of individual objects

In unit C, we studied how the three great conservation laws that lie behind all of physics constrain the behavior of isolated *systems* of objects in *spite* of the detailed characteristics of their interactions. In this unit, we will narrow our focus from systems of objects to single objects. Our goal is to answer the following question: *How do interactions affect the motion of an individual object?*

We have already learned in unit C most of what we need to answer this question. But before we dig into the details, I'd like to briefly review the history of humanity's work on this topic and underline its intellectual impact.

Aristotelian physics

As mentioned in unit C, the Greek philosopher Aristotle (384–322 B.C.) was one of the first people to think about the laws of motion in any systematic way. In his treatises *Physica* ("Physics") and *De Caelo* ("On the Heavens") Aristotle described laws of motion for both terrestrial and celestial motion. His ingenious and careful work on the subject provided the foundation for Western thought on the subject until well into the Renaissance.

Aristotle's fundamental and ambitious goal in these works was to understand motion in the context of a general philosophy of *change* and how it is related to an object's nature. His fundamental model of change was actually more biological than what we might think of as mechanical. According to Aristotle, just as it is in an acorn's intrinsic nature to grow into an oak, different kinds of objects move in different ways according to their intrinsic natures. Heavy objects have a natural tendency to move toward the center of the universe (the earth) and remain at rest there. Fire, by its nature, yearns to move away from the earth toward the heavens, where fiery objects naturally reside. Celestial bodies, because of their natures, move endlessly in the heavens in perfect circles.

Since a heavy object's nature is to be at rest on the earth, if it is moving, there must be some kind of special cause for that motion. When that cause ceases, the object again comes to rest. Celestial objects, in contrast, naturally move in perfect circles. Any deviation from *that* motion would require a special cause. The important thing to note here is that *different laws of motion apply to terrestrial and celestial objects* in Aristotle's thinking, because the natures of celestial and terrestrial objects are fundamentally different.

The fact that Aristotle's ideas provided the *starting point* for Western thought about motion until the 1500s did not mean that scholars unquestioningly accepted his ideas. Aristotle's simple vision of celestial objects moving in perfect circles was quickly seen to be inadequate. In 140 A.D., the Alexandrian astronomer Ptolemy published a much more sophisticated geocentric model of celestial motion that involved nested circular motions. Ptolemy's book was so encyclopedic and brilliantly written that it became the accepted text on astronomy until the 1600s (9th-century Arab astronomers referred to this book as the *Almagest*, "The Greatest"). Medieval Western philosophers also noted problems with Aristotle's models of motion for terrestrial objects and came up with improvements. However, these were just incremental adjustments to the model of motion that treated celestial physics as fundamentally distinct from terrestrial physics.

The Copernican model of the solar system

The Aristotelian consensus began to unravel in the mid-1500s, when the Polish scholar Nicolaus Copernicus proposed a model that put the sun, instead of the earth, at the center of the universe. Copernicus hoped that this model would make it possible to explain the observed behavior of the planets in terms of perfectly circular orbits without the complicated nesting of circular motions required by Ptolemy's model. While this hope ultimately proved vain, astronomers found his model so handy in doing calculations that Copernicus's book was widely circulated even as opposition mounted from other scholars. These scholars clearly recognized that Copernicus's model contradicted not only Ptolemy's great work but also the very foundation of Aristotle's ideas of motion. Aristotle's scheme of "natural" motion toward the earth or toward the heavens rested firmly on the assumption that the earth was the fixed center of the universe. Copernicus's scholarly foes were unwilling to overturn a consensus that had lasted nearly 2000 years just to make a few astronomical calculations simpler, particularly as no reasonable replacement for Aristotle's scheme was available.

However, the Italian physicist Galileo Galilei became convinced that Copernicus's model was too simple and beautiful to be wrong. In the late 1500s and early 1600s, he took the first steps toward a new way of looking at motion consistent with this model, developing nascent forms of Newton's first law and the theory of relativity to ingeniously rebut arguments in the *Almagest* purporting to show that the earth must be at rest. Galileo also was on the leading edge of several hot trends in the "natural philosophy" of his time: (1) the desire to make science *quantitatively predictive* as well as qualitatively descriptive and (2) the development of an experimental outlook that sought to *test* scientific hypotheses instead of relying purely upon the power of reason. These values prompted Galileo to do the experiments that enabled him to make the first *quantitative* description of the motion of falling objects.

Several other crucial developments laid the foundation for Newton's work. In 1609, Johannes Kepler showed (to his surprise and some shock) that Copernicus's model was quantitatively consistent with observations only if the planets move in *ellipses* instead of perfect circles. By 1618, Kepler published three empirical laws that provided a complete description of planetary motion in Copernicus's model, finally making that model both simpler and more accurate than Ptolemy's. About the same time, observation of variation in the size and phases of Venus using the newly invented telescope enabled Galileo and others to argue convincingly that Venus (at least) must orbit the sun. Finally, the philosopher René Descartes began in the mid-1600s to express the radical idea that it should be possible to explain *all* phenomena in terms of moving and colliding particles. The stage was set for revolution.

In 1661, the 19-year-old Isaac Newton arrived in Cambridge to begin his college education. In his classes, Newton learned Aristotle, but the ideas of Galileo and Descartes circulated unofficially. In 1665, the university closed because of the Black Plague, and Newton spent the next two years at his home in rural Lincolnshire. During this time, Newton essentially invented calculus and applied it to the motion of the moon and planets. Starting with the idea that *all* objects (including planets) naturally move in straight lines at constant velocities, Newton was able to show that the planets behaved as if they were attracted to the sun by a force that varied as the inverse square of their distance from the sun, a force that Newton interpreted as being the same as the gravitational force that causes an apple to drop from a tree. In spite of these successes, Newton did not publish any of this work at the time.

In 1679, after he had returned to Cambridge as a professor, correspondence with his rival Robert Hooke goaded Newton to renew his explorations of planetary motion. In 1687, Newton published a work entitled *Philosophiae Naturalis Principia Mathematica* ("Mathematical Principles of Natural Philosophy") that used his techniques of calculus, three simple laws of motion, and his law of inverse-square gravitation to explain all Kepler's empirical laws of planetary motion as well as a variety of terrestrial phenomena, including the moon's influence on the tides and the empirical work of Galileo on falling bodies.

Galileo and Kepler lay the foundations for Newton

Newton invents a model that explains *both* terrestrial and celestial physics

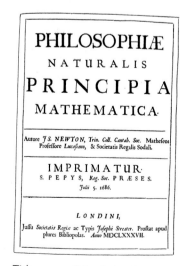

Title page of the first edition of Newton's *Principia*.

The newtonian synthesis
makes physics a science

This turned out to be one of the single most important books in the history of science. Newton's theory of motion was the first that was able to explain both terrestrial *and* celestial phenomena using the same theoretical framework. This unification, which is called the **newtonian synthesis,** stands as one of the greatest intellectual achievements of all time. Indeed, the *Principia* was so impressive that nearly everyone doing physics at the time dropped what they were doing and began to explore the rich implications of this new theory. The *Principia* forged from a diversity of competing schools and perspectives a *single* community of scholars dedicated to a single overarching theory, transforming physics from a branch of natural philosophy into a quantitative science.

In the next few sections, we will discuss the three laws of motion that provide the foundation for Newton's astonishing model of mechanics. The remainder of this unit is devoted to exploring the implications of these laws (particularly Newton's second law) for a variety of terrestrial and celestial phenomena. Our journey will culminate in chapters N12 and N13 with a demonstration that these laws indeed explain Kepler's empirical laws of planetary motion, allowing us to experience for ourselves the crowning achievement of the newtonian synthesis.

Newton's First Law

An important implication of
the definition of a system's
center of mass

In chapter C4, we saw that the total momentum of any system of particles is equal to the system's total mass times the velocity of its center of mass:

$$M\vec{v}_{CM} = \vec{p}_1 + \vec{p}_2 + \cdots + \vec{p}_N \equiv \vec{p}_{tot} \tag{N1.1}$$

This law tells us that we can calculate the total momentum of an extended object (or even a system of objects!) by treating it as if it were a point particle of equal total mass located at the center of mass.

Conservation of momentum implies that *the total momentum of an isolated system is conserved.* Since equation N1.1 implies that $\vec{v}_{CM} = \vec{p}_{tot}/M$, and M and $\vec{p}_{tot}$ have fixed values for an isolated object, this means that

> In the absence of external interactions, an object's (or system's) center of mass moves at a constant velocity.

A statement of Newton's
first law

This is **Newton's first law** as it applies to everyday macroscopic objects.

Should we take this law or
conservation of momentum
to be fundamental?

At present, we consider this law to be a consequence of the law of conservation of momentum, which we now think of as being more fundamental. Newton, however, used this law as one of the basic assumptions of his model, and he essentially derived conservation of momentum as a *consequence* of this and his other two laws of motion. Newton's approach is not wrong. *Every* physical model is based on assumptions, and it is possible to build the same basic model on *different* sets of assumptions. The choice between two different sets of assumptions that logically lead to the same physical model is more a matter of taste and/or rhetorical aims than anything else.

Newton had a very important rhetorical reason for stating the first law as he did: the first law clearly and vividly contradicts the previously accepted Aristotelian assumptions about both terrestrial and celestial motion. Newton wanted to make it clear that the *natural* state of a terrestrial object is *not* to be at rest as close to the center of the earth as possible, but rather is constant, unending motion in a straight line. Similarly, the *natural* state of a celestial object is not unending motion in a *circle*, but rather unending motion in a *straight line*. Given the context of the time, making this point clearly and vividly was essential.

In the context of contemporary physics, in addition to expressing an important and useful consequence of the law of conservation of momentum, Newton's first law plays an important role in the definition of what we call an *inertial reference frame*. We will explore this issue in greater depth in chapter N9.

N1.3   Newton's Third Law

According to the basic model of interactions presented in chapter C3, an interaction between two objects A and B *transfers* an impulse $[d\vec{p}]$ out of one object and into the other during a short interval of time dt (rather than creating or destroying momentum). Now, a tiny impulse $[d\vec{p}]$ flowing *out* of object A is the same as $-[d\vec{p}]$ flowing *into* A, so we could just as easily say that the interaction puts an impulse $[d\vec{p}]$ into object B and puts an impulse $-[d\vec{p}]$ into A (conserving the objects' total momentum). Since the force $\vec{F}$ exerted by an interaction on an object is defined to be the *rate* at which the interaction transfers momentum into the object, we have

$$\vec{F}_B \equiv \frac{[d\vec{p}]}{dt} = -\left(\frac{-[d\vec{p}]}{dt}\right) \equiv -\vec{F}_A \qquad \text{(N1.2)}$$

A mathematical statement of Newton's third law

That is, the rate at which a given interaction transfers momentum into object B is the negative of the rate at which it transfers momentum into object A.

We might express the implication of equation N1.2 in words as follows:

When objects A and B interact, the force the interaction exerts on A is equal in magnitude and opposite in direction to the force that it exerts on B.

A verbal statement of Newton's third law

This is a modern statement of **Newton's third law.**

We see that this law directly follows from the idea that interactions *transfer* momentum. Newton, on the other hand, again took this law as being fundamental and derived the momentum transfer model from it. Since either approach yields the same final physical model, the argument over which is better is again a matter of rhetorical utility in a given context.

While this law seems logical and straightforward in the context of the momentum transfer model, when we apply this law to realistic situations, we often get counterintuitive results. For example, consider a 10,000-kg truck traveling at 65 mi/h (30 m/s). Imagine that it hits a parked 500-kg Volkswagen Beetle. Which vehicle exerts the stronger force on the other during the collision? According to the third law, the magnitudes of the forces that the contact interaction exerts on either vehicle *must be the same:* any momentum that the interaction transfers out of the truck must go into the Beetle at the same rate!

This law is often counterintuitive

But this seems absurd! We know that the Beetle will be totally destroyed in this interaction (probably simultaneously crushed flat and thrown many meters), while the truck will hardly be fazed. How can we possibly *imagine* that the magnitude of the force the interaction exerts on the truck will be the same as the magnitude of the force it exerts on the Beetle?

The answer is that the fact that the *forces* are the same on each doesn't mean that each has to *respond* in the same way. Because the Beetle is 20 times less massive than the truck, a given amount of momentum flowing into the Beetle will lead to a change in its velocity whose magnitude is 20 times larger than the change in the truck's velocity from the same momentum flow. Thus forces of the same magnitude can have entirely different effects.

Which exerts the stronger force on the other?

Even so, this law is true and really works!

Part of the trick in successfully understanding and using Newton's third law is to recognize that (1) it does indeed lead to superficially counterintuitive results sometimes, but (2) it is a *necessary* consequence of the momentum flow model, and (3) it really does *correctly* describe interactions between objects (even if they have very different masses) when the situation is carefully analyzed.[†]

We will discuss the consequences and implications of Newton's third law in greater detail later (primarily in chapters N3 and N7).

Exercise N1X.1

Imagine that the truck is gently pushing the Beetle at a constant speed of 10 m/s. Which exerts the larger force on the other now?

Exercise N1X.2

Imagine now a small car pushing a large disabled truck in such a way that the pair slowly gains speed. Which exerts the larger force on the other in this circumstance?

Exercise N1X.3

Imagine that in a schoolyard fight, 75-lb Sal shoves a passive 90-lb Terry to the ground. Who exerts the larger force on the other?

N1.4   Newton's Second Law

Derivation of Newton's second law for a particle

Consider now a single point particle of mass m participating in a number of interactions with external objects. During a given short interval of time dt, the first interaction delivers to the object a certain impulse $[d\vec{p}]_1$, the second transfers a certain amount of momentum $[d\vec{p}]_2$, and so on. The net change in the particle's momentum due to all these interactions is

$$d\vec{p} = [d\vec{p}]_1 + [d\vec{p}]_2 + \cdots \qquad \text{(N1.3)}$$

Dividing both sides of this equation by dt, we get

$$\frac{d\vec{p}}{dt} = \frac{[d\vec{p}]_1}{dt} + \frac{[d\vec{p}]_2}{dt} + \cdots \equiv \vec{F}_1 + \vec{F}_2 + \cdots \equiv \vec{F}_{\text{net}} \qquad \text{(N1.4)}$$

where $\vec{F}_1 \equiv [d\vec{p}]_1/dt$ is the force that the first interaction exerts on the particle, $\vec{F}_2 \equiv [d\vec{p}]_2/dt$ is the force that the second exerts, and so on.

We see that the rate at which the particle's *total* momentum changes is equal to the *vector sum* $\vec{F}_{\text{net}}$ of *all* the forces its interactions exert on it. Now, $\vec{p} = m\vec{v}$ and $d\vec{p} \equiv \vec{p}_f - \vec{p}_i$ (for the very short time interval dt) $= mv_f - mv_i = m(v_f - v_i) \equiv m\,d\vec{v}$, since a particle's mass m is fixed. Therefore

Newton's second law for a point particle

$$\vec{F}_{\text{net}} = \frac{d\vec{p}}{dt} = m\frac{d\vec{v}}{dt} \equiv m\vec{a} \qquad \text{(N1.5)}$$

where I am defining the particle's **acceleration** $\vec{a} \equiv d\vec{v}/dt$ to be the *rate at which its velocity changes*. We will explore the meaning of acceleration in

---

[†]The only qualification is that for very long-range interactions, it can take a finite time for the impulse to flow through the interaction "conduit." Modern theories of such interactions handle this by treating the interaction essentially as a third object that can have its own momentum.

depth in chapter N2: for now, this qualitative definition is all you need to know. Equation N1.5 says that a particle's acceleration in response to the forces its interactions exert on it is equal to the **net force** $\vec{F}_{net}$ (that is, the *vector sum* of all the forces) acting on it, divided by its mass. Equation N1.5 is **Newton's second law** for a point particle: it links an aspect of the particle's *motion* (its acceleration) with the *forces* acting on it.

Consider now a *system* of particles (such as an extended object) that interacts with its surroundings (it is thus *not* isolated). Each particle in the system may participate in *internal* interactions with other particles *inside* the system and/or *external* interactions with things *outside* the system. However (as we discussed in chapter C4), internal interactions only transfer momentum from particle to particle *inside* the system, so they will not change the system's *total* momentum. Therefore, the rate at which the system's *total* momentum changes is the vector sum of the rates at which *external* interactions with particles in the system supply momentum to that system. This implies that

$$\frac{d\vec{p}_{tot}}{dt} = \vec{F}_{net,ext} \qquad (N1.6)$$

where $\vec{F}_{net,ext}$ is the vector sum of all *external* forces acting on all the particles in the system. Using equation N1.1 and assuming that the total mass M of the system is fixed (i.e., assuming that it doesn't lose or gain particles), we have

$$\vec{F}_{net,ext} = \frac{d\vec{p}_{tot}}{dt} = M\frac{d\vec{v}_{CM}}{dt} \equiv M\vec{a}_{CM} \qquad (N1.7)$$

Newton's second law for extended objects

Purpose: This equation links the acceleration $\vec{a}_{CM}$ of an object's center of mass to its total mass M and the vector sum $\vec{F}_{net,ext}$ of the external forces acting on it.

Limitations: The object must have a fixed mass for the last two steps to be true. The object must also have a speed that is small compared to that of light (as we will see in unit R).

Note: Read this equation verbally as follows: the net external force on an object *causes* its mass to accelerate.

Equation N1.7 is **Newton's second law** for *systems* of particles (including extended objects). *This is one of the most important equations of physics:* we will spend the rest of the unit exploring its rich implications!

Equation N1.5 is really a special case of equation N1.7: all interactions are external to a point particle, and such a particle's position is automatically its center of mass. However, comparing these equations reminds us of a very important point we first encountered in chapter C4: *an extended object's center of mass responds to external forces acting on an object exactly as if it were a point particle responding to those forces.* Therefore, we can always *model* an extended object by treating it as if it were a point particle, completely ignoring all complexities associated with the object's *internal* interactions. Therefore, we will usually express Newton's second law as in equation N1.5, even when we are applying it to extended objects: we *model* the extended object as a point particle.

An extended object's CM acts as a point particle

Let me emphasize again how wonderful this result is! When one object pushes or pulls on another, we do not need to understand in *detail* why and how each atom of one object interacts with the atoms of the other, or how the atoms inside each object are interacting. To understand how an object's center of mass moves, we only need to be able to quantify the *net external force* acting on that object (however that force is applied). Everything else is irrelevant!

Newton's second law contradicts subconscious notions about force and motion in some important ways that are worth making explicit. First, it asserts that forces *cause* acceleration. At the literal level, equation N1.7 expresses a quantitative equality between two sides of a mathematical equation. At the more important conceptual level, though, it expresses a *causal* relationship. The net external force is not in any sense "equivalent" to a mass accelerating (these are completely distinct ideas); rather, the net external force on an object *causes* its center of mass to accelerate. This is what the final note in the box reminds us.

Moreover (contrary to the Aristotelian world view), forces do not cause *motion*, they cause *acceleration*. Newton's first and second laws address this issue from two directions, the first law declaring that rectilinear motion is *natural* and needs no explanatory cause and the second proclaiming that forces cause an object's motion to *change* away from its natural motion.

Finally, Newton's second law tells us that it is the *net* force (the *vector sum* of *all* external forces) on the object that causes this acceleration, not the strongest force or the most recently applied force. A force does not *overcome* another force to cause an object to move in a certain way: rather all forces acting on the object act in concert to direct the object's acceleration.

The second law makes three important statements

Example N1.1

Problem The forward speed of a coasting bicycle and rider with a total mass of 62 kg is observed to decrease from 12 m/s to 9.0 m/s during a 6.0-s interval. What must be the average net external force acting on the bike and rider?

Model/Solution Assuming that the bike is moving in a straight line, and assuming that the speeds given refer to the motion of the system's center of mass, its change in velocity is 3.0 m/s rearward in 6.0 s, implying that its average acceleration is (3.0 m/s)/(6.0 s) = 0.50 m/s² rearward. Therefore, the average external force on the system must be $\vec{F}_{net,ext} = M\vec{a}_{CM} =$ (62 kg) (0.50 m/s² rearward) = 31 kg·m/s² rearward = 31 N rearward (since $1\,N \equiv 1\,kg·m/s^2$: see chapter C8).

Exercise N1X.4

A car having a mass of 2200 kg is traveling due west at a constant speed of 28 m/s. What is the magnitude of the net force acting on this car? Explain your response.

Exercise N1X.5

A small jet whose mass is 14,000 kg accelerates for takeoff, going from rest to a speed of 120 m/s in 45 s. What is the magnitude of the average net external force that must be exerted on the jet?

N1.5   Classification of Forces

In this unit we will use Newton's second law to link an object's motion to the forces that act on it. It will help us, as we do this, to have a clear and agreed-upon scheme for describing and naming these forces.

Review of the basic macroscopic force categories

Forces ultimately reflect *interactions*. As discussed in chapter C1, it is helpful at the macroscopic level to divide interactions into **long-range interactions** (gravitational and electromagnetic interactions) that can act between objects even when they are physically separated, and **contact interactions** that arise from the microscopic electromagnetic interactions between the atoms of the objects when the objects are in physical contact.

Figure N1.1 shows a useful scheme for further subdividing these categories. Let me make a few brief comments about this scheme. First, it is important to understand that *all* known long-range interactions are either gravitational or electromagnetic: there are no other possibilities. While *every* object exerts a gravitational force on every other object, unless at least one object in the interacting pair has a mass of planetary proportions, these gravitational forces will be utterly negligible. Loosely speaking, an electromagnetic interaction is *electrostatic* if it involves at least one electrically charged object, and *magnetic* if it involves at least one magnet or flowing electric current. We will explore the distinction between these categories more fully in unit E.

Subcategories of long-range forces

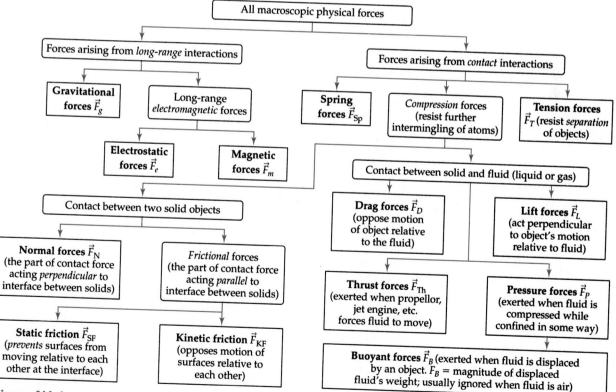

Figure N1.1

A scheme for categorizing forces. The names of the force categories are shown in boldface adjacent to the symbol that we will use for such a force.

Subcategories of contact forces

All forces that are not long-range gravitational or electromagnetic forces arise from contact interactions. The subcategories of contact forces, however, are neither completely exhaustive nor even completely logical: these are simply categories that people have found *useful* in dealing with common situations. Broadly speaking, contact forces can be divided into two subcategories: *compression* forces that arise when the atoms in the contacting surfaces oppose further intermingling or penetration, and *tension* forces, where the atoms (e.g., in a string, cable, or chain) resist being pulled apart. I have set aside the force exerted by a spring as a third subcategory because it can be either a compression force (if the spring is squeezed) *or* a tension force (if it is stretched).

The distinction between normal and frictional forces

The primary distinction between *normal* and *frictional* compression forces between two solid objects is the direction in which they act. In a certain sense, the basic compression force exerted by the contact interaction on a solid is divided somewhat artificially into a part that is *perpendicular* to the interface between the solids (the normal force) and a part that is *parallel* to this interface (the frictional force), as discussed in chapter C8. (Remember that *normal* here does not mean "typical" but "perpendicular.") A similar distinction distinguishes *lift* forces from *drag* forces on an object moving relative to a fluid: a lift force is the part of the force on the object that acts *perpendicular* to its motion relative to the fluid, and a drag force is the part that acts *opposite* the object's motion.

It turns out to be convenient to divide the frictional forces acting on solid objects in contact into two further categories depending on whether the interacting surfaces are actually sliding relative to each other (kinetic friction) or not (static friction). A *static* friction force acts to *prevent* the surfaces from starting to slide. For example, imagine that you push horizontally on a crate, but it doesn't move. An object at rest is not accelerating, so according to Newton's second law it must have zero net force on it, and therefore *some* force must be opposing your push. What is happening is that the crate and floor atoms are intertwined and locked together, and until you supply enough force to break these bonds, they will oppose any sideward push that you might apply.

If it bothers you to call such a force a friction force (because *friction* to you connotes "rubbing" and thus relative motion of the surfaces), you can think of the subscript SF as meaning *sticking force* instead of *static friction*.

The categories can be somewhat fuzzy

The distinctions between the varieties of fluid forces can be somewhat artificial. (For example, should we think of the force exerted by the air on a helicopter rotor a *lift* force or a *thrust* force? A case could be made for either!) Even so, these categories are reasonably helpful in most common contexts.

Buoyancy

The most important things to know about *buoyancy* are that (1) buoyant forces arise from height-dependent pressure *differences* in a fluid body (e.g., the ocean, the atmosphere) confined by gravity; (2) the buoyant force on an object generally acts opposite to its weight and is equal in magnitude to the weight of the fluid it displaces; and (3) buoyant forces on an object in air are generally very small compared to its weight and are usually ignored unless *explicitly* stated.

Conventional notation for forces

Note that in this text, I will *always* use the symbol $\vec{F}$ for a force (this sharply distinguishes forces from other vector quantities). I indicate the various categories of force by attaching subscripts, as shown in figure N1.1.

Since contact interactions (with the exception of the spring interaction) are not described by potential energy functions, we cannot find nice force laws for contact interactions. The strength of such interactions is usually determined in other ways, as we will discover in chapter N3.

Exercise N1X.6

A motorboat moves at a constant speed across a lake. Classify the forces acting on the boat according to the categories in figure N1.1.

N1.6   Free-Body Diagrams

Newton's second law says that an object's acceleration is determined by the external forces acting on it. A **free-body diagram** is a graphical tool that helps us recognize and display these external forces. Drawing a free-body diagram is a very useful first step in almost any problem involving Newton's second law: it helps us clarify for ourselves exactly what forces act on the object and how they are oriented relative to one another.

The purpose of a free-body diagram

To draw a free-body diagram of an object, we do the following:

Steps to follow in drawing a free-body diagram

1. *Imagine the object in its context*, thinking of the things that interact with it. Be especially careful to look along the object's boundary in your mental image for places where it touches things outside itself.
2. *Draw a sketch of the object alone*, as if it were floating in space. Part of the purpose of a free-body diagram is to direct our attention to the object itself, reducing its surroundings merely to abstract forces that act on it.
3. For each force that an external interaction exerts on the object, *draw a dot* that indicates roughly where the force is applied, and then *draw an arrow* that depicts (at least qualitatively) the magnitude and direction of that force, attaching to the dot either the arrow's tail end or tip (preferably the tail end, but whichever makes the diagram clearer). *Label each arrow* with an appropriate symbol (in some cases you may also need to provide some explanation to help the reader interpret your symbol).

In the last step, you can depict long-range forces as if they were applied to the object's center of mass, and a contact force acting on a surface as if it were applied to the surface's center. To make a diagram clearer, you may also displace force arrows slightly if they would otherwise overlap.

Here are some general "dos and don'ts" about drawing free-body diagrams:

General dos and don'ts

1. *Do not draw any arrows on the diagram depicting quantities that are* not *forces.* The main point of a free-body diagram is to display the *forces* that act on an object. Do not confuse the issue by also drawing arrows displaying the object's velocity or acceleration.
2. *Draw arrows for* only *those forces that act directly on the object in question.* As stated above, part of the purpose of a free-body diagram is to direct our attention to a *single* object and display the forces acting on it. Do not confuse the issue by drawing forces that act on other objects.
3. *Make sure that every force arrow you draw reflects an interaction* with some other object. A common error is to draw arrows on a free-body diagram for alleged forces associated with an object's velocity and/or acceleration that have no place in the newtonian model. If an alleged force is not obviously associated with an interaction involving an external object, it is not a real force in the newtonian model.

In this chapter, we will use such diagrams only as a device for qualitatively recognizing and displaying the forces that act on an object, and so we will not pay much attention to the lengths of the force arrows. (In chapter N3, we will be much more interested in displaying the force magnitudes correctly.)

Example N1.2

Problem   Draw a free-body diagram of a skier sliding down a hill (see figure N1.2a).

Model   The skier participates in a gravitational interaction with the earth, which exerts a downward force on the skier's center of mass. The skier also participates in contact interactions with the snow and the air. The force due to the interaction with the snow can be divided into normal and frictional parts. The interaction with the air exerts buoyant and drag forces (we will ignore the former). There is nothing else that the skier is in contact with, so these will be the only forces that act.

Solution   Figure N1.2b shows the finished free-body diagram.

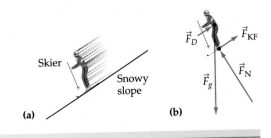

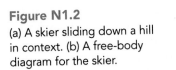

Figure N1.2
(a) A skier sliding down a hill in context. (b) A free-body diagram for the skier.

Exercise N1X.7

Someone takes a rock at the end of string and then whirls it around her or his head in a horizontal circle. Draw a free-body diagram for the rock, ignoring drag. (*Hint:* If you think there is an outward force on the rock, can you describe the interaction that gives rise to it? If not, is it real?)

TWO-MINUTE PROBLEMS

N1T.1   Do the following statements about the physical world express a primarily Aristotelian viewpoint (A), do they express a primarily newtonian viewpoint (B), or are they consistent with either viewpoint (C)?
a. If you push on something, it moves. If you push twice as hard, it moves twice as fast.
b. A heavy object falls faster than a light object.
c. The moon is held in its orbit around the earth by the force of the earth's gravity.
d. If you push on an object, it moves. After you release it, it gradually comes to rest because friction drains away the force of the initial push.
e. The speed of a falling object increases as it falls.

N1T.2   A 6-kg bowling ball moving at 3 m/s collides with a 1.4-kg bowling pin at rest. How do the magnitudes of the forces exerted by the collision on each object compare?

A. The force on the pin is larger than the force on the ball.

B. The force on the ball is larger than the force on the pin.

C. Both objects experience the same magnitude of force.

D. The collision exerts no force on the ball.

E. There is no way to tell.

N1T.3 A large car drags a small trailer in such a way that their common speed increases rapidly. Which tugs harder on the other?

A. The car tugs harder on the trailer than vice versa.

B. The trailer tugs harder on the car than vice versa.

C. Both tug equally on each other.

D. The trailer exerts no force on the car at all.

E. There is no way to tell.

N1T.4 A parent pushes a small child on a swing so that the child moves rapidly away while the parent remains at rest. How does the magnitude of the force that the child exerts on the parent compare to the magnitude of the force that the parent exerts on the child?

A. The force on the child is larger in magnitude.

B. The force on the parent is larger in magnitude.

C. These forces have equal magnitudes.

D. The child exerts zero force on the parent.

E. There is no way to tell.

N1T.5 A spaceship with a mass of 24,000 kg is traveling in a straight line at a constant speed of 320 km/s in deep space. What is the magnitude of the net thrust force acting on this spaceship?

A. 7.7 MN

B. 7.7 GN

C. 75 N

D. 0.075 N

E. 0

F. Other (specify)

N1T.6 Arrows representing four forces having equal magnitudes are shown below. What combinations of these forces, acting together on the same object, will allow that object to move with a constant velocity?

$\vec{F}_1$ $\vec{F}_2$ $\vec{F}_3$ $\vec{F}_4$ $\vec{F}_5$

A. $\vec{F}_2$ and $\vec{F}_3$

B. $\vec{F}_3$ and $\vec{F}_4$

C. $\vec{F}_1$, $\vec{F}_2$, and $\vec{F}_4$

D. $\vec{F}_1$, $\vec{F}_3$, and $\vec{F}_4$

E. A and D

F. None of the above

N1T.7 The free-body diagram at the right is supposed to represent a box sliding at a constant speed toward the right along a tabletop as it is pulled by a string. How should we label the leftward force?

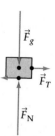

A. $\vec{F}_D$

B. $\vec{F}_{KF}$

C. $\vec{F}_{SF}$

D. The diagram is wrong: the left force should not be there at all.

E. Some other force (specify).

N1T.8 The free-body diagram below is supposed to represent a rock at the end of a string which is being whirled *clockwise* in a vertical circle. The rock is at the top of its circular path at the instant shown. How should we label the rightward force?

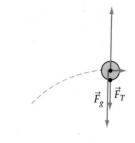

A. $\vec{F}_M$ (force of motion)

B. $\vec{F}_I$ (force of inertia)

C. $\vec{F}_D$

D. The diagram is wrong: the right force should not be there at all.

E. The diagram is wrong: there should be a leftward force instead of a rightward force.

F. Some other force (specify).

N1T.9 Consider again the free-body diagram in problem N1T.8. How should we label the upward force?

A. $\vec{F}_C$ (centrifugal force)

B. $\vec{F}_C$ (centripetal force)

C. $m\vec{a}$

D. $\vec{F}_I$ (force of inertia)

E. The diagram is wrong: there is no upward force.

N1T.10 Imagine that you serve a tennis ball directly forward and parallel to the ground. *Ignoring* air resistance, what is the approximate direction of the *net* force acting on the ball after it leaves your racquet?

A. Directly forward.

B. Directly downward.

C. Zero.

D. Forward and a bit downward.

E. Downward and a bit backward.

F. Some other direction (specify).

HOMEWORK PROBLEMS

Basic Skills

N1B.1 Imagine a cannon that is free to roll on wheels. According to Newton's third law, the interaction between the cannon and a cannonball as the latter is fired should exert equal forces on both. If this is so, why does the cannonball fly away at a very high speed and the cannon only recoils modestly? If the cannon is bolted to the ground, it does not recoil at all. How is this possible if the interaction with the cannonball exerts a force on the cannon? Please explain carefully.

N1B.2 The moon is about 1/81 times that of the earth. How does the gravitational force that the moon exerts on the earth compare to the gravitational force the earth exerts on the moon? Explain your answer.

N1B.3 A car with a mass of 1300 kg decreases its forward speed from 22 m/s to 14 m/s in 4.0 s. What are the magnitude and direction of the net force on the car? Show how you arrived at your result.

N1B.4 The magnitude of the net external force on a flying 0.25-kg ball is 2.5 N. What is the magnitude of the ball's acceleration? Explain how you arrived at your result.

N1B.5 A batter hits a baseball vertically in the air. Ignoring air resistance, draw a free-body diagram of the ball after it has left the bat but while it is still moving upward.

N1B.6 A crane lifts a crate at a constant velocity upward. Draw a free-body diagram of the crate.

N1B.7 An airplane is flying horizontally at a constant velocity. Draw a free-body diagram for this airplane.

N1B.8 A baseball player slides into second base. Draw a free-body diagram of the player.

Synthetic

N1S.1 Two friends push on a 2200-kg car. The car is initially at rest, but after pushing on it for about 2.0 s, they manage to get it to a final speed of 0.80 m/s. What is the *minimum* average magnitude of the total force the friends exert on the car? Show your work, and explain why this is the *minimum* average force.

N1S.2 A 52-kg skater sliding due north at an initial speed of 2.5 m/s runs into a snowbank and comes to rest in a time interval of 0.25 s. What are the magnitude and direction of the average force exerted on the skater by the snowbank?

N1S.3 You push on a 32-kg box with a force of 75 lb forward. The sliding friction between the box and the floor is 60 lb backward. If the box starts from rest, how long will it take you to get it moving at a speed of 5 mi/h? (Remember that 1 lb = 4.45 N and 1 m/s = 2.2 mi/h.)

N1S.4 Two people, one with a mass of 70 kg and one with a mass of 45 kg, sit on 5-kg frictionless carts that are initially at rest on a flat surface. The lighter person pushes on the more massive person, causing the latter to accelerate at a rate of 1.0 m/s² toward the east. What are the magnitude and direction of the lighter person's acceleration vector?

N1S.5 (a) Draw a free-body diagram for a box that is sliding down an incline at a constant speed.
(b) Draw a free-body diagram for a box that is being pulled up an incline at a constant speed by a rope.

N1S.6 (a) Draw a free-body diagram for a motorboat at rest on a still lake. How should we classify the vertically upward force on the boat in this case?
(b) Now draw a free-body diagram for the boat when it is moving at a constant velocity. There is an additional upward force in this case (that causes the boat to rise somewhat in the water). How should this force be classified? Explain in both cases.

Rich-Context

N1R.1 If one looks at the planet Venus through even a small telescope, one sees that sometimes it looks like a tiny crescent moon and at other times like a tiny full moon. Galileo noted that whenever Venus looks like a crescent, it also looks significantly *larger* than it does when it looks like a full moon. Argue (with the help of some diagrams) that this makes sense if Venus goes around the *sun*, but not if it goes around the earth. Contrast this situation with that of the moon, which *does* go around the earth.

N1R.2 One of the arguments against the concept that the earth is moving that is described (and refuted) by Galileo is this: Imagine dropping an object from a tall tower. If the earth is moving in a certain direction through space at a speed of many kilometers per second (as is required by the Copernican model), then it will move forward a considerable distance even during the short time before the object hits the ground. This means that since the earth moves forward under the object as it falls, the object will hit the ground at a point a considerable distance away from the tower in the opposite direction to the earth's motion. Since observations clearly show that an object dropped from a tower hits the ground directly below the drop point, the earth must be at rest. Carefully construct an argument based on momentum flow (see chapter C3) in a reference frame at rest with respect to the sun to explain why, according to Newton's model, the object will hit the ground directly below the drop point even if the earth *does* move.

N1R.3 Imagine that you have discovered a way to create a repulsive force field and you are trying to use it to make a device to protect against car crashes. Your device is mounted in a 1000-kg car and is capable of exerting a constant repulsive force of 200,000 N on another car until the cars come to rest with respect to each other. Generating the force field creates a lot of electric power, though, and you are trying to decide how big a battery you need to use. You want the battery to be able to supply power long enough to bring the cars to rest with respect to each other (not necessarily the road) when the protected car is involved in a head-on collision with a 2000-kg car when both are traveling at 30 m/s on frictionless ice. How long must your battery be able to supply power to the force field?

ANSWERS TO EXERCISES

N1X.1 They both exert the same force on each other.

N1X.2 They both exert the same force on each other. (The car/truck system accelerates *not* because the car exerts more force on the truck than the truck exerts on the car, but because the forward force exerted on the small car by its wheels exceeds the friction forces acting on both the car and the truck. These are the only *external* horizontal forces acting on the system, and for the system to gain momentum, it needs *external* forces.)

N1X.3 Each exerts the same force on the other, even though Terry does nothing to actively exert a force on Sal. Terry falls to the ground and Sal does not, probably because Terry was surprised and did not brace against the push, but Sal was braced to exert the push (and thus receive a passive push in return).

N1X.4 The car is not accelerating, so the net force on it must be *zero*.

N1X.5 The plane's change in velocity is 120 m/s during the 45-s run, so the plane's average acceleration is $(120 \text{ m/s})/(45 \text{ s}) = 2.7 \text{ m/s}^2$ forward. The average thrust force required to produce this average acceleration will be $(14,000 \text{ kg})(2.7 \text{ m/s}^2) = 37,000 \text{ kg·m/s}^2 = 37,000$ N.

N1X.6 The boat experiences a gravitational force downward, a buoyant force upward, a thrust force forward, and a drag force backward.

N1X.7 Ignoring drag from the air, the string is the only thing touching the rock, and the string exerts an inward tension force as shown. Otherwise, the only force that can act on the rock is gravity. An outward force is neither possible nor necessary (as we will see in chapter N9).

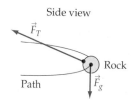

N2

Vector Calculus

Chapter Overview

Introduction

To proceed further with Newton's second law $\vec{F}_{net} = m\vec{a}$, we need to use calculus to define velocity $\vec{v}$ and acceleration $\vec{a}$ precisely.

Section N2.1:   The Time Derivative of a Vector

By analogy to the derivative of a simple function $f(x)$, if an arbitrary vector $\vec{q}$ depends on time, we define its **time derivative** to be the following vector:

$$\frac{d\vec{q}}{dt} \equiv \lim_{\Delta t \to 0} \frac{\vec{q}(t + \Delta t) - \vec{q}(t)}{\Delta t} \tag{N2.5}$$

Section N2.2:   The Definition of Velocity

Using this idea, we can precisely define the **instantaneous velocity** of a particle (or an object's center of mass) at a given instant of time t to be

$$\vec{v} = \frac{d\vec{r}}{dt} \equiv \lim_{\Delta t \to 0} \frac{\vec{r}(t + \Delta t) - \vec{r}(t)}{\Delta t} = \begin{bmatrix} \lim_{\Delta t \to 0} \dfrac{x(t + \Delta t) - x(t)}{\Delta t} \\ \lim_{\Delta t \to 0} \dfrac{y(t + \Delta t) - y(t)}{\Delta t} \\ \lim_{\Delta t \to 0} \dfrac{z(t + \Delta t) - z(t)}{\Delta t} \end{bmatrix} \equiv \begin{bmatrix} \dfrac{dx}{dt} \\ \dfrac{dy}{dt} \\ \dfrac{dz}{dt} \end{bmatrix} \tag{N2.9}$$

Purpose: This equation defines an object's velocity $\vec{v}(t)$ at an instant t.
Symbols: $\vec{r}(t) = [x(t), y(t), z(t)]$ is the position of the object's CM at time t, and $\vec{r}(t + \Delta t) = [x(t + \Delta t), y(t + \Delta t), z(t + \Delta t)]$ is the same evaluated at time $t + \Delta t$, where Δt is some finite interval of time. The symbols $d\vec{r}/dt, dx/dt,$ $dy/dt,$ and dz/dt now officially represent time derivatives.
Limitations: There are none: this is a definition.

This is a precisely defined vector at every instant of time. The object's **average velocity** $\vec{v}_{\Delta t} \equiv \Delta \vec{r}/\Delta t$ during a *finite* time interval Δt (without taking the limit) turns out to be a good approximation to the object's instantaneous velocity at an instant halfway through the interval.

Section N2.3:   The Definition of Acceleration

In physics, an object's **acceleration** at an instant of time t is a vector that expresses *how rapidly* and *in what direction its velocity vector is changing* at that instant. More mathematically, an object's **instantaneous acceleration** is

$$\vec{a}(t) \equiv \frac{d\vec{v}}{dt} \equiv \lim_{\Delta t \to 0} \frac{\vec{v}(t + \Delta t) - \vec{v}(t)}{\Delta t}$$

$$= \begin{bmatrix} a_x \\ a_y \\ a_z \end{bmatrix} = \begin{bmatrix} \lim\limits_{\Delta t \to 0} \dfrac{v_x(t + \Delta t) - v_x(t)}{\Delta t} \\ \lim\limits_{\Delta t \to 0} \dfrac{v_y(t + \Delta t) - v_y(t)}{\Delta t} \\ \lim\limits_{\Delta t \to 0} \dfrac{v_z(t + \Delta t) - v_z(t)}{\Delta t} \end{bmatrix} \equiv \begin{bmatrix} \dfrac{dv_x}{dt} \\ \dfrac{dv_y}{dt} \\ \dfrac{dv_z}{dt} \end{bmatrix} \qquad (N2.11)$$

Purpose: This equation defines the acceleration $\vec{a}(t)$ of an object's center of mass at an instant of time t.

Symbols: $\vec{v}(t)$, with components $v_x(t)$, $v_y(t)$, and $v_z(t)$, is the velocity of the object's center of mass at time t; $\vec{v}(t + \Delta t)$ (and analogously for its components) is the same evaluated at time $t + \Delta t$, where Δt is some finite interval of time. The vector $d\vec{v}/dt$ and components dv_x/dt, dv_y/dt, and dv_z/dt now represent formal time derivatives of the corresponding functions.

Limitations: There are none: this is a definition.

Note: The SI units of acceleration are $(\text{m/s})/\text{s} = \text{m/s}^2$.

Again, an object's **average acceleration** $\vec{a}_{\Delta t} \equiv \Delta \vec{v}/\Delta t$ during a *finite* time interval Δt (without taking the limit) turns out to be a good approximation to the object's instantaneous velocity at an instant halfway through the interval.

Section N2.4: Motion Diagrams

A **motion diagram** is a tool for depicting and determining an object's acceleration as it moves in a specified way. A complete motion diagram includes

1. A single image of the object in question.
2. A set of dots that represent the positions of the object's center of mass at equally spaced instants of time Δt apart.
3. Arrows between adjacent dots that represent average velocities $\vec{v}_{\Delta t} \Delta t$.
4. Acceleration arrows centered on each dot (except the first and last).

One constructs an acceleration arrow by constructing the difference vector $\Delta \vec{v} \, \Delta t$ between two adjacent average velocity arrows. This arrow closely approximates $\vec{a} \, \Delta t^2$ as the object passes the dot between the velocity arrows. As long as Δt is fixed, the arrows $\vec{v}_{\Delta t} \, \Delta t$ and $\vec{a} \, \Delta t^2$ are proportional to, and thus visually represent, the object's velocity and acceleration at selected instants.

Section N2.5: Numerical Results from Motion Diagrams

This section illustrates that as long as one knows the scale of the diagram and the value of Δt, one can calculate numerical estimates of the object's velocity and acceleration at various times.

Section N2.6: Uniform Circular Motion

This section illustrates how one can use a quantitatively accurate motion diagram to show that an object executing **uniform circular motion** (i.e., whose trajectory is circular and whose speed is constant) has an acceleration vector such that

$$a \equiv \text{mag}(\vec{a}) = \frac{\text{mag}(\vec{v})^2}{R} = \frac{v^2}{R} \qquad (N2.19)$$

Purpose: This equation specifies the magnitude of the acceleration $\vec{a}$ of an object in uniform circular motion.

Symbols: R is the radius of the object's circular path; v is the object's speed along that path.

Limitations: The object's path must be circular and its speed constant.

Note: The object's acceleration $\vec{a}$ points toward the center of the circle.

N2.1   The Time Derivative of a Vector

Our general goal in this unit is to understand applications of Newton's second law $\vec{F}_{net,ext} = M\vec{a}_{CM}$. I *qualitatively* defined the acceleration $\vec{a}_{CM}$ of an object's center of mass in chapter N1 to be "the rate $d\vec{v}_{CM}/dt$ at which its velocity changes." To use Newton's second law effectively, we need to understand this definition much more deeply and mathematically.

Calculus is the key to a better mathematical understanding of such rates of change. Indeed, Newton had to *invent* calculus to work out the implications of his theory, as there are only a handful of applications that can be studied in depth *without* calculus. In this chapter, we will explore how calculus helps us define both velocity and acceleration more precisely.

The time derivative of an ordinary mathematical function

Consider a quantity $f(t)$ that varies with time t. We define the **time derivative** df/dt of such a quantity at a given instant of time t as follows:

$$\frac{df}{dt} \equiv \lim_{\Delta t \to 0} \frac{f(t + \Delta t) - f(t)}{\Delta t} \tag{N2.1}$$

In words, this equation says that we compute the change $\Delta f = f(t + \Delta t) - f(t)$ in the quantity's value during a time interval Δt starting at time t, calculate the value of the ratio $\Delta f/\Delta t$, and then take the limiting value of this ratio as Δt approaches zero. The derivatives of ordinary numerical functions of a single variable are typically described in an introductory course in calculus. Appendix NA on differential calculus at the end of this volume provides a more detailed review if you need it.

The time derivative of a vector

We discovered in unit C that many important quantities in physics (such as force, velocity, momentum, and the like) are described not by ordinary numbers but by *vectors*. Let $\vec{q}$ be some vector quantity that varies with time t. In analogy to equation N2.1, we define the time derivative $d\vec{q}/dt$ of $\vec{q}$ to be

$$\frac{d\vec{q}}{dt} \equiv \lim_{\Delta t \to 0} \frac{\vec{q}(t + \Delta t) - \vec{q}(t)}{\Delta t} \tag{N2.2}$$

In words, this says that we compute the *change* $\Delta\vec{q} = \vec{q}(t + \Delta t) - \vec{q}(t)$ in the vector $\vec{q}$ during the time interval Δt starting at time t, calculate the ratio $\Delta\vec{q}/\Delta t$ (which is a vector), and determine this ratio's limiting value as Δt approaches zero.

The components of a vector's time derivative

Using the component definition of the difference between two vectors, we can write the change $\Delta\vec{q} = \vec{q}(t + \Delta t) - \vec{q}(t)$ in $\vec{q}$ as

$$\Delta\vec{q} = \begin{bmatrix} q_x(t + \Delta t) \\ q_y(t + \Delta t) \\ q_z(t + \Delta t) \end{bmatrix} - \begin{bmatrix} q_x(t) \\ q_y(t) \\ q_z(t) \end{bmatrix} = \begin{bmatrix} q_x(t + \Delta t) - q_x(t) \\ q_y(t + \Delta t) - q_y(t) \\ q_z(t + \Delta t) - q_z(t) \end{bmatrix} \tag{N2.3}$$

Therefore,

$$\frac{d\vec{q}}{dt} \equiv \lim_{\Delta t \to 0} \frac{\Delta\vec{q}}{\Delta t} = \lim_{\Delta t \to 0} \frac{1}{\Delta t} \begin{bmatrix} q_x(t + \Delta t) - q_x(t) \\ q_y(t + \Delta t) - q_y(t) \\ q_z(t + \Delta t) - q_z(t) \end{bmatrix} \tag{N2.4}$$

If we apply the definition of multiplication of a vector by a scalar (in this case $\Delta\vec{q}$ by $1/\Delta t$) and note that the limit applies to each component individually, we get

$$\frac{d\vec{q}}{dt} \equiv \begin{bmatrix} \lim_{\Delta t \to 0} \frac{1}{\Delta t}[q_x(t + \Delta t) - q_x(t)] \\ \lim_{\Delta t \to 0} \frac{1}{\Delta t}[q_y(t + \Delta t) - q_y(t)] \\ \lim_{\Delta t \to 0} \frac{1}{\Delta t}[q_z(t + \Delta t) - q_z(t)] \end{bmatrix} \equiv \begin{bmatrix} \dfrac{dq_x}{dt} \\ \dfrac{dq_y}{dt} \\ \dfrac{dq_z}{dt} \end{bmatrix} \tag{N2.5}$$

In words, this says that the components of the time derivative vector $d\vec{q}/dt$ are just the ordinary time derivatives of the components of $\vec{q}$.

Exercise N2X.1

Imagine that the position of a certain object's center of mass is

$$\vec{r}(t) = \begin{bmatrix} (3.0 \text{ m/s}^2)t^2 \\ 1.2 \text{ m} \\ (2.5 \text{ m/s})t \end{bmatrix} \qquad (N2.6)$$

Find the components of the time derivative $d\vec{r}/dt$ of this vector, and evaluate these components at time $t = 2.0$ s. (See appendix NA for a review of how to take the time derivative of simple powers of t.)

N2.2   The Definition of Velocity

In unit C we defined an object's velocity vector $\vec{v}$ at a time t to be

$$\vec{v} \equiv \frac{d\vec{r}}{dt} \equiv \frac{\text{small displacement}}{\text{short time interval}} \qquad (N2.7)$$

The definition of velocity we used in unit C

where $d\vec{r}$ is the object's displacement during the time interval dt, which in turn (1) encloses the instant t in question and (2) is "sufficiently short" that the velocity doesn't change significantly during the interval.

This definition is reasonably intuitive and was satisfactory for our purposes in unit C, but really is a bit fuzzy. What constitutes a "sufficiently short" dt is not clearly defined, partly because the criterion refers to the quantity (velocity) that we are trying to define! While we may have an *intuitive* sense of what this means, it is not rigorous enough to be used as a basis for a mathematically precise definition of velocity. The lack of a precise definition of velocity at an *instant* was a significant stumbling block for Western physicists until Newton.

The concept of the *time derivative of position* provides a natural way to put the concept of velocity on a mathematically firm foundation. An object's **instantaneous velocity** vector at a given instant of time t is defined to be

$$\vec{v} \equiv \frac{d\vec{r}}{dt}$$

The formal definition of instantaneous velocity

$$\equiv \lim_{\Delta t \to 0} \frac{\Delta \vec{r}}{\Delta t}$$

$$\equiv \lim_{\Delta t \to 0} \frac{\vec{r}(t + \Delta t) - \vec{r}(t)}{\Delta t} \qquad (N2.8)$$

This definition is essentially the same as that given by equation N2.7 except that *no interval Δt is considered "sufficiently short"*: we define $\vec{v}$ by the *limiting value* that $\Delta \vec{r}/\Delta t$ approaches as Δt goes to *zero*. This definition thus entirely avoids the problem of defining when an interval is "sufficiently short." It also makes it mathematically clear what we can possibly mean by an object's velocity *at an instant*, an idea that superficially seems to be incompatible with the definition of velocity as the object's *displacement* during a (nonzero!) *interval* of time divided by the duration of that interval.

According to equation N2.5, this definition implies that

$$\vec{v} \equiv \frac{d\vec{r}}{dt} \equiv \begin{bmatrix} \lim\limits_{\Delta t \to 0} \frac{1}{\Delta t}[x(t+\Delta t) - x(t)] \\ \lim\limits_{\Delta t \to 0} \frac{1}{\Delta t}[y(t+\Delta t) - y(t)] \\ \lim\limits_{\Delta t \to 0} \frac{1}{\Delta t}[z(t+\Delta t) - z(t)] \end{bmatrix} \equiv \begin{bmatrix} \dfrac{dx}{dt} \\ \dfrac{dy}{dt} \\ \dfrac{dz}{dt} \end{bmatrix} \quad \text{(N2.9)}$$

Purpose: This equation formally defines an object's velocity $\vec{v}(t)$ at an instant t.

Symbols: $\vec{r}(t) = [x(t), y(t), z(t)]$ is the position of the object's CM at time t, and $\vec{r}(t+\Delta t) = [x(t+\Delta t), y(t+\Delta t), z(t+\Delta t)]$ is the same evaluated at time $t + \Delta t$, where Δt is some finite interval of time. The symbols $d\vec{r}/dt, dx/dt, dy/dt$, and dz/dt now officially represent time derivatives.

Limitations: There are none: this is a definition.

We will still call the components $v_x \equiv dx/dt$, $v_y \equiv dy/dt$, and $v_z \equiv dz/dt$ the object's **x-velocity**, **y-velocity**, and **z-velocity** at time t, respectively. The object's instantaneous speed at time t is

$$v \equiv \text{mag}(\vec{v}) \equiv \sqrt{v_x^2 + v_y^2 + v_z^2} = \sqrt{\left(\frac{dx}{dt}\right)^2 + \left(\frac{dy}{dt}\right)^2 + \left(\frac{dz}{dt}\right)^2} \quad \text{(N2.10)}$$

Example N2.1

Problem The components of an object's position as a function of time are given by $x(t) = at^2 + b$, $y(t) = ct$, and $z(t) = 0$, where $a = 2.0$ m/s^2, $b = 5.2$ m, and $c = 1.6$ m/s. What is the object's speed at time $t = 0$?

Solution Taking the time derivative of each of these position components (using the methods discussed in appendix NA), we find that the components of the object's velocity (as functions of time) are $v_x(t) = dx/dt = 2at$, $v_y(t) = dy/dt = c$, and $v_z(t) = dz/dt = 0$. Evaluating these at time $t = 0$, we find that $v_x(0) = 2a \cdot 0 = 0$, $v_y(0) = c$, and $v_z(0) = 0$. Therefore, the object's speed at time $t = 0$ is $v(0) = [v_x^2 + v_y^2 + v_z^2]^{1/2} = [0 + c^2 + 0]^{1/2} = c = 1.6$ m/s.

Exercise N2X.2

The components of an object's position vector as functions of time are given by $x(t) = at^2 + b$, $y(t) = -ct$, and $z(t) = -at^2 + ct$, where $a = 1.5$ m/s^2, $b = 3.0$ m, and $c = 4.0$ m/s. Find the components of this object's velocity and its speed at $t = 0$ and $t = 2.0$ s.

Average velocity as an approximation to instantaneous velocity

As figure N2.1 illustrates, the ratio $\Delta\vec{r}/\Delta t$ computed for any nonzero time interval Δt starting at time t is only *approximately* equal to the object's instantaneous velocity $\vec{v}$ at time t (the approximation gets better as Δt gets smaller). We will call $\vec{v}_{\Delta t} \equiv \Delta\vec{r}/\Delta t$ the object's **average velocity** during the

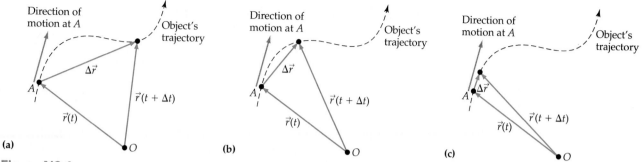

Figure N2.1

The shorter the time interval Δt, the closer that the direction of $\Delta \vec{r}/\Delta t$ (which is the same as that of $\Delta \vec{r}$) becomes to the direction of the object's velocity $\vec{v}$ at time t (which is the direction of the object's motion as it passes point A).

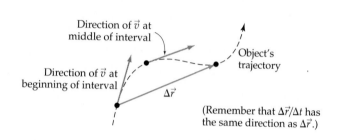

Figure N2.2
The direction of the average velocity $\Delta \vec{r}/\Delta t$ of an object during an interval Δt is generally closer to the direction of its instantaneous velocity $\vec{v}$ at the *middle* of the interval than at the beginning of the interval (the same statement also applies to the *magnitudes* of the vectors).

time interval Δt to distinguish it from the object's instantaneous velocity. The subscript on the symbol $\vec{v}_{\Delta t}$ specifies the interval used to calculate it and distinguishes it from the symbol $\vec{v}$, which from now on we will use exclusively for *instantaneous* velocity.

As figure N2.2 illustrates, $\vec{v}_{\Delta t} \equiv \Delta \vec{r}/\Delta t$ for a nonzero Δt generally most closely approximates the object's instantaneous velocity at an instant t halfway through the interval Δt. We will use this idea extensively from now on.

Example N2.2

Problem   The diagram below is a top view of a ball rolling toward the right along an inclined track, showing the ball's position every 0.1 s. Is the ball's speed increasing, decreasing, or staying constant? Estimate its speed at $t = 0.3$ s.

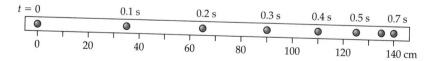

Model   Since the ball's displacement during successive intervals gets smaller, the ball is slowing down. Its speed at $t = 0.3$ s will be most closely approximated by the magnitude of $\Delta \vec{r}/\Delta t$ for the shortest interval having that instant as its midpoint.

Solution   Between $t = 0.2$ s and $t = 0.4$ s, the ball travels 45 cm, so $v(0.3 \text{ s}) \approx$ mag$(\Delta \vec{r}/\Delta t)$ during this interval $\approx (45 \text{ cm})/(0.2 \text{ s}) = 225 \text{ cm/s}$.

N2.3   The Definition of Acceleration

The word *acceleration* in everyday English carries a connotation of "speeding up." Its meaning in *physics* is both more general and more precise:

> An object's **acceleration** at an instant is a vector that expresses *how rapidly* and *in what direction* its velocity vector is changing at that instant.

In physics, then, we use this term to describe *any* change in the magnitude or direction of an object's velocity. A car that is speeding up is indeed accelerating (because the magnitude of its velocity is increasing); but according to the physics definition of the word, a car that is slowing down is *also* accelerating, because the magnitude of its velocity vector is changing (in this case decreasing). Even a car moving at a constant speed can be accelerating if it is going around a bend in the road (in this case because the *direction* of its velocity is changing). If an object's velocity vector changes in *any* way, the object is accelerating.

Mathematically, we define an object's **instantaneous acceleration** vector $\vec{a}(t)$ at an instant of time t as follows:

$$\vec{a}(t) \equiv \frac{d\vec{v}}{dt} \equiv \lim_{\Delta t \to 0} \frac{\vec{v}(t + \Delta t) - \vec{v}(t)}{\Delta t}$$

$$= \begin{bmatrix} a_x \\ a_y \\ a_z \end{bmatrix} = \begin{bmatrix} \lim_{\Delta t \to 0} \dfrac{v_x(t + \Delta t) - v_x(t)}{\Delta t} \\ \lim_{\Delta t \to 0} \dfrac{v_y(t + \Delta t) - v_y(t)}{\Delta t} \\ \lim_{\Delta t \to 0} \dfrac{v_z(t + \Delta t) - v_z(t)}{\Delta t} \end{bmatrix} \equiv \begin{bmatrix} \dfrac{dv_x}{dt} \\ \dfrac{dv_y}{dt} \\ \dfrac{dv_z}{dt} \end{bmatrix} \quad \text{(N2.11)}$$

Purpose: This equation defines the acceleration $\vec{a}(t)$ of an object's center of mass at an instant of time t.

Symbols: $\vec{v}(t)$, with components $v_x(t)$, $v_y(t)$, and $v_z(t)$, is the velocity of the object's center of mass at time t; $\vec{v}(t + \Delta t)$ (and analogously for its components) is the same evaluated at time $t + \Delta t$, where Δt is some finite interval of time. The quantity $d\vec{v}/dt$ and components dv_x/dt, dv_y/dt, and dv_z/dt represent formal time derivatives of the corresponding functions.

Limitations: There are none: this is a definition.

In words, this equation says that an object's acceleration at time t is the limiting value as Δt goes to 0 of the *change* $\Delta \vec{v}$ in the object's instantaneous velocity between times t and $t + \Delta t$, divided by the duration Δt of that interval. Since velocity is measured in meters per second, the SI units of acceleration are meters per second squared (m/s^2).

We also see that if we know the components of an object's velocity vector as a function of time, we can compute the components of its acceleration simply by taking the time derivatives of the corresponding velocity components.

Exercise N2X.3

Imagine that the components of an object's velocity vector are $v_x(t) = q$, $v_y(t) = 0$, and $v_z(t) = bt + q$, where $q = 5.0$ m/s and $b = -10$ m/s^2. Find the components of the object's acceleration at time $t = 0$ and $t = 2$ s.

Just as an object's average velocity $\vec{v}_{\Delta t} \equiv \Delta \vec{r}/\Delta t$ for a given nonzero (but fairly short) interval of time Δt is a good approximation for its instantaneous velocity midway through the interval, so an object's **average acceleration** $\vec{a}_{\Delta t} \equiv \Delta \vec{v}/dt$ during a given nonzero (reasonably short) interval Δt is a good approximation for its instantaneous acceleration midway through the interval:

The definition of the average acceleration

$$\vec{a}\left(t + \tfrac{1}{2}\Delta t\right) \approx \vec{a}_{\Delta t} \equiv \frac{\Delta \vec{v}}{\Delta t} \equiv \frac{\vec{v}(t + \Delta t) - \vec{v}(t)}{\Delta t} \qquad \text{(N2.12)}$$

So if we know an object's velocity at two instants of time, we can *estimate* its acceleration halfway through the interval by computing the difference $\Delta \vec{v}$ between the velocities and dividing by Δt. This approximation improves as Δt becomes smaller compared to the time required for the acceleration to change significantly. If the object's acceleration is *constant* during Δt, then the instantaneous acceleration throughout the interval is the *same* as this average acceleration.

In situations where we need only to *estimate* an object's acceleration, equation N2.12 provides a quick way to connect an object's acceleration to the change in its velocity *without* doing any calculus. Examples N2.3 and N2.4 illustrate the application of this idea in several contexts.

Example N2.3

Problem A person driving eastward on a straight road at 20 m/s (44 mi/h) sees the brake lights of the car in front go on and so applies the brakes for 1.5 s. This slows the car to 14 m/s. What were the magnitude and direction of the car's average acceleration during this period?

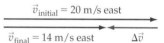

Model/Solution The car's initial velocity vector is 20 m/s eastward; its final velocity vector is only 14 m/s eastward. As the vector construction in figure N2.3 illustrates, this means that the car's change in velocity is $\Delta \vec{v} = 6.0$ m/s *westward*. (Remember that difference $\Delta \vec{v}$ between the final and initial velocities is the vector that we would have to add to the initial velocity to get the final velocity.) The car's average acceleration in this case is thus $\vec{a}_{\Delta t} = (6.0 \text{ m/s west})/(1.5 \text{ s}) = 4.0 \text{ m/s}^2$ westward.

Figure N2.3
The car's *change* in velocity in this example points westward, according to the definition of the vector difference.

Example N2.4

Problem A bicyclist travels at a constant speed of 12 m/s around a bend in the road during a 34-s interval of time. If the bicyclist was traveling northward at the beginning of the bend and westward at the end, what are the magnitude and direction of the bicyclist's average acceleration during this interval?

Solution Figure N2.4 shows that $\Delta \vec{v}$ in this case is a vector pointing *southwest*. According to the pythagorean theorem, the magnitude of this vector is

$$\text{mag}(\Delta \vec{v}) = \sqrt{(12 \text{ m/s})^2 + (12 \text{ m/s})^2} = 17 \text{ m/s} \qquad \text{(N2.13)}$$

The magnitude of the bicyclist's average acceleration is therefore $\text{mag}(\Delta \vec{v})/\Delta t = (17 \text{ m/s})/(34 \text{ s}) = 0.50 \text{ m/s}^2$. So the cyclist's average acceleration during this interval is 0.5 m/s² southwest (even though the cyclist's speed is constant!).

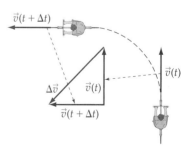

Figure N2.4
The bicyclist's change in velocity.

Exercise N2X.4

A car starting from rest travels forward in a straight line with an average acceleration of 3.0 m/s² for 8.0 s. What is the car's final speed?

Exercise N2X.5

Imagine that you throw a ball vertically into the air. The ball leaves your hand, traveling upward at a speed of 15 m/s. Three seconds later, the ball passes you at the same speed on its way downward. What are the magnitude and direction of the ball's average acceleration during this time interval?

N2.4   Motion Diagrams

Why motion diagrams are useful

When we apply the newtonian model in many practical situations (as we will see), it is often helpful to know the *direction of an object's acceleration vector* before we can even start. A **motion diagram** is a powerful tool that both vividly illustrates an object's motion in one or two dimensions and provides a way to visually determine the direction of an object's acceleration.

Steps in drawing a motion diagram

Before we draw such a diagram, it helps to visualize what a *strobe photograph* of the object would look like (such a photograph is made by using a pulsing flash to record on a single picture multiple images of a moving object at equally spaced instants of time). Figure N2.5a illustrates such a "strobe photograph" of a car that is braking to a stop. Note that the distance between successive images of the car gets smaller as time passes, because as the car's speed decreases, its displacement between equally spaced instants of time also decreases.

The first step in drawing a motion diagram is to draw a *single* image for the object (to show what object we are considering and how it is oriented). Then we draw a sequence of dots, starting below (or perhaps alongside) that image that represents the positions of an object's *center of mass* at equally spaced instants of time, as shown in figure N2.5b. Let's label these dots 1, 2, 3, . . . in sequence.

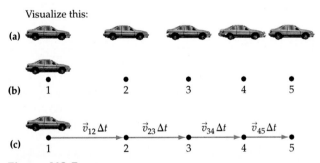

Figure N2.5

How to draw a motion diagram of a car that is braking to a stop. (a) Visualize how a strobe photograph of the moving object might look. (b) Draw a single image of the object, and then draw dots to indicate the position of the object's center of mass at equally spaced instants of time. (c) Draw arrows from each dot to the next. These arrows are proportional to the object's average velocity between the instants depicted.

We then draw arrows from one dot to the next, as shown in figure N2.5c. Each such arrow corresponds to the *displacement* $\Delta\vec{r}$ of the object's center of mass as it moves from the dot at the arrow's tail to the dot at its head. The object's average velocity during this time interval is $\vec{v}_{\Delta t} = \Delta\vec{r}/\Delta t$, so the arrow is *also* equivalent to the vector $\vec{v}_{\Delta t}\,\Delta t$. We conventionally give the arrow that we draw between dots 1 and 2 the label $\vec{v}_{12}\,\Delta t$, the arrow that we draw between dots 2 and 3 the label $\vec{v}_{23}\,\Delta t$, and so on (see figure N2.5c). The subscripts 12, 23, and so on take the place of the Δt subscript that we have been using to indicate the average velocity, but serve the same purpose of specifying the time interval over which we are determining the average velocity.

Note that if the time interval Δt between adjacent dots is fixed, then the arrows $\vec{v}_{12}\,\Delta t$, $\vec{v}_{23}\,\Delta t$, ... have the same direction as, and are proportional in magnitude to, the object's actual average velocity vectors $\vec{v}_{12}$, $\vec{v}_{23}$, ..., respectively. These arrows therefore vividly depict the object's changing velocity as it moves.

Indeed, the object's average velocity $\vec{v}_{12}$ between instants 1 and 2 approximates the object's *instantaneous* velocity most closely at an instant halfway between 1 and 2, an instant we might call "1.5." Similarly, the object's average velocity $\vec{v}_{23}$ approximates its instantaneous velocity at the instant halfway between 2 and 3 (instant "2.5"). So the velocity arrows drawn in figure N2.5c also (approximately) depict the directions and the relative magnitudes of the object's *instantaneous* velocities at instants 1.5, 2.5, and so on.

The final step in drawing a motion diagram is to use these velocity arrows to construct arrows representing the object's *acceleration* as it passes the numbered dots. Since instants 1.5 and 2.5 are also separated by a time interval of Δt, the object's average *acceleration* between instants 1.5 and 2.5 is

$$\vec{a}_{\Delta t} = \frac{\Delta\vec{v}}{\Delta t} = \frac{\vec{v}(2.5) - \vec{v}(1.5)}{\Delta t} \approx \frac{\vec{v}_{23} - \vec{v}_{12}}{\Delta t} \qquad (N2.14)$$

The arrows drawn on the diagram *actually* correspond to $\vec{v}_{12}\,\Delta t$ and $\vec{v}_{23}\,\Delta t$. If we multiply both sides of equation N2.14 by Δt^2, we see that

$$\vec{a}_{\Delta t}\,\Delta t^2 \approx \vec{v}_{23}\,\Delta t - \vec{v}_{12}\,\Delta t \qquad (N2.15)$$

The average acceleration $\vec{a}_{\Delta t}$ in turn approximates the object's *instantaneous* acceleration most closely at the instant halfway between instants 1.5 and 2.5, which is the instant that the object's center of mass passes point 2. Therefore, the arrow representing the vector *difference* $\vec{v}_{23}\,\Delta t - \vec{v}_{12}\,\Delta t$ between adjacent velocity arrows on the diagram is a good approximation to the vector $\vec{a}_2\,\Delta t^2$, where $\vec{a}_2$ is the object's instantaneous acceleration as it passes point 2:

$$\vec{a}_2\,\Delta t^2 \approx \vec{a}_{\Delta t}\,\Delta t^2 \approx \vec{v}_{23}\,\Delta t - \vec{v}_{12}\,\Delta t \qquad (N2.16)$$

Figure N2.6 shows how we can construct this arrow on a motion diagram (for an object moving in one dimension).

1. Draw dashed vertical lines down from each dot,
2. Redraw the arrow $\vec{v}_{23}\,\Delta t$ slightly below the arrow $\vec{v}_{12}\,\Delta t$ so that the tail of $\vec{v}_{23}\,\Delta t$ lines up with the vertical line coming down from dot 1.

Figure N2.6

How to construct an arrow that represents the object's (approximate) acceleration at the instant when the object passes position 2. The curved dotted arrow shows how we move $\vec{v}_{23}\,\Delta t$ to perform the subtraction.

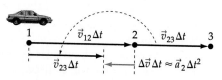

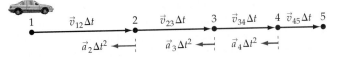

Figure N2.7
A complete motion diagram for a braking car (with optional labels).

3. Then draw an arrow from the tip of the initial velocity arrow $\vec{v}_{12}\,\Delta t$ to the tip of the final velocity $\vec{v}_{23}\,\Delta t$: this constructed arrow corresponds to the vector difference $\vec{v}_{23}\,\Delta t - \vec{v}_{12}\,\Delta t = \Delta\vec{v}\,\Delta t \approx \vec{a}_2\,\Delta t^2$.

Using the dashed lines as guides for displacing the arrows a bit away from each other (instead of piling them on top of each other) makes the diagram clearer. (Note that this process conveniently puts the tail end of $\vec{a}_2\,\Delta t^2$ at point 2.)

We can construct the acceleration arrows $\vec{a}_3\,\Delta t^2$, $\vec{a}_4\,\Delta t^2$, and so on analogously. Again, since Δt^2 is a scalar and has the same value for all such arrows on a given diagram, these acceleration arrows have the same direction and are proportional in length to the object's instantaneous acceleration $\vec{a}$ as it passes the numbered dots on the diagram.

A complete motion diagram thus shows the following items:

A summary of the items that should appear in a motion diagram

1. A single image of the object in question.
2. A set of dots that represent the positions of the object's center of mass at equally spaced instants of time.
3. (Average) velocity arrows drawn between these dots.
4. Acceleration arrows centered on each dot (except the first and last).

Figure N2.7 illustrates a complete motion diagram for the braking car. In a motion diagram that you are using for purely *qualitative* purposes, you do not need to include labels for the points or arrows. (Note that you can tell the difference between the velocity and acceleration arrows even without labels, because the velocity arrows always stretch between the dots while the acceleration arrows generally do not.) You also do not *need* to show how you constructed the acceleration arrows (if you can do the construction in your head); but if you include such diagrams showing how you constructed the acceleration arrows, draw them below or alongside the motion diagram, as shown in figure N2.6.

Example N2.5

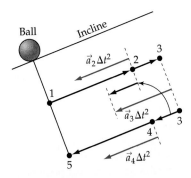

Figure N2.8
Motion diagram of a ball rolling first up and then down an incline.

Problem   Draw a motion diagram of a ball that rolls up an incline and then back down.

Model/Solution   Figure N2.8 shows an appropriate motion diagram for this situation, with the ball slowing down as it rolls up the incline and speeding up as it rolls back down the incline. The tricky thing here is that the ball's positions as it rolls up the incline overlap with its positions as it rolls down. This would be very confusing unless we offset the position points for the up and down motions as shown in the diagram. The two points labeled 3 at the right of the diagram both represent the ball's position at the *single* instant when it is at rest at the top of the incline (this instant is common to both the up and down motions).

The diagram explicitly shows the construction of the ball's acceleration arrow $\vec{a}_3\,\Delta t^2$ as the ball passes that position, which is also when its instantaneous velocity passes through zero as it changes direction. (The solid thin curved line shows how I have moved the arrow $\vec{v}_{34}\,\Delta t$ to make its tail coincide with the tail of the arrow $\vec{v}_{23}\,\Delta t$ so that I can do the subtraction yielding $\vec{a}_3\,\Delta t^2$.) This construction clearly shows that the object's acceleration is nonzero even at that instant. This may seem counterintuitive to you, but the fact that an object's velocity is *passing through* zero at a certain instant doesn't mean that the *rate of change* of its velocity has to be zero at that instant! Indeed, the ball's acceleration seems to be nonzero and directed down the incline throughout its motion.

Exercise N2X.6

Draw a qualitative motion diagram of a basketball falling toward the floor and then rebounding from it.

One can often determine the direction (and even estimate the length) of the acceleration arrows by eye on a *one*-dimensional motion diagram. When drawing a motion diagram for an object moving in *two* dimensions, though, you should always show the construction of the acceleration arrow explicitly. This is usually pretty easy, since the arrows to be subtracted usually do *not* lie on top of each other (and so do not have to be displaced in complicated ways, as in figure N2.8). To construct the acceleration arrow at a given point (say, point 2), do the following:

Drawing a motion diagram for two-dimensional motion

1. Take the velocity arrow following the point in question (point 2 here) and move it (without changing its direction) until its tail end coincides with the tail end of the previous velocity arrow.
2. Then draw an arrow from the head of the previous velocity arrow to the head of the arrow you just moved: by definition of vector subtraction, the arrow that you have just constructed is $\vec{v}_{23}\,\Delta t - \vec{v}_{12}\,\Delta t = \Delta\vec{v}\,\Delta t \approx \vec{a}_2\,\Delta t^2$.

(See Figure N2.9.) Note that this method also automatically leaves the acceleration arrow with its tail end at the point to which it applies.

Exercise N2X.7

Draw a qualitative motion diagram of a car going over the crest of a hill at a constant speed. Is its acceleration nonzero?

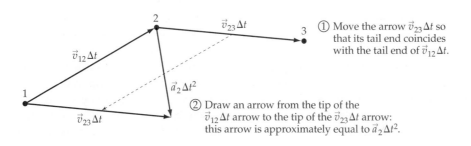

① Move the arrow $\vec{v}_{23}\,\Delta t$ so that its tail end coincides with the tail end of $\vec{v}_{12}\,\Delta t$.

② Draw an arrow from the tip of the $\vec{v}_{12}\,\Delta t$ arrow to the tip of the $\vec{v}_{23}\,\Delta t$ arrow: this arrow is approximately equal to $\vec{a}_2\,\Delta t^2$.

Figure N2.9
How to construct an object's acceleration arrow on a motion diagram for an object moving in two dimensions.

N2.5   Numerical Results from Motion Diagrams

Calculating the magnitudes of velocities and accelerations on a motion diagram

Even a hastily sketched motion diagram can give us important *qualitative* information about the direction and approximate relative magnitudes of an object's acceleration as it follows the motion depicted. However, a carefully drawn motion diagram can additionally allow us to compute the approximate *quantitative magnitudes* of the object's velocity and acceleration at various points along its trajectory by measuring the lengths of the arrows on the diagram.

For example, if we know that the time interval Δt between dots on a motion diagram is 0.10 s and we measure the velocity arrow $\vec{v}_{12}\,\Delta t$ on the diagram to have a length of 2.5 cm, then the actual magnitude of the object's average velocity during the time interval between dots 1 and 2 is

$$v_{12} \equiv \mathrm{mag}(\vec{v}_{12}) = \frac{\mathrm{mag}(\vec{v}_{12}\,\Delta t)}{\Delta t} = \frac{2.5 \text{ cm}}{0.10 \text{ s}} = 25 \text{ cm/s} \qquad \text{(N2.17a)}$$

Similarly, if we measure the acceleration arrow $\vec{a}_2\,\Delta t^2$ on the diagram to have a length of 1.2 cm, the magnitude of the object's acceleration at point 2 is roughly

$$a_2 \equiv \mathrm{mag}(\vec{a}_2) = \frac{\mathrm{mag}(\vec{a}_2\,\Delta t^2)}{\Delta t^2} = \frac{1.2 \text{ cm}}{(0.10 \text{ s})^2} = 120 \text{ cm/s}^2 \qquad \text{(N2.17b)}$$

Example N2.6

Problem   Assume that the time between dots in figure N2.9 is 0.20 s. What is the magnitude of the object's average velocity between points 1 and 2? What is the magnitude of the object's acceleration at point 2?

Solution   I measure the arrow $\vec{v}_{12}\,\Delta t$ on figure N2.9 to be 4.1 cm long. (You may get a different result if the drawing's size is not exactly preserved during printing.) According to equation N2.8a, this means that $v_{12} = (4.1 \text{ cm})/(0.20 \text{ s}) = 20.5$ cm/s. I also measure the arrow $\vec{a}_2\,\Delta t^2$ to be ≈ 2.4 cm long, so $a_2 = (2.4 \text{ cm})/(0.20 \text{ s})^2 = 60 \text{ cm/s}^2$.

Exercise N2X.8

The drawing below shows an unfinished motion diagram for a button sliding on a tabletop. If $\Delta t = 0.05$ s, what is v_{23}? What is a_3?

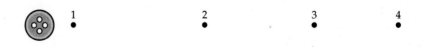

Handling scale diagrams

Of course, the examples that we have considered so far assume that the motion diagram has been drawn to actual size. Many of the motion diagrams that we will draw (if they are quantitatively accurate at all) will be *scale* drawings of the object's motion. For example, the diagram for the braking car shown in figure N2.10 is drawn to a scale such that 1 cm on the diagram

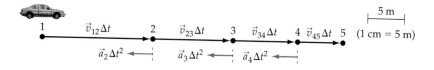

corresponds to 5 m of actual distance (as the scale marker on that diagram indicates). In such a case, we have to convert the distances we measure on the diagram to actual physical distances before we can compute velocities or accelerations.

For example, I measure the drawn arrow representing $\vec{v}_{12}\,\Delta t$ in figure N2.10 to be 3.0 cm long, corresponding to

$$3.0 \text{ cm}\left(\frac{5 \text{ m}}{1 \text{ cm}}\right) = 15 \text{ m} \qquad \text{(N2.18)}$$

of actual displacement for the car. If the time interval between position dots is 0.50 s, then the magnitude of the car's average velocity between points 1 and 2 is (15 m)/(0.50 s) = 30 m/s.

Exercise N2X.9

Similarly, find the magnitude of $\vec{v}_{23}$ and $\vec{a}_2$ for the car shown in figure N2.10 (assuming that the time between dots is still 0.5 s).

N2.6   Uniform Circular Motion

Circular motion (or approximately circular motion) is a very common kind of motion, applicable to a wide variety of physical situations such as atomic orbitals, bicycle wheels, hurricanes, satellite orbits, and galactic rotation. Historically, one of the ideas that helped Newton build his theory was growing recognition in the middle 1600s that even an object that is moving along a circular path (even if it has a constant speed) is accelerating, because the *direction* of its velocity is changing. In this section we will use motion diagrams to learn about the acceleration of an object moving in a circular path.

Consider an object moving at a constant speed around a circle of radius R: this kind of motion is called **uniform circular motion.** What are the direction and magnitude of the object's acceleration in this case?

The definition of uniform circular motion

Figure N2.11a shows a motion diagram of an object traveling around a circle with a radius of 7.7 cm. For the sake of argument, let's assume that the time interval between position dots is 0.10 s. Note that the position dots are equally spaced, consistent with the idea that the object is moving with a constant speed. Using the techniques discussed in sections N2.4 and N2.5, I have constructed arrows representing the average velocities $\vec{v}_{12}\,\Delta t$ and $\vec{v}_{23}\,\Delta t$ and the arrow representing the object's acceleration as it passes point 2.

Note that when this arrow is attached to point 2, it points toward the center of the circle. Since there is nothing special about point 2 (all points on the circle are equivalent), this result applies generally: *the acceleration of an object moving at a constant speed around a circle points directly toward the center of the circle at every instant.* Note that this applies *only* if the object is moving at a constant speed: you can see from the diagram that if either velocity arrow is longer than the other, the acceleration arrow will *not* point toward the center.

The direction of the object's acceleration

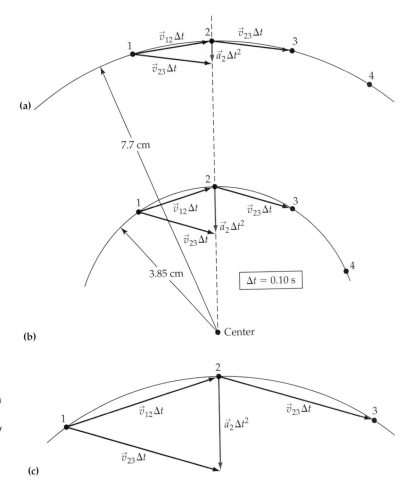

Figure N2.11

(a) An actual-size motion diagram for an object moving at about 22 cm/s in a circle of 7.7 cm in radius. Note that the object's acceleration vector as it passes point 2 points directly at the center of the circle. (b) If we decrease the radius by a factor of 2 but keep the object's speed the same, the magnitude of its acceleration doubles. This is because the velocity vectors change direction twice as fast as the object goes around the smaller circle. (c) If we keep the radius the same but double the speed, not only does the direction of the velocity vector change twice as fast but also each velocity vector is twice as long. This has the effect of quadrupling the acceleration.

What happens if we vary R and v?

For future reference, let's determine the actual magnitude of the velocities and acceleration in this case. I measure the arrows $\vec{v}_{12}\,\Delta t$ and $\vec{v}_{23}\,\Delta t$ in figure N2.11a to be 2.2 cm long on the diagram. Since $\Delta t = 0.10$ s here, the actual magnitude of the velocities is $(2.2 \text{ cm})/(0.10 \text{ s}) = 22$ cm/s. I measure the $\vec{a}_2\,\Delta t^2$ arrow to be about 0.6 cm long on the diagram, corresponding to an actual acceleration of $(0.6 \text{ cm})/(0.10 \text{ s})^2 = 60$ cm/s^2.

Now, the *magnitude* of the object's acceleration can only depend on the radius of the circle and the object's speed as it travels around its circular path: these two quantities completely determine the object's motion. What happens if we vary the values of these two quantities?

Figure N2.11b shows that if we keep the object's speed unchanged while we decrease R to *one-half* of its original value, the lengths of the velocity arrows are unchanged, but the angle between them *doubles* (since the object covers twice as great a fraction of the circle during each time interval Δt). This has the effect of *doubling* the length of the acceleration arrow.[†] We see that the magnitude of an object's acceleration in uniform circular motion must therefore depend inversely on the radius of its circular path: $a = \text{mag}(\vec{a}) \propto 1/R$.

Figure N2.11c shows that if we keep the radius of the circle unchanged while we double the object's speed, not only does the *angle* between the

---

[†]Technically, the acceleration arrow exactly doubles in size only in the limit that Δt (and thus the angle between velocity arrows) goes to zero.

velocity vectors double (since the object goes around the circle twice as fast as before) but their length doubles, too. The result is an acceleration vector that is 4 times as large as it was in figure N2.11a. We see that the magnitude of an object's acceleration in uniform circular motion depends on the square of its speed: $a \propto v^2$.

The simplest formula for the magnitude of the object's acceleration in uniform circular motion that embraces both the observation that $a \propto 1/R$ and the observation that $a \propto v^2$ is the following:

$$a \equiv \mathrm{mag}(\vec{a}) = \frac{\mathrm{mag}(\vec{v})^2}{R} = \frac{v^2}{R} \qquad (N2.19)$$

The magnitude of the acceleration of an object in uniform circular motion

Purpose: This equation specifies the magnitude of the acceleration $\vec{a}$ of an object in uniform circular motion.
Symbols: R is the radius of the object's circular path, and v is the object's speed along that path.
Limitations: The object's path must be circular, and its speed must be constant.
Note: The object's acceleration vector $\vec{a}$ points toward the center of the circle.

We could multiply the right side of this equation by any constant and get another equation that is *still* consistent with the constraints $a \propto 1/R$ and $a \propto v^2$, but it happens that equation N2.19 is correct as it stands. We can check this as follows. We saw earlier that the object in figure N2.11a moves with a speed of 22 cm/s. According to equation N2.19, the object in that figure should have an acceleration of magnitude $a = v^2/R = (22 \text{ cm/s})^2/(7.7 \text{ cm}) = 63 \text{ cm/s}^2$. This compares reasonably well with the result of 60 cm/s^2 that we found earlier by measuring the length of the acceleration arrow on the diagram (particularly considering the uncertainties associated with the various measurements and in the construction of the arrow).

To summarize, we have learned (by studying a quantitatively accurate motion diagram) that if an object moves at a constant speed v around a circle of radius R (that is in uniform circular motion), its acceleration always points toward the center of the circle and has a magnitude $a = v^2/R$. This is, as we will see, a very important and useful result! Perhaps you can see that motion diagrams can be a powerful tool for analyzing and understanding acceleration, particularly when objects move in two dimensions.

Exercise N2X.10

Construct an arrow on figure N2.11a that represents the object's acceleration as it passes point 3. When this arrow is attached to that point, does it point toward the circle's center?

Exercise N2X.11

A car traveling at a constant speed of 20 m/s ($\approx$ 44 mi/h) goes around a circular curve to the left in the road whose radius is 600 m. At an instant when the car is going due north, what are the magnitude and direction of the car's acceleration?

TWO-MINUTE PROBLEMS

N2T.1 Two marbles are rolling along parallel tracks (which may or may not be inclined). A stroboscopic photograph showing a top view of the positions of the marbles at equally spaced instants of time looks like this:

t_A t_B t_C t_D t_E t_F t_G

○   ○   ○   ○   ○   ○   ○

○ ○ ○   ○   ○   ○   ○
t_A t_B t_C t_D t_E t_F t_G

At what instant(s) of time do the marbles have the same instantaneous velocity?
A. At time t_B.
B. At time t_G.
C. At both t_B and t_G.
D. Some instant between t_C and t_D.
E. Some instant between t_D and t_E.
F. Roughly time t_D.

N2T.2 Two marbles are rolling along parallel tracks (which may or may not be inclined). A stroboscopic photograph showing a top view of the positions of the marbles at equally spaced instants of time looks like this:

t_A t_B t_C t_D t_E t_F t_G

○   ○   ○   ○ ○ ○ ○

○ ○ ○   ○   ○   ○   ○
t_A t_B t_C t_D t_E t_F t_G

At what instant(s) of time do the marbles have the same instantaneous velocity?
A. At time t_B.
B. At time t_F.
C. At both t_B and t_F.
D. Some instant between t_C and t_D.
E. Some instant between t_D and t_E.
F. Roughly time t_D.

N2T.3 An object travels halfway around a circle at a constant instantaneous speed v. What is the magnitude of its average velocity during this time interval?
A. v
B. $1.41v$
C. $1.57v$
D. $v/1.41$
E. $v/1.57$
F. Some other multiple of v (specify).
T. We do not have enough information to answer.

N2T.4 An object can have a constant speed and still be accelerating, true (T) or false (F)?

N2T.5 An object's acceleration vector always points in the direction that it is moving, T or F?

N2T.6 An object can have zero velocity (at an instant of time) and still be accelerating, T or F?

N2T.7 The x-velocity of an object can be positive at the same time that its x-acceleration is negative, T or F?

N2T.8 An object falling vertically at a speed of 20 m/s lands in a snowbank and comes to rest 0.5 s later. What is the object's average acceleration during this interval?
A. 10 m/s^2 up
B. 10 m/s^2 down
C. 40 m/s^2 up
D. 40 m/s^2 down
E. 5 m/s^2 up
F. 80 m/s^2 down
T. Other (specify)

N2T.9 A boat hits a sandbar and slides some distance before coming to rest. Which of the arrows shown below best represents the direction of the boat's acceleration as it is sliding? (*Hint:* Draw a motion diagram.)

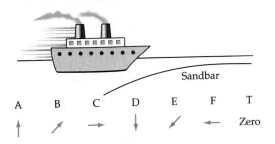

A   B   C   D   E   F   T

↑   ↗   →   ↓   ↙   ←   Zero

N2T.10 A car moving at a constant speed travels past a valley in the road, as shown. Which of the arrows shown most closely approximates the direction of the car's acceleration at the instant that it is at the position shown? (*Hint:* Draw a motion diagram.)

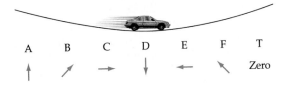

A   B   C   D   E   F   T

↑   ↗   →   ↓   ←   ↘   Zero

N2T.11 A bike (shown in a top view in the diagram) travels around a curve with its brakes on, so that it is constantly slowing down. Which of the arrows shown below most closely approximates the direction of its acceleration at the instant that it is at the position shown? (*Hint:* Draw a motion diagram.)

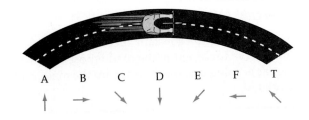

A   B   C   D   E   F   T

↑   →   ↘   ↓   ↙   ←   ↖

Basic Skills

N2B.1 What are the derivatives of the following functions of time (where a, b, and c are constants)?

    (a) $at + c$

    (b) $at^2 + (bt)^{-2}$

    (c) $(at + b)^2$

N2B.2 What are the derivatives of the following functions of time (where a, b, and c are constants)?

    (a) $ct(at^2 + b)$

    (b) $(at + b)^{-2}$

    (c) $\dfrac{at^3 + b}{t}$

N2B.3 An object's position as a function of time is given by $\vec{r}(t) = [-a/t, bt^2 + c, qt]$, where a, b, c, and q are constants with values $a = 2.0$ m·s, $b = 1.0$ m/s², $c = 5.0$ m, and $q = 2.0$ m/s. Calculate the components of this object's instantaneous velocity at time $t = 2.0$ s.

N2B.4 An object travels exactly once around a circle of radius 5.0 m at a constant instantaneous speed of 3.0 m/s. What is its average velocity during this interval?

N2B.5 The driver of a car moving at 24 m/s due east sees something on the road and applies the brakes. The car comes to rest 4.0 s later. What are the magnitude and direction of the car's average acceleration during this interval?

N2B.6 A person hits a trampoline while moving downward with a speed of 5.0 m/s and rebounds a short time later with roughly the same speed upward. If the person is in contact with the trampoline for about 1.8 s, what are the magnitude and direction of the person's average acceleration during this time interval?

N2B.7 A spaceship accelerates from rest with a constant average acceleration of 15 m/s² (about the maximum that a human being can tolerate indefinitely). About how long will it take the ship to reach one-half the speed of light? (*Hint:* 1 Ms = 10^6 s $\approx$ 12 days.)

N2B.8 Consider the unfinished motion diagram shown. What is the direction of the object's acceleration at the instant that it passes point 4? Show how you arrived at your answer.

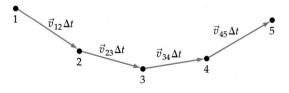

N2B.9 Consider the unfinished motion diagram shown. What is the general direction of the object's acceleration? Explain your reasoning.

N2B.10 A jet plane flies at a constant speed of 120 m/s (260 mi/h) clockwise in a circular holding pattern of radius 8.0 km. What are the magnitude and direction of the plane's acceleration at an instant when it is traveling due east?

Synthetic

N2S.1 An object's position as a function of time is given by $\vec{r}(t) = [c(bt + 1)^{-1}, cbt, at^2]$, where a, b, and c are constants.

    (a) What are the SI units of a, b, and c?

    (b) Find an expression for the object's *speed* as a function of time.

N2S.2 Between $t = 3.0$ s and $t = 4.0$ s, an object moving along a straight line moves a distance of 2.0 m in the $+x$ direction. Between $t = 4.0$ s and $t = 5.0$ s, the object moves a distance of 4.0 m in this same direction. *Estimate* the object's instantaneous speed at $t = 4.0$ s, and explain how you arrived at this estimate and *why* it is an estimate.

N2S.3 A certain object's velocity vector has components $v_x(t) = bt^2 + c$, $v_y(t) = qt$, and $v_z(t) = 0$, where $b = 10$ m/s³, $c = 5$ m/s, and $q = -2.0$ m/s². What is the magnitude of the object's acceleration at time $t = 0$? At time $t = 3.0$ s?

N2S.4 A spaceship initially traveling at a speed of $0.1c$ (where $c = 3.0 \times 10^8$ m/s = the speed of light) decreases its speed smoothly to zero in 30 min as it approaches a space station. Assuming that the spaceship travels in a straight line, what is the magnitude of its average acceleration during this time interval? Compare to 60 m/s², which is roughly the maximum acceleration that a human being can tolerate for a short period of time.

N2S.5 A car traveling at a constant speed of 20 m/s (about 44 mi/h) that is initially traveling due northwest rounds a corner so that after 10 s, the car is traveling due northeast. What are the magnitude and direction of the car's average acceleration during this interval of time? (*Hint:* It really helps if you draw a picture.)

N2S.6 Figure N2.12 shows a set of dots indicating a certain object's position every 0.10 s. After tracing these dots to your own sheet of paper, draw a complete motion diagram for this object and compute the magnitude of its acceleration at point 3.

N2S.7 Figure N2.13 shows a set of dots indicating a certain object's position every 0.05 s. After tracing these

•
1

•
2

•
3

•
4

• •
5 6

Figure N2.12
(For problem N2S.6.)

•
4

•
3

•
5

• 2

•
6

1 •

Figure N2.13
(For problem N2S.7.)

dots to your own sheet of paper, draw a complete motion diagram for this object and compute the magnitude of its acceleration at point 4.

N2S.8   Figure N2.14 shows a set of dots indicating a certain object's position every 0.10 s (note that 1 cm on this diagram corresponds to 0.1 m of actual distance). After tracing these dots to your own sheet of paper, draw a complete motion diagram for this object and compute the magnitude of its acceleration at points 2, 3, and 4.

N2S.9   Consider the motion of an object swinging back and forth on the end of a string. Draw a qualitatively accurate motion diagram (with at least seven position points) of a single swing (left to right) of such a pendulum.

N2S.10   Draw a qualitatively accurate motion diagram of a ball thrown vertically into the air. (Your diagram should include at least seven position points.)

Rich-Context

N2R.1   Imagine that you are driving on a large, dark parking lot in the middle of the night. Your headlights suddenly illuminate a wall directly ahead of you that is perpendicular to your direction of travel and stretching away on both sides as far as you can see. To avoid hitting the wall, is it better to turn your car to the right or left without braking or to brake as hard as you can while moving in a straight line toward the wall? (*Hints:* There is a limit to the force the interaction between your car's tires and the road can exert on your car before you begin to skid. Also remember Newton's second law.)

Advanced

N2A.1   (Do this problem only if you are familiar with partial derivatives.) One can show that Newton's second law follows from the time derivative of the law of conservation of energy. For the sake of simplicity, we will assume the particles are constrained to move in one dimension, but it is fairly easy to generalize the proof to three dimensions. Assume that two particles participate in a long-range interaction that can be described by some potential energy function $V(r)$, where $r = x_2 - x_1$ and x_1 and x_2 are the x-positions of both particles (we are assuming for the sake of argument that $x_1 > x_2$).

•
2

•
3

•
1

|— 0.1 m —|

•
4

•
5

Figure N2.14
(For problem N2S.8.)

(a) Show that we can write the time derivative of the particles' kinetic energies as

$$\frac{dK_1}{dt} = m_1 a_{1x} v_{1x} \quad \text{and} \quad \frac{dK_2}{dt} = m_2 a_{2x} v_{2x}$$
(N2.20)

(b) Show that the time derivative of the potential energy can be written as

$$\frac{dV}{dt} = -F_{1x} v_{1x} + (-F_{2x} v_{2x})$$
(N2.21)

if we define

$$F_{1x} = -\frac{\partial V}{\partial x_1} \quad \text{and} \quad F_{2x} = -\frac{\partial V}{\partial x_2}$$
(N2.22)

(see equation C11.3).

(c) Use these results to compute the time derivative of the law of conservation of energy, and argue that this equation will be satisfied for all possible particle velocities if and only if

$$F_{1x} = m_1 a_{1x} \quad \text{and} \quad F_{2x} = m_2 a_{2x}$$
(N2.23)

Since these equations are the one-dimensional expression of Newton's second law for each particle, we see that Newton's second law does indeed follow from the time derivative of the law of conservation of energy!

ANSWERS TO EXERCISES

N2X.1 $d\vec{r}/dt = [12 \text{ m/s}, 0, 2.5 \text{ m/s}]$

N2X.2 At an arbitrary time t, the components of this object's velocity are

$$\vec{v} = \begin{bmatrix} dx/dt \\ dy/dt \\ dz/dt \end{bmatrix} = \begin{bmatrix} 2at \\ -c \\ -2at + c \end{bmatrix}$$
(N2.24)

At $t = 0$, the velocity is $\vec{v}(0) = [0, -4 \text{ m/s}, 4 \text{ m/s}]$, and the speed is $v = 5.7 \text{ m/s}$. At $t = 2.0$ s, the velocity is $\vec{v}(2\text{ s}) = [6 \text{ m/s}, -4 \text{ m/s}, -2 \text{ m/s}]$, and the speed is $v = 7.5 \text{ m/s}$.

N2X.3 $a_x = dv_x/dt = 0$, $a_y = dv_y/dt = 0$, and $a_z = dv_z/dt = b$. So $a_x = a_y = 0$ and $a_z = b = -10 \text{ m/s}^2$ at both $t = 0$ and $t = 2$ s.

N2X.4 24 m/s

N2X.5 The ball's change in velocity during this interval is 30 m/s downward, so the ball's average acceleration in this case must be 10 m/s^2 downward.

N2X.6 The motion diagram for the ball is shown below. As in the case of the ball rolling up the incline, we have to separate the downward and upward portions of the motion. The ball falls with increasing speed toward the ground, suddenly and violently

(The colored arrows depict $\vec{a}\,\Delta t^2$.)

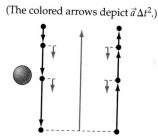

reverses direction, and then moves upward with *decreasing* speed.

N2X.7 The motion diagram looks like this.

I have only drawn three position points here: five would probably be better (but the diagram would have to be bigger). The direction of the car's velocity is changing, so the car is indeed accelerating, at least at the crest of the hill.

N2X.8 I measure the distance between points 2 and 3 to be about 2.95 cm, so $v_{23} \approx (2.95 \text{ cm})/(0.05 \text{ s}) = 59$ cm/s. I measure the distance between points 3 and 4 to be 2.3 cm, so the *difference* in length between the arrows representing $\vec{v}_{23}\,\Delta t$ and $\vec{v}_{34}\,\Delta t$ is 0.65 cm. This would be the length of the acceleration arrow $\vec{a}_3\,\Delta t$ that we would draw for point 3, so $a_3 = (0.65 \text{ cm})/(0.05 \text{ s})^2 = 260 \text{ cm/s}^2$.

N2X.9 I find the length of the arrow between points 2 and 3 to be 2.2 cm, which scales up to $(2.2 \text{ cm})(5 \text{ m/cm}) = 11$ m in real space. Therefore, $v_{23} = (11 \text{ m})/(0.50 \text{ s}) = 22$ m/s. The difference between the lengths of the arrows for $\vec{v}_{12}\,\Delta t$ and $\vec{v}_{23}\,\Delta t$ is $15 \text{ m} - 11 \text{ m} = 4.0$ m in real space, so the arrow representing the *change* in the object's velocity (and thus the acceleration) will have this length in real space. Thus $a_2 = (4.0 \text{ m})/(0.50 \text{ s})^2 = 16 \text{ m/s}^2$.

N2X.10 The acceleration arrow should indeed point toward the circle's center when it is moved back to point 3.

N2X.11 $\text{mag}(\vec{a}) = v^2/R = (20 \text{ m/s})^2/600 \text{ m} \approx 0.67 \text{ m/s}^2$. When the car is traveling north, the center of the circle is to the west, so $\vec{a}$ is 0.67 m/s^2 west.

N3

Forces from Motion

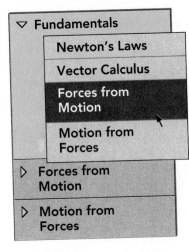

- ▽ Fundamentals
  - Newton's Laws
  - Vector Calculus
  - Forces from Motion
  - Motion from Forces
- ▷ Forces from Motion
- ▷ Motion from Forces

Chapter Overview

Introduction

There are two fundamentally different ways to use Newton's second law. If we know how an object moves (and therefore what its acceleration is), we can use Newton's second law to determine things about the forces acting on an object. On the other hand, if we know the forces that act on an object, we can use Newton's second law to determine its acceleration and thus how it moves. In this chapter, we will explore qualitative examples of the first approach so as to lay solid conceptual foundations for the deeper study of that approach in chapters N5 through N9. Chapter N4 will similarly introduce the second approach, as preparation for chapters N10 through N13.

Section N3.1: The Kinematic Chain

According to chapter N2, an object's position, velocity, and acceleration are linked in the following chainlike relationship:

$$\vec{r}(t) \ \begin{array}{c}\text{time}\\ \text{derivative}\end{array} \longrightarrow \ \vec{v}(t) \ \begin{array}{c}\text{time}\\ \text{derivative}\end{array} \longrightarrow \ \vec{a}(t) \qquad (N3.1)$$

We call this the **kinematic chain** (note that **kinematics** is the mathematical study of motion).

Equation N3.1 implies that if we know either $\vec{r}(t)$ or $\vec{v}(t)$, we can determine the object's acceleration. We can then use Newton's second law to determine the net force acting on the object, which in turn allows us to infer things about the characteristics of the individual forces acting on that object.

Section N3.2: Net-Force Diagrams

A **net-force diagram** is a tool for clearly displaying the net force on an object. To construct a net-force diagram, do the following:

1. Construct a free-body diagram.
2. Copy the force arrows from the diagram in sequence (i.e., place the tail of each new arrow at the tip of the previous one).
3. Draw a double-line arrow to represent the vector sum $\equiv \vec{F}_{\text{net}}$.

Section N3.3: Qualitative Examples

To solve a qualitative motion-to-force problem, you will generally do the following:

1. Determine the object's acceleration (perhaps by using a motion diagram).
2. Draw a free-body diagram and a net-force diagram.
3. Adjust the forces in these diagrams until the net force points in the direction of the object's known acceleration.

The examples in this section illustrate that (1) $\text{mag}(\vec{F}_{\text{N,tot}}) = \text{mag}(\vec{F}_g)$ for an object on a flat, level surface, but (2) this is *not* true if the object is not on a level surface or has a nonzero vertical component of acceleration. The examples also illustrate how static friction forces make it possible for a car to speed up, slow down, round corners, or maintain its speed against an opposing drag force.

Section N3.4:   Third-Law and Second-Law Pairs

Newton's third law states:

> When objects A and B interact, the force the interaction exerts on A is equal in magnitude and opposite in direction to the force it exerts on B.

This implies that any pair of forces connected by Newton's third law must have the following characteristics:

1. Each force in the pair must act on *different* objects.
2. Each must reflect the *same interaction* between those objects.

We will call such a pair of forces **third-law partners.**

By contrast, two opposing forces that we know are equal in magnitude because of Newton's *second* law **(second-law partners)** act on the *same* object and reflect *different* interactions. Note that the second law implies that two opposing forces on an object are equal *only* if (1) no other forces have components in the same direction and (2) the object is not accelerating in that direction. Third-law partners, on the other hand, are *always* equal in magnitude, no matter how the interacting objects move.

Section N3.5:   Coupling Forces

This section illustrates how we can use Newton's second and third laws together to determine the magnitudes of forces produced by an interaction that couples two objects together.

Section N3.6:   Graphs of One-Dimensional Motion

When an object moves along a straight line (which we can define to be the x axis), its position, velocity, and acceleration are completely described by its x components $x(t)$, $v_x(t)$, and $a_x(t)$ (which are simple signed numbers). A graph of $a_x(t)$ usefully displays how the net force on an object changes with time.

Conventionally, we stack such graphs vertically so that we always place a graph of $x(t)$ above a graph of $v_x(t)$, which in turn we always place above a graph of $a_x(t)$. Given this convention, we can always construct a given graph from the graph above it by remembering that at any given instant of time, the *slope* on the upper graph in the pair corresponds to the value displayed on the lower graph: *the slope above equals the value below.*

Section N3.7:   A Few Quantitative Examples

The *qualitative* tools we discuss in this chapter make it easier to analyze *quantitative* problems correctly, as the examples in these sections illustrate. If an object moves along the x axis, then $\mathrm{mag}(\vec{a}_{\Delta t}) = |a_{\Delta t,x}| = |\Delta v_x/\Delta t|$ provides a handy estimate of the object's instantaneous acceleration during a time interval.

N3.1   The Kinematic Chain

As we saw in chapter N2, an object's position $\vec{r}(t)$, its velocity $\vec{v}(t)$, and its acceleration $\vec{a}(t)$ as a function of time are connected by the following chain of derivative relationships:

The kinematic chain

$$\vec{r}(t) \quad \left[\begin{array}{c} \text{time} \\ \text{derivative} \end{array} \right] \rightarrow \vec{v}(t) \quad \left[\begin{array}{c} \text{time} \\ \text{derivative} \end{array} \right] \rightarrow \vec{a}(t) \qquad (N3.1)$$

We will call this chain of relationships the **kinematic chain** (note that **kinematics** is the mathematical study of motion). The chain of vector derivatives in equation N3.1 is equivalent to three separate chains of component derivatives:

$$x(t) \quad \left[\begin{array}{c} \text{time} \\ \text{derivative} \end{array} \right] \rightarrow v_x(t) = \frac{dx}{dt} \quad \left[\begin{array}{c} \text{time} \\ \text{derivative} \end{array} \right] \rightarrow a_x(t) = \frac{dv_x}{dt} \quad (N3.2a)$$

$$y(t) \quad \left[\begin{array}{c} \text{time} \\ \text{derivative} \end{array} \right] \rightarrow v_y(t) = \frac{dy}{dt} \quad \left[\begin{array}{c} \text{time} \\ \text{derivative} \end{array} \right] \rightarrow a_y(t) = \frac{dv_y}{dt} \quad (N3.2b)$$

$$z(t) \quad \left[\begin{array}{c} \text{time} \\ \text{derivative} \end{array} \right] \rightarrow v_z(t) = \frac{dz}{dt} \quad \left[\begin{array}{c} \text{time} \\ \text{derivative} \end{array} \right] \rightarrow a_z(t) = \frac{dv_z}{dt} \quad (N3.2c)$$

Finding forces from motion: an overview

In many situations, there are one or more forces acting on an object whose magnitude and/or direction we do not know. The kinematic chain implies that if we know how an object *moves* (i.e., we know either its position or its velocity as a function of time), then we can compute the object's *acceleration* as a function of time. Newton's second law in turn links an object's acceleration to the *net force* acting on the object, and knowing the net force often is sufficient information to determine the unknown forces. This is how we determine *forces from motion*. We will talk about this in fairly general terms in the remainder of this chapter and in much greater detail in chapters N5 through N9.

Exercise N3X.1

Say that we know that $x(t) = \frac{1}{2}bt^2 + c$, where b and c are constants. What are the units of b and c? What are $v_x(t)$ and $a_x(t)$ in this case? Do the units come out right for these quantities?

N3.2   Net-Force Diagrams

In chapter N1, we learned how to draw a *free-body diagram* of an object. Free-body diagrams are very useful for helping us to recognize what forces might act on an object and to display these forces in a vivid way. In situations where we are trying to determine force magnitudes more precisely, it helps to *supplement* a free-body diagram with a different kind of diagram that we call a **net-force diagram** which more vividly displays the quantitative relationship between these forces.

How to draw a net-force diagram

Newton's second law tells us that the acceleration of an object's center of mass is determined by the *net external force* acting on the object, that is, the vector sum of the external forces acting on the object. The most important

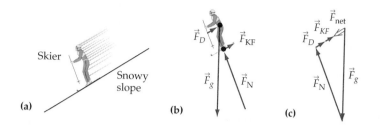

Figure N3.1
(a) A skier sliding down a snowy slope. (b) A free-body diagram of the same skier. (c) A net-force diagram showing the vector sum of the forces acting on the skier.

purpose of a net-force diagram is to display this vector sum. To draw a net-force diagram, do the following:

1. *Draw a free-body diagram first.* A net-force diagram is essentially a re-arrangement of a free-body diagram to make the quantitative relationships between forces clearer. A net-force diagram rarely makes sense without an accompanying free-body diagram.
2. *Copy the force arrows* (and labels) from the free-body diagram to the net-force diagram, *arranging them in sequence* by drawing each new force arrow with its tail end starting at the tip of the previously drawn arrow. If two force arrows point in opposite directions, draw them right next to each other (displaced by a small distance).
3. *Draw the arrow that represents the vector sum* of these force arrows. To distinguish it visually from the other arrows, make its body a double line instead of a single line. If the force arrows add up to zero, you can simply write $\vec{F}_{net} = 0$.

Figure N3.1 displays both a free-body diagram and a net-force diagram for a skier sliding down a hill. Note that if the forces on the skier really have the magnitudes shown, the net force on the skier is a vector pointing down the incline, and thus the skier will be *accelerating* down the incline.

N3.3   Qualitative Examples

In this section, we will work a number of qualitative example problems that illustrate how we can use motion diagrams, free-body diagrams, net-force diagrams, and Newton's second law to determine the magnitude, direction, and/or even the very existence of certain individual forces acting on an object. The basic process for doing such qualitative problems is as follows:

1. *Determine the direction of the object's acceleration (or determine that it is zero),* using information from the problem statement. You often will find a motion diagram useful in establishing the direction of the acceleration when the acceleration is nonzero.
2. *Draw a free-body diagram* of the object, displaying the directions and possible magnitudes of the forces that you know act on the object.
3. *Draw a net-force diagram,* using these force arrows. If the net force arrow that you construct does not have the same direction as the acceleration found in the first step (as required by Newton's second law), adjust the magnitudes of the force arrows and/or add new forces until the net force and acceleration are consistent. Be sure to update your free-body diagram to keep it consistent with the net-force diagram.

An outline of the steps that we will follow in the following examples

In this section, we will apply this process explicitly and in detail to make some basic qualitative statements about the forces that must be acting on an object to make it move in the manner stated. When we do more quantitative problems of this type in chapters N5 through N8, we will have to follow this same process (in more compressed form) before we can do the calculations. What we are doing here is thus necessary background for doing those problems correctly.

Example N3.1

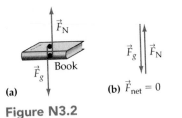

(a)

(b) $\vec{F}_{net} = 0$

Figure N3.2
(a) A free-body diagram of the book. (b) A net-force diagram for the book.

Problem   A book sits at rest on a table. Does the contact interaction between the book and the table exert a force on the book? If so, what is the magnitude of this force compared to that of the gravitational force $\vec{F}_g$ acting on the book?

Model/Solution   A book sitting at rest has a constant velocity of zero, so it has no acceleration. Newton's second law then implies that the net force exerted on the book must be zero. Since gravity clearly exerts a downward gravitational force $\vec{F}_g$ on the book, *something* must be exerting an upward force on the book to cancel the gravitational force. This force must be exerted by the *table*, since it is the only substantial thing in contact with the book. (The surrounding air does not significantly support the book, as you can vividly illustrate by removing the table suddenly.) This force must be equal and opposite to $\vec{F}_g$ to cancel it. This means that the force exerted by the contact interaction must be perpendicular to the book-table interface, so it is a *normal* force $\vec{F}_N$ (see figure N3.2).

Example N3.2

(a)

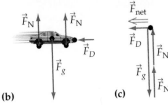

(b)   **(c)**

Figure N3.3
(a) A moving car. (b) A first draft for a free-body diagram for the car. (c) The net-force diagram shows that if the free-body diagram is correct, there will be a nonzero net force on the car.

Problem   A car travels at a constant velocity along a level, straight highway. Draw an accurate free-body diagram for the car. (Do not ignore air drag.)

Model/Solution   As usual, the car experiences a gravitational force $\vec{F}_g$ directed toward the center of the earth. The car touches only the road and the air that rushes by. The air exerts a rearward drag force $\vec{F}_D$ on the car. The road plausibly also exerts an upward normal force $\vec{F}_N$ on each of the car's tires for reasons similar to those discussed in Example N3.1. A plausible first guess for a free-body diagram is shown in figure N3.3b. (This diagram assumes that the normal forces on each tire are equal in magnitude and direction.)

However, we are told that the car is moving at a constant velocity, which means that its acceleration is zero. Therefore, by Newton's second law, the net force on the car must also be zero. However, the net-force diagram of figure N3.3c makes it clear that the forces drawn in figure N3.3b do not add up to zero. Therefore, some other force must be acting on the car in the forward direction to cancel the rearward drag force. What is this force?

A natural first guess would be an "engine force." The engine *must* be involved somehow, since without the engine, the car would not be able to maintain its forward velocity. However, an "engine force" does not appear in our list of categories in chapter N1. Moreover, the force that moves the car forward must be an *external* force, and there is no way that an engine that is *inside* the car could exert an external force on the car. A second guess might be that the force is a *thrust* force, but the engine (in a normal car at least) does not propel the car forward by pushing a fluid backward.

Example N3.5

Problem   A car travels around a bend in the road at a constant speed. Draw a free-body diagram of the car. (Do *not* ignore air resistance.)

Model/Solution   The situation is almost exactly the same as in example N3.4, except that the car is going around a horizontal bend instead of over a hill, as shown in figure N3.7a. The car thus stays in the horizontal plane (and so does not accelerate vertically), but as the motion diagram in figure N3.7b shows, the car *is* accelerating *horizontally* (toward the center of the circle). According to Newton's second law, this means that the net force on the car has to be toward the center of the circle, so there has to be *some* force on the car that has a component toward the circle's center (perpendicular to the car's motion!). Where does this force come from?

The clues are that (1) the road is the only solid thing that touches the car and (2) if the road were very slippery, the car could not negotiate the curve. Thus the *road* must exert the force on the car (through its tires). Moreover, since this force acts parallel to the tire-road interface, and the tires are not sliding relative to the road, it must be a *static friction force*.

As shown in figure N3.7c, the *total* static friction force exerted by the road on the tires must therefore be tipped at an angle, so that it has both a forward component and a component toward the center of the curve. (Note that in the top view of this diagram, I have added up the static friction force exerted on each tire and simply presented the *total* static friction force as if it acted on the car's center.) The net-force diagram in figure N3.7d shows that the vector sum of all the forces on the car will then (and only then) yield a net force directed horizontally toward the center of the road's curve, which is what is necessary to be consistent with the car's observed motion around the curve.

Note that in this case, we need two different views to completely illustrate this intrinsically three-dimensional free-body diagram!

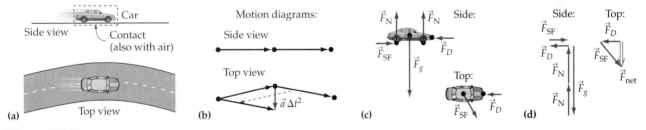

Figure N3.7
(a) Side and top views of the car in its context. (b) Side and top views of the car's motion diagram. (c) Side and top views of a free-body diagram for the car. (d) Side and top views of the net-force diagram for the car. Note how the interaction with the road must exert an angled total static friction force to keep the car in its circular path.

Exercise N3X.2

Draw free-body and net-force diagrams for a box sitting at rest on an incline.

Exercise N3X.3

Draw motion, free-body, and net-force diagrams for a car skidding toward a stop (with its wheels locked).

Exercise N3X.4

Draw motion, free-body, and net-force diagrams for an upward-bound elevator just as it begins to pull away from the ground floor.

N3.4   Third-Law and Second-Law Pairs

We have seen now a number of situations where we have determined that certain pairs of opposing forces acting on an object (e.g., the gravitational and normal force) must have equal magnitudes. Newton's third law talks about forces that are equal in magnitude but opposite in direction. How are the cases we have considered so far related to Newton's third law?

The short answer is that the cases we have considered so far of equal and opposite force pairs acting on an object have *nothing* to do with Newton's third law: it is Newton's *second* law that tells us that these opposing forces have equal magnitude. The purpose of this section is to clarify when forces are related by Newton's third law and when they are related by Newton's second law.

Newton's *third* law asserts that

> When objects A and B interact, the force that the interaction exerts on A is equal in magnitude and opposite in direction to the force that it exerts on B.

This law is *always* true because all known interactions simply transfer momentum from one object to the other (see the discussion in chapter C3).

This statement implies that if a given pair of forces is connected by Newton's third law, they must have the following characteristics:

1. Each force in the pair must act on a *different* object.
2. Both forces must reflect the *same interaction* between the objects.

We will call a pair of forces fitting these conditions **third-law partners.** For example, the gravitational forces that two objects interacting gravitationally exert on each other are third-law partners.

Since third-law partners must act on *different* objects, it follows that two forces that act on the same object (and thus appear on the same free-body diagram) *cannot* be third-law partners. For example, the normal and gravitational forces on the book in figure N3.2 are indeed equal in magnitude and opposite in direction; but they both act on the *same* object and they do not reflect the same interaction (one reflects a gravitational interaction and the other a contact interaction). So they *cannot* possibly be third-law partners.

It is in fact Newton's *second* law that tells us that the particular forces in figure N3.2 have equal magnitudes: since the book does not accelerate vertically, the net vertical force on the book must be zero; and since these two forces are the only ones acting in the vertical direction, they must be equal and opposite, so that they cancel. In general, a pair of opposing forces whose magnitudes are constrained to be equal by Newton's *second* law has the following characteristics:

1. Both forces act on the *same* object.
2. The forces oppose each other along a certain axis.
3. No other forces have any component along that axis.
4. The component of the object's acceleration along that axis is *zero*.

Newton's third law

Characteristics of third-law partners

Forces appearing on a single object's free-body diagram *cannot* be third-law partners

Characteristics of second-law partners

If these things are true, then Newton's second law implies that the net force along that axis must be zero, which implies that the two forces must be equal in magnitude (so that they cancel). Such forces are **second-law partners.**

Exercise N3X.5

Are the forces in each of the following pairs third-law partners or second-law partners?

(a) The drag force and static friction force in figure N3.6.
(b) A helicopter's rotor exerts a downward force on the air; the air exerts an upward opposing force on the helicopter.
(c) A car's tires push backward on the road; the road pushes forward on the tires.
(d) You exert an upward force on a ball sitting on your hand; gravity exerts an opposing downward force on the ball.

N3.5   Coupling Forces

Example N3.6 illustrates how we can use Newton's second and third laws *together* to determine the magnitude and direction of the forces that two coupled objects exert on each other.

Example N3.6

Problem   A small car of mass m_1 is pushing a disabled truck of mass m_2 along a level road in such a way that both accelerate forward with a certain acceleration $\vec{a}$. Assuming that both the car and the truck are moving slowly enough that frictional forces are negligible, draw free-body diagrams of both the car and truck, describe which pair (or pairs) of forces on these diagrams are linked by Newton's third law, and determine the magnitude of the force the truck exerts on the car.

Model   Figure N3.8a shows the situation described in the problem. Figure N3.8b shows free-body diagrams for both the car and the truck. The car must exert a forward contact force on the truck to accelerate the latter. Since this force acts perpendicular to the bumper of the car that is exerting it, it is a

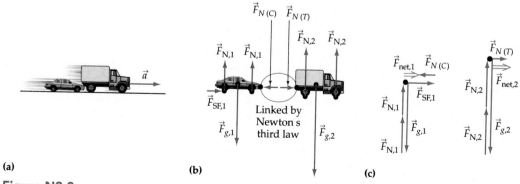

Figure N3.8

(a) The car and truck in context. (b) Free-body diagrams for the car and truck. (c) Net-force diagrams for the car and truck.

normal force, so I have labeled this force $\vec{F}_{N(C)}$ (the subscript indicates that it is exerted *by* the car) to distinguish it from the normal forces that the road exerts on the tires of each. Newton's *third* law then implies that the truck must exert an opposing force of equal magnitude on the car, which I have given the label $\vec{F}_{N(T)}$. It is these forces that are equal by Newton's third law.

The net force on the car must point to the right if the car is to accelerate in that direction, so the contact interaction between the car and the road must exert a forward static friction force $\vec{F}_{SF,1}$ on the car that exceeds the rearward force applied by the truck. This is shown in both the free-body and net-force diagrams for the car in figure N3.8b and N3.8c.

Solution   The net-force diagram for the truck in figure N3.8c shows that the net force on the truck is simply equal to the force $\vec{F}_{N(C)}$ the car exerts on it, so by Newton's *second* law we must have

$$\vec{F}_{N(C)} = \vec{F}_{net,2} = m_2\vec{a} \tag{N3.3}$$

Newton's *third* law, in turn, implies that the forces which the car and truck exert on each other must be equal in magnitude, so

$$F_{N(T)} = F_{N(C)} = m_2 a \tag{N3.4}$$

(where I have taken the magnitude of equation N3.3 to arrive at the second equality). If we know the truck's mass and the magnitude of its acceleration, therefore, we can determine the magnitudes of both the force that the car exerts on the truck and the force that the truck exerts on the car.

Note that the net force diagram for the car implies that net force on the car has the magnitude $F_{SF,1} - F_{N(T)}$. From this and equation N3.4, you should be able to use Newton's second law to show that

$$F_{SF,1} = (m_1 + m_2)a \tag{N3.5}$$

Note that this is the result we would get if we were to consider the truck and car to be a single unit acted on by the single horizontal external force $\vec{F}_{SF,1}$.

Exercise N3X.6

Verify equation N3.5.

N3.6   Graphs of One-Dimensional Motion

Free-body diagrams help us see how force and motion are related at an instant of time: they are like snapshots of the force on an object. To understand more about how position, velocity, acceleration, and force are related in time-*dependent* situations, it helps to draw graphs of these quantities versus time.

In the case of one-dimensional motion, vectors are represented by signed numbers

In cases where an object's motion is confined to a line, we can generally choose to orient our reference frame so that the line of motion coincides with the x axis (or we might choose the z axis if the motion is vertical). If we do this, then only the x component of an object's position vector $\vec{r}(t)$ is ever nonzero, and this in turn means that only the x component of its velocity vector $\vec{v}(t)$ is nonzero:

$$\vec{r}(t) = \begin{bmatrix} x(t) \\ 0 \\ 0 \end{bmatrix} \quad \Rightarrow \quad \vec{v}(t) \equiv \frac{d\vec{r}}{dt} = \begin{bmatrix} dx/dt \\ d(0)/dt \\ d(0)/dt \end{bmatrix} = \begin{bmatrix} dx/dt \\ 0 \\ 0 \end{bmatrix} \tag{N3.6}$$

Similarly, the object's acceleration $\vec{a}(t)$ only has one nonzero component.

$$\vec{a}(t) \equiv \frac{d\vec{v}}{dt} = \begin{bmatrix} dv_x/dt \\ 0 \\ 0 \end{bmatrix} \qquad (N3.7)$$

Therefore, in this very special case of one-dimensional motion (and *only* in this case), we can represent an object's position, velocity, and acceleration vectors each by a *single signed number* considered to be a function of time [these numbers are $x(t)$, $v_x(t)$, and $a_x(t)$, respectively]. These signed numbers provide all that we need to know about the vectors $\vec{r}(t)$, $\vec{v}(t)$, and $\vec{a}(t)$ in this case: the *absolute value* of the number is the same as the *magnitude* of the vector [for example, mag$(\vec{a}) = |a_x|$], and the *sign* of the number tells us the *direction* of the corresponding vector (a plus sign indicates that the vector points in the $+x$ direction, whereas a minus sign indicates that it points in the $-x$ direction).

Now, according to equation N3.7, $a_x(t) = dv_x/dt$. This means that (as discussed in appendix NA), the *value* of the object's x-acceleration at every instant of time will correspond to the *slope* of $v_x(t)$ when the latter is plotted versus t on a graph. Therefore, given a graph of $v_x(t)$, in principle we can construct a graph of $a_x(t)$. Figure N3.9a illustrates the process of constructing such a graph.

Similarly, because $v_x = dx/dt$, the *value* of v_x at every instant of time will correspond to the *slope* of a graph of $x(t)$ versus t (see figure N3.9b). It is conventional when drawing pairs of graphs like these to put the graph of $x(t)$ above the paired graph of $v_x(t)$ and a graph of $v_x(t)$ above a paired graph of $a_x(t)$ so that the slope of the upper graph always corresponds to the value on the lower graph (always remember: *slope* above equals *value* below).

In this text, to keep the distinction between the components and the magnitude of a vector clear, I will *always* use the symbols x, v_x, and a_x to refer to these signed numbers (because they really are just components of the corresponding vectors) and *always* use r, v, and a to refer to the (nonnegative) magnitudes of the corresponding vectors. I strongly advise you to do the same: blurring the distinction between components and magnitudes is a very common source of sign errors.

According to Newton's second law, an object's acceleration is proportional to the net force acting on the object, so a graph of $a_x(t)$ essentially tells us how $F_{\text{net},x}$ depends on time. This can give us insight into the forces acting on the object, as example N3.7 illustrates.

Constructing a graph of $a_x(t)$ from one of $v_x(t)$

Constructing a graph of $v_x(t)$ from one of $x(t)$

Keep components and magnitudes distinct!

Inferring forces

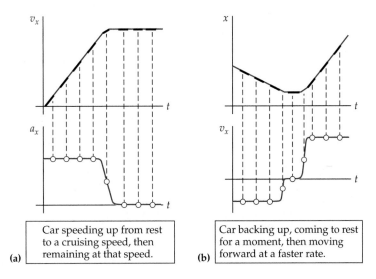

(a) Car speeding up from rest to a cruising speed, then remaining at that speed.

(b) Car backing up, coming to rest for a moment, then moving forward at a faster rate.

Figure N3.9

(a) How to construct a graph of $a_x(t)$ from a graph of $v_x(t)$ for a car speeding up to and then holding at a certain cruising speed. Each little black line segment on the top graph indicates the *slope* of that graph of $v_x(t)$ at a certain instant, while the circle on the bottom graph indicates the value of a_x at that instant. Note that at first the car's velocity is steadily increasing, so its slope is a constant positive number. When the car reaches its cruising speed, the slope quickly falls to zero and remains zero. (b) How to construct a graph of $v_x(t)$ from a graph of $x(t)$ using the same technique.

Example N3.7

Problem   Imagine that you hold a basketball some distance above the floor and then release it from rest. It falls toward the floor, hits the floor, and then rebounds upward at about the same speed. Draw graphs of the ball's vertical velocity and acceleration components as functions of time.

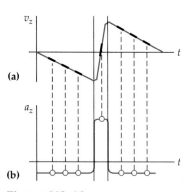

Figure N3.10
(a) A graph of $v_z(t)$ and (b) a graph of $a_z(t)$ for a basketball dropped on the floor. The ball is in contact with the floor during the time interval bracketed by the thin vertical lines.

Model/Solution   The first thing that we have to do is to define our reference frame. Let us define our z axis to be vertically upward and take $z = 0$ to correspond to the floor. (In the case of vertical motion, it is more natural to align the z axis with the object's motions than to use the x axis.) Let us also define $t = 0$ to be the instant when the ball is dropped.

According to the description of the situation, the ball is released from rest ($v_z = 0$). As the ball falls, its unopposed gravitational interaction with the earth transfers downward momentum to the ball at a constant rate, and thus the ball's z-velocity should steadily become more negative as time decreases. When the ball hits the floor, its z-velocity rapidly changes from being negative to positive (downward to upward). After the ball leaves the floor, its z-velocity again becomes steadily more negative.

From this information, we can construct the graph of v_z versus t shown in figure N3.10a. Then we can use the methods discussed in this section to construct the graphs of $a_z(t)$ below this graph (figure N3.10b). Note how in each case the slope of the upper graph at a given t is equal to the value of the graph below it (as indicated at selected places).

We see from figure N3.10b that the net force on the ball when it is *not* in contact with the floor must be constant and downward: this is the constant force of gravity. When the ball is in contact with the floor, however, the graph indicates that the net force on the ball must become very large and upward. The gravitational force on the ball always has a constant (downward) value, so we must have a new upward force acting on the ball. This must be a normal force arising from the ball's interaction with the ground. The graph clearly indicates that this force must be larger than the gravitational force, and indeed, the briefer the bounce is, the larger this normal force has to be.

Drawing such sets of graphs is an excellent way to visualize (and thus think more carefully about) the time dependence of these motion functions. *Drawing such graphs is a skill that you should practice and master:* becoming adept at both constructing and interpreting these graphs is one of the surest ways to gain a firm understanding of the mathematical relationship between position, velocity, and acceleration. Note that it is not critical that your graphs be *quantitatively* accurate: the important thing is that they be *qualitatively* accurate.

Exercise N3X.7

How does the net x-force on the car described in figure N3.9a change as time passes? What force provides the forward force on the car, and why does it change with time? (Ignore air resistance.)

Exercise N3X.8

A car traveling along a straight road at a constant speed suddenly brakes to a stop to avoid an animal and remains at rest thereafter. Draw graphs of $v_x(t)$ and $a_x(t)$ for this situation.

N3.7   A Few Quantitative Examples

So far in this chapter, I have focused exclusively on examples in which we *qualitatively* determine the existence and/or direction of forces by analyzing an object's motion. These techniques of qualitative analysis are the bedrock on which we can construct more detailed quantitative analyses: without the guidance that the qualitative tools provide, one can quickly become lost in complicated problems. We will examine a number of progressively more complex problems in chapters N5 through N9, but I would like to close this chapter with a few simple examples of quantitative analysis designed to give you a foretaste of what we can do.

A simple way to get a quick *estimate* of the magnitude of the net force on an object that is accelerating along a straight line during a certain time interval Δt is to assume that the object's instantaneous acceleration during the interval is approximately equal to its average acceleration over that interval. If the motion is along the x direction, for example, this assumption means that

How to *estimate* the magnitude of the net force on an object moving in one dimension

$$a \approx a_{\Delta t} = |a_{\Delta t,x}| = \left| \frac{\Delta v_x}{\Delta t} \right| \qquad (N3.8)$$

where $\Delta v_x \equiv v_{f,x} - v_{i,x}$ is the change in the object's x-velocity during the interval. This is a useful (if often crude) first model for a variety of situations. As example N3.8 illustrates, once we know the magnitude of the acceleration, we can use the magnitude of Newton's second law to determine the magnitude of the net force.

Example N3.8

Problem   Imagine that in example N3.3 the box being carried by the pickup truck has a mass $m = 10$ kg. If the truck goes from 0 to 60 mi/h in 10 s, what is the magnitude of the average static friction force exerted on the box? How does this compare to the magnitude of the object's weight?

Model   We can see from our previous qualitative analysis (see particularly the net-force diagram in figure N3.5c) that the net force on the box in this case is equal to the static friction force. Let's define the $+x$ direction to be the direction of the truck's motion. The truck is initially at rest ($v_{i,x} = 0$) and is moving with x-velocity $v_{f,x} = +60$ mi/h: if the box remains at rest with respect to the truck, this will be its initial and final x-velocities as well.

Solution   The magnitude of Newton's second law applied to the box then implies that

$$F_{SF} = F_{net} = ma = m|a_x| \approx m\frac{|\Delta v_x|}{\Delta t}$$

$$= \frac{10\text{ kg}}{10\text{ s}}|+60\text{ mi/h}| \left(\frac{1\text{ m/s}}{2.24\text{ mi/h}} \right)\left(\frac{1\text{ N}}{1\text{ kg·m/s}^2} \right) = 27\text{ N} \qquad (N3.9)$$

If an object has mass m, the magnitude of its weight is $F_g = mg$, which in this case is $F_g = (10\text{ kg})(9.8\text{ m/s}^2) = 98\text{ kg·m/s}^2 = 98$ N. So the magnitude of the static friction force acting on the box in this case is a bit less than one-third of the magnitude of its weight.

Exercise N3X.9

If the box's interaction with the truck bed can exert no more than 45 N on the box before the box begins to slip, what is the minimum time that the driver should allow to go from 0 to 60 mi/h?

Exercise N3X.10

If the basketball in example N3.7 is traveling at 3.0 m/s when it hits the ground and the bounce lasts 0.10 s, how does the magnitude of the normal force exerted by the floor compare to that of the basketball's weight?

An example involving circular motion

When an object is moving in a circular path of radius R at a constant speed v, we saw in chapter N2 that the magnitude of its acceleration is $a = v^2/R$. As example N3.9 illustrates, we can use this in conjunction with Newton's second law to determine the magnitude of forces acting on the object.

Example N3.9

Problem Imagine that in example N3.4 the car going over the hill is traveling at a constant speed of 30 m/s and (near the top of the hill, anyway) the road curves vertically as if it were a part of a circle whose radius is 450 m. What is the ratio of the magnitudes of the total normal force and gravitational forces acting on the car in this situation?

Model The net-force diagram in figure N3.6d makes it clear that the magnitude of the net force acting on the car is $F_{net} = F_g - F_{N,tot} = mg - F_{N,tot}$. This means that the total normal force acting on the car is $F_{N,tot} = mg - F_{net}$. The magnitude of Newton's second law implies that $F_{net} = ma$. The car's acceleration in this case is given by $a = v^2/R$.

Solution Putting all this together, we find that the requested ratio is equal to

$$\frac{F_{N,tot}}{F_g} = \frac{mg - ma}{mg} = 1 - \frac{a}{g} = 1 - \frac{v^2}{Rg} = 1 - \frac{(30 \text{ m/s})^2}{(450 \text{ m})(9.8 \text{ m/s}^2)} = 0.80$$

(N3.10)

So the total normal force acting on the car in this case is only 80% as strong as the gravitational force on the car as it goes over the top of the hill.

TWO-MINUTE PROBLEMS

N3T.1 A car passes a dip in the road, going first down, then up. At the very bottom of the dip, when the car's instantaneous velocity is passing through horizontal, how does the magnitude of the total normal force on the car compare to the magnitude of the car's weight?

A. $F_{N,tot} < F_g$
B. $F_{N,tot} = F_g$
C. $F_{N,tot} > F_g$
D. $F_{N,tot} = 0$
E. We do not have enough information to answer.

N3T.2 A crate sits on the ground. You push as hard as you can on it, but you cannot move it. At any given time when you are pushing, what is the magnitude of the static friction force exerted on the crate by its contact interaction with the ground compared to the magnitude of your push (which is a *normal* force)?

A. $F_{SF} < F_N$
B. $F_{SF} = F_N$
C. $F_{SF} > F_N$
D. $F_{SF} = 0!$
E. We do not have enough information to answer.

N3T.3 A box sits at rest on an inclined plank. How do the magnitudes of the normal force and the gravitational force exerted on the box compare? (*Hint:* Draw a picture!)

A. $F_N < F_g$
B. $F_N = F_g$
C. $F_N > F_g$
D. $F_N = 0!$
E. We do not have enough information to answer.

N3T.4 The drawing below is supposed to be a free-body diagram of a box that sits without slipping on the back of a truck that is moving to the right but is slowing down. Is the diagram correct?

A. Yes.
B. No: $\vec{F}_{SF}$ should point leftward.
C. No: the $\vec{F}_{SF}$ label should be $\vec{F}_{KF}$.
D. No: there should be a leftward drag force.
E. No: F_N should not be equal to F_g.
F. No: there is some other problem (specify).

N3T.5 The drawing shown is supposed to be a free-body diagram of a crate that is being lowered by a crane and is speeding up as it is being lowered. Is the diagram correct? (Ignore air resistance.)

A. Yes.
B. No: $\vec{F}_T$ should be labeled $\vec{F}_N$.
C. No: $\vec{F}_T$ should be equal to F_g.
D. No: $\vec{F}_T$ should be greater than F_g.
E. No: there should be an upward $\vec{F}_D$.
F. No: there is some other problem (specify).

N3T.6 Which of the following are third-law partners? Answer T if the two forces described are third-law partners, F if they are not.

a. A thrust force from its propeller pulls a plane forward; a drag force pushes it backward.
b. A car exerts a forward force on a trailer; the trailer tugs backward on the car.
c. A motorboat propeller pushes backward on the water; the water pushes forward on the propeller.
d. Gravity pulls down on a person sitting in a chair; the chair pushes back up on the person.

N3T.7 A large truck with mass m_1 pushes a disabled small car with mass $m_2 < m_1$, giving it a forward acceleration a. Each vehicle exerts a force on the other as a result of their contact interaction. Which vehicle exerts the *greater* force on the other?

A. The car.
B. The truck.
C. Both forces have the same magnitude.
D. The car does not exert any force on the truck.
E. We do not have enough information to answer.

N3T.8 In the situation described in problem N3T.7, what is the *magnitude* of the force $\vec{F}$ that the car exerts on the truck?

A. $F = m_1 a$
B. $F = m_2 a$
C. $F = (m_1 + m_2)a$
D. $F = 0$
E. $F = -m_2 a$
F. Other (specify)

N3T.9 An object's x-position $x(t)$ is shown in the boxed graph of the following set of graphs. Which of the other graphs in the set most correctly describes its x-velocity?

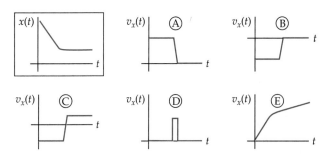

N3T.10 Which graph best describes the x-acceleration of the object described in problem N3T.9?

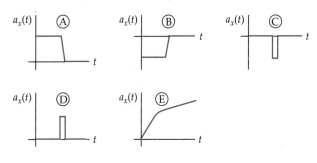

Basic Skills

N3B.1 Draw a motion diagram, a free-body diagram, and a net-force diagram for a moving motorboat whose motor has just run out of gas.

N3B.2 Draw a motion diagram, a free-body diagram, and a net-force diagram for a child bouncing on a trampoline (at the instant that the child is at the lowest part of the bounce).

(See problem N3B.2.)

N3B.3 Draw a motion diagram, a free-body diagram, and a net-force diagram for a box sitting in an elevator whose downward speed is increasing as it begins its descent from the top floor.

N3B.4 Draw a motion diagram, a free-body diagram, and a net-force diagram for a child holding on for dear life (with feet in the air!) to a rapidly spinning merry-go-round.

N3B.5 A car travels at a constant speed through a dip in the road that takes the car first down and then up. Draw a motion diagram, a free-body diagram, and a net-force diagram for the car as it passes the *bottom* of the dip, paying special attention to correctly indicating the relative magnitudes of the vertical forces on the car.

N3B.6 Imagine a person hanging onto the bottom of a helicopter as it accelerates upward. Draw free-body diagrams for both the person and the helicopter. Indicate which pairs of forces (if any) are third-law partners, and explain your reasoning.

N3B.7 Imagine that a person jumps off the floor. Draw free-body diagrams of the earth and the person at some instant while the person is beginning the jump (but has not yet left the floor). Indicate on your diagram which pairs of forces (if any) are third-law partners, and explain your reasoning.

(See problem N3B.6.)

N3B.8 Construct a graph of *x*-acceleration as a function of time for an object whose *x*-velocity is as shown.

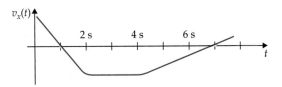

N3B.9 Construct a graph of *x*-velocity as a function of time for an object whose *x*-position is as shown.

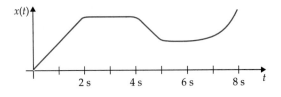

Synthetic

N3S.1 An airplane travels at a constant speed in a horizontal circular path around an airport. Draw a *top-view* motion diagram and a *rear-view* free-body diagram of the plane. (*Hint:* A plane has to "bank," that is, lower the inner wing and raise the outer wing, when it flies in a circular turn to orient the lift force exerted on the plane by the wings away from vertical. Why is this so?)

N3S.2 A box sitting on the floor of a van slides toward the front of the van when the van suddenly brakes to a stop.

(a) Draw a motion diagram (as viewed from the *ground*, not the van), a free-body diagram, and a net force diagram of the box.

(b) Explain qualitatively why the box moves forward relative to the van.

N3S.3 Draw a motion diagram and a free-body diagram for a roller-coaster car at the top of a "loop-de-loop" (i.e., when the car is upside down). Do not ignore air resistance, and assume that the car is moving rapidly enough to remain in firm contact with the rails. In particular, explain why there is no *outward* force on the car.

(See problem N3S.3.)

N3S.4 Imagine that in a movie chase scene, the director wants a 1000-kg car traveling at 15 m/s to run head-on into a brick wall without knocking over the wall. If the wall brings the car to rest in about 0.2 s, roughly what force will the car exert on the wall during the collision? (This calculation would give the director's assistant some advice about how strong the wall has to be.)

N3S.5 Imagine that an unpowered go-cart (with a child rider) traveling at an initial speed of 5 m/s coasts on a level road for about 50 m before coming to rest. If the cart and rider have a mass of 40 kg, about

what force would you have to apply to keep the cart moving at a constant speed? Assume that the friction forces acting on the cart are essentially independent of speed. (*Hint:* Use the concept of work to compute the magnitude of the total friction/drag force acting on the cart, then link this force to your force.)

N3S.6 Imagine that the car in example N3.5 must convert chemical energy to other forms at a rate of 10,000 W to keep moving at a constant speed of 25 m/s. If the car's mass is $m = 1100$ kg, its constant speed is $v = 25$ m/s, and the radius of the curve is $R = 120$ m, what angle does the static friction force on the car make with the car's direction of travel as it rounds the curve? Be sure to describe your reasoning. (*Hint:* Argue that the rate at which drag extracts energy from the car's kinetic energy is given by $[dk]/dt = \vec{F}_D \cdot \vec{v} = -F_D v$.)

N3S.7 A balloon of mass M is floating at rest a distance H above the earth's surface. The rider throws out some ballast, and the balloon begins to rise with an initial acceleration a.

(a) What is the mass m of the ballast the rider throws out?

(b) Why is the word *initial* important in the description of the problem? What will cause this acceleration to change as time passes?

N3S.8 Imagine that a certain jet plane weighing 25 tons must be moving at 200 mi/h to take off (1 ton = 2000 lb).

(a) What is the *minimum* thrust that the jet's exhaust must exert on its engines if the plane is to successfully take off from a 500-ft airstrip? (*Hint:* If you assume that the jet's acceleration is constant, then its average speed during the takeoff run is 100 mi/h.)

(b) Explain carefully why your estimate is the *minimum* thrust required.

N3S.9 Imagine that during the time interval $0 \le t \le 2$ s, the x-velocity of a car with a mass $m = 1200$ kg is given by $v_x(t) = v_0 - bt^2$, where $v_0 = 20$ m/s and $b = 5$ m/s^3. Sketch graphs of $v_x(t)$ and $a_x(t)$ for this situation. Find an expression (in terms of m, b, t, and whatever else you need) for the x component of the combined static friction and drag forces acting on the car during that time interval, and calculate the magnitude of this force at time $t = 0.5$ s.

Rich-Context

N3R.1 The top of a small hill in a certain highway has a circular (vertical) cross section with an approximate radius of 57 m. A car going over this hill too fast might leave the ground and thus lose control. What speed limit should be posted? (*Hint:* Note that the magnitude of the normal force acting on the car as it passes over the crest of the hill is *not* equal to the car's weight. What must it be, approximately, if the car's tires just barely maintain contact with the road?)

N3R.2   You are designing an ejector seat for an automobile. Your seat contains two rocket engines that will burn

for no more than 0.5 s (so as not to fry passers-by) and yet must throw the seat and occupant at least 150 m in the air after the engines shut off (to allow the parachute to deploy). Estimate the combined thrust that your rockets will have to exert on the seat, and check whether the seat's acceleration will exceed the safe limit of about $10g$.

ANSWERS TO EXERCISES

N3X.1   The units of b are meters per second squared, and the units of c are meters. The x-velocity and x-acceleration in this case are $v_x = bt$ and $a_x = b$. The units of these expressions are correct (meters per second and meters per second squared, respectively).

N3X.2   Figure N3.11 shows a drawing of the situation, a free-body diagram of the box, and a net-force diagram, respectively.

N3X.3   Figure N3.12 shows a drawing of the situation, a motion diagram, a free-body diagram, and a net-force diagram for the car, respectively.

N3X.4   Figure N3.13 shows a drawing of the situation and motion diagram, a free-body diagram, and a net-force diagram for the elevator, respectively.

N3X.5   (a) These forces are second-law partners.
(b) These forces are third-law partners.
(c) These forces are third-law partners.
(d) These forces are second-law partners.

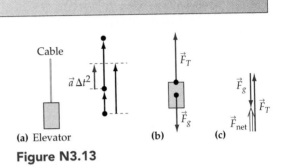

Figure N3.13

N3X.6   According to Newton's second law, the magnitude of the net force on the car is linked to the magnitude of its acceleration as follows:

$$F_{SF,1} - F_{N(T)} = F_{net} = m_1 a \qquad (N3.11)$$

But according to equation N3.4, $F_{N(T)} = F_{N(C)} = m_2 a$. Plugging this into the equation above, we find that

$$F_{SF,1} - m_2 a = m_1 a \quad \Rightarrow \quad F_{SF,1} = (m_1 + m_2)a \qquad (N3.12)$$

N3X.7   Initially the net x-force is positive and it remains relatively constant, but then it drops to zero. This force is a static friction force exerted by the road on the wheels (as a result of the engine rotating the wheels against the road). Presumably, this force drops to whatever value is equal in magnitude to the drag force when the driver decides the car has reached cruising speed and lets up on the gas pedal.

N3X.8   Graphs of $v_x(t)$ and $a_x(t)$ look like this:

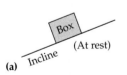

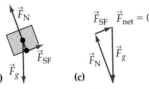

Figure N3.11

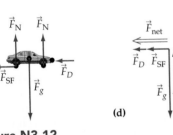

Figure N3.12

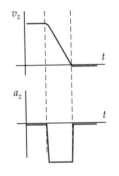

N3X.9 In this case the maximum possible deceleration that the box can have before it begins to slip is given by $a_{max} = F_{net,max}/m$. If we approximate this acceleration by the average acceleration, then we have

$$\frac{F_{net,max}}{m} = a_{max} \geq a \approx \frac{|\Delta v_x|}{\Delta t} \qquad \text{(N3.13)}$$

Since $|\Delta v_x|$ is given, the acceleration will get larger when Δt is smaller, so this equation puts a lower limit on Δt:

$$\Delta t \geq \frac{m|\Delta v_x|}{F_{net,max}} = \frac{(10 \text{ kg})(60 \text{ mi/h})}{45 \text{ N}} \left(\frac{1 \text{ m/s}}{2.24 \text{ mi/h}} \right)$$

$$= 6.0 \frac{\text{kg·m/s}}{\text{N}} \left(\frac{1 \text{ N}}{1 \text{ kg·m/s}^2} \right) = 6.0 \text{ s} \qquad \text{(N3.14)}$$

N3X.10 If we define the x direction to be positive upward, the change in the basketball's x-velocity is given by $\Delta v_x = v_{f,x} - v_{i,x} = +3.0 \text{ m/s} - (-3.0 \text{ m/s}) = 6.0 \text{ m/s}$. If this takes place in 0.10 s and we approximate the ball's acceleration by its average acceleration during this interval, we find that the magnitude of the ball's acceleration is $a = |\Delta v_x|/\Delta t = 60 \text{ m/s}^2$. The net upward force on the ball while it touches the floor has a magnitude of

$$F_{net} = F_N - F_g = F_N - mg \qquad \text{(N3.15)}$$

since the ball's weight opposes the normal force that is driving the ball upward. The magnitude of Newton's second law then implies that

$$F_N - mg = F_{net} = ma \qquad \Rightarrow \qquad F_N = m(a + g)$$

$$\Rightarrow \qquad \frac{F_N}{mg} = \frac{m(a + g)}{mg}$$

$$= \frac{a}{g} + 1 = \frac{60 \text{ m/s}^2}{10 \text{ m/s}^2} + 1 = 7 \qquad \text{(N3.16)}$$

So the magnitude of the normal force is about 7 times larger than that of the ball's weight.

N4

Motion from Forces

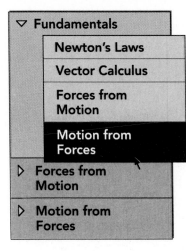

Chapter Overview

Introduction

In chapter N3, we explored a number of examples in which we use information about an object's motion to learn about the forces acting on that object. In this chapter, we will discuss examples in which we use knowledge of the net force acting on an object to determine the object's motion. This will lay important foundations for exploring more complicated problems of this type in chapters N10 through N13.

Section N4.1: The Reverse Kinematic Chain

We can reverse the kinematic chain discussed in chapter N3 as follows:

$$\vec{a}(t) \quad \boxed{\begin{array}{c} \text{time} \\ \text{antiderivative} \end{array}} \rightarrow \quad \vec{v}(t) \quad \boxed{\begin{array}{c} \text{time} \\ \text{antiderivative} \end{array}} \rightarrow \quad \vec{r}(t) \qquad \text{(N4.2)}$$

If we know the net force acting on an object, we can use Newton's second law to determine the object's acceleration. The chain of relationships in equation N4.1 means that we can then predict the object's motion by computing **antiderivatives** of $\vec{a}(t)$. The only problem with doing this is that computing antiderivatives is *hard*. In this chapter we will look at four different approaches to computing these antiderivatives.

Section N4.2: Graphical Antiderivatives

If the object moves only along the x axis, then we can use graphs of $a_x(t)$, $v_x(t)$, and $x(t)$ to describe an object's motion. If we stack such graphs vertically so that a graph of $x(t)$ is above a graph of $v_x(t)$ above a graph of $a_x(t)$, we can construct an antiderivative graph from the graph below it by using one of two methods.

The **slope method** uses the idea that the *slope above equals the value below.* To use this method:

1. Draw a short line segment on the upper graph of the pair whose *slope* reflects the *value* on lower graph at a given instant of time.
2. Align successively drawn segments so that they sketch out a continuous curve.

The **area method** uses the fact that the antiderivative of a function is related to the area under a graph of that function. To use this method, draw the value at a given instant of time on the upper graph of a pair of graphs so that it corresponds to the total area under the curve of the lower graph between the given instant and $t = 0$.

In either method, we must use separately stated **initial conditions** $x(0)$ and $v_x(0)$ to determine where to start on the upper graph.

Section N4.3: Integrals for One-Dimensional Motion

The mathematical definitions of velocity and acceleration imply that

$$v_x(t) = \int a_x(t)\,dt + C_1 \qquad \text{and} \qquad x(t) = \int v_x(t)\,dt + C_2 \qquad \text{(N4.4)}$$

where $\int f(t)\,dt$ is any antiderivative of $f(t)$ and C_1 and C_2 are constants of integration. We can determine the values of C_1 and C_2 by plugging $t = 0$ into the expressions above and setting the values C_1 and C_2 so that the values of $v_x(0)$ and $x(0)$ match the specified initial conditions.

We can evaluate these constants of integration automatically (without separate steps) by using definite integrals:

$$v_x(t) - v_x(0) = \int_0^t a_x(t)\, dt \quad \text{and} \quad x(t) - x(0) = \int_0^t v_x(t)\, dt \quad \text{(N4.10)}$$

where the **definite integral** of $f(t)$ is any of its possible antiderivative functions evaluated at the integral's upper limit minus the same function evaluated at the lower limit.

When an object's x-acceleration is *constant*, these integrals imply that

$$v_x(t) = a_x t + v_{0x} \quad \text{and} \quad x(t) = \tfrac{1}{2} a_x t^2 + v_{0x} t + x_0 \quad \text{(N4.8)}$$

Purpose: These equations specify an object's x-velocity $v_x(t)$ and x-position $x(t)$ as functions of time t whenever its x-acceleration a_x is constant.

Symbols: $v_{0x} \equiv v_x(0)$ is the object's initial x-velocity, and $x_0 \equiv x(0)$ is its initial x-position.

Limitations: *Important!* This equation *only* applies if object's x-acceleration is constant (or we can reasonably model it as constant).

Section N4.4: Free Fall in One Dimension

An object is freely falling if the only significant force acting on it is its weight. Newton's second law then says that $m\vec{a} = \vec{F}_{net} = m\vec{g}$, which implies that the object's acceleration is $\vec{a} = \vec{g}$, where $\vec{g}$ is the local **gravitational field vector** (sometimes called the **acceleration of gravity**). If the object moves only vertically, we can adapt equations N4.8 by replacing the x subscripts with z subscripts and using $a_z = -g$.

Section N4.5: Integrals in Three Dimensions

In general, the definitions of velocity and acceleration imply that

$$\vec{v}(t) - \vec{v}(0) = \int_0^t \vec{a}(t)\, dt \quad \text{and} \quad \vec{r}(t) - \vec{r}(0) = \int_0^t \vec{v}(t)\, dt \quad \text{(N4.22)}$$

Purpose: These equations describe how to calculate an object's velocity $\vec{v}(t)$ and its position $\vec{r}(t)$ as functions of time t given its acceleration $\vec{a}(t)$, its initial velocity $\vec{v}(0)$, and its initial position $\vec{r}(0)$.

Limitations: The acceleration and velocity functions must be well defined.

Note: These equations compactly express three independent component equations that look like equations N4.10.

Section N4.6: Constructing Trajectory Diagrams

When an object's motion is two-dimensional, one can calculate its motion graphically by constructing a **trajectory diagram** as follows:

1. Draw an arrow $\vec{v}_0\,\Delta t$ that is *centered* on the object's initial position $\vec{r}_0$. This arrow's endpoints are approximately the object's positions at $t_1 = -\tfrac{1}{2}\Delta t$ and $t_2 = +\tfrac{1}{2}\Delta t$. Label these points 1 and 2.
2. Draw the arrow $\vec{a}_2\,\Delta t^2$ with its tail attached to point 2.
3. Construct the arrow $\vec{v}_{23}\,\Delta t = \vec{v}_{12}\,\Delta t + \vec{a}_2\,\Delta t^2$.
4. Move $\vec{v}_{23}\,\Delta t$ so that its tail is at point 2: its tip then is the object's position at time $t_3 = t_2 + \Delta t$.
5. Repeat steps 2 through 4 to find the object's positions at times t_4, t_5, and so on.

This process works in *all* circumstances (as long as Δt is sufficiently small).

Section N4.7: The Newton Program

Trajectory diagrams are very useful, but tedious to construct by hand. This section describes a computer program that can construct trajectory diagrams very rapidly.

N4.1   The Reverse Kinematic Chain

The kinematic chain

In chapter N3 we discussed the *kinematic chain* of relationships that link an object's position to its velocity and its acceleration:

$$\vec{r}(t) \quad \left[\begin{array}{c} \text{time} \\ \text{derivative} \end{array} \right] \rightarrow \quad \vec{v}(t) \quad \left[\begin{array}{c} \text{time} \\ \text{derivative} \end{array} \right] \rightarrow \quad \vec{a}(t) \qquad \text{(N4.1)}$$

This chain of relationships means that given an object's position or velocity as a function of time, we can determine its acceleration as a function of time. As we demonstrated in chapter N3, we can then use Newton's second law to link the object's acceleration to the net force on the object, and use that information to learn things about the forces that go into that net force.

We can reverse the chain by taking *antiderivatives*. The **antiderivative** of a function $f(t)$ is simply any function $F(t)$ such that $dF/dt = f(t)$. For example, $\vec{v}(t)$ is the antiderivative of $\vec{a}(t)$, since $d\vec{v}/dt = \vec{a}(t)$. Therefore

Reversing the kinematic chain using antiderivatives

$$\vec{a}(t) \quad \left[\begin{array}{c} \text{time} \\ \text{antiderivative} \end{array} \right] \rightarrow \quad \vec{v}(t) \quad \left[\begin{array}{c} \text{time} \\ \text{antiderivative} \end{array} \right] \rightarrow \quad \vec{r}(t) \quad \text{(N4.2)}$$

So, if we know how to compute a function's antiderivative and we know an object's acceleration as a function of time, we can find its velocity and its position as functions of time as well. (For more information about antiderivatives, see appendix NB on integral calculus at the end of this volume.)

Finding motion from forces

If we know all the forces that act on an object, we can find the net force on the object. Newton's second law then allows us to find the object's acceleration at all times, and the reversed kinematic chain allows us to find the object's velocity and position at all times. This is how we determine *motion from forces* (at least in principle).

The problem here is not the physics but the mathematics. By applying only a handful of rules, one can pretty easily calculate the derivative of almost any function. Finding the *antiderivative* of an arbitrary function is generally not nearly so easy.

In this chapter, we will discuss *four* different approaches to finding the antiderivatives of acceleration—two approaches that work for one-dimensional motion and two for multidimensional motion. When an object moves in only one dimension, we can essentially reverse the graphical construction techniques we learned in the chapter N3 to construct graphs of $v_x(t)$ and $x(t)$ from a graph of $a_x(t)$. Alternatively, we can sometimes use the techniques of integral calculus to find these functions mathematically.

If the object moves in three dimensions, we can still use the mathematical approach: we simply have to keep track of three vector components instead of one. We can also reverse the approach we used to construct motion diagrams in chapter N3: whereas before we used information about the object's position at equally spaced instants of time to construct arrows representing the object's acceleration, we now use known acceleration arrows to construct the object's trajectory. This latter technique turns out to be especially easy to turn into a computer algorithm that computes trajectories. (It is also essentially the approach that Newton himself used in his *Principia*.)

N4.2   Graphical Antiderivatives

We begin with the graphical approach to one-dimensional motion. In chapter N3, we learned how to use the "slope above equals value below" method to construct a graph of $v_x(t)$ below a graph of $x(t)$ and/or a graph of $a_x(t)$

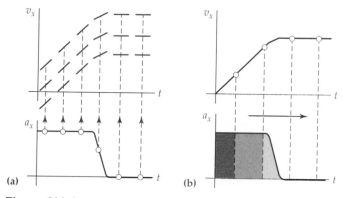

Figure N4.1

Two methods for constructing a graph of $v_x(t)$ from a graph of $a_x(t)$. (a) The antiderivative method involves drawing short line segments on the upper graph whose *slopes* are consistent with the *value* of the lower graph at the same time. We can only choose between the various possible upper curves we generate in this way by actually choosing a value for, say, $v_x(0)$. (b) The integral method involves linking the value on the upper graph with the area under the lower graph. We still have to choose a value for $v_x(0)$: the single curve I have drawn corresponds to the choice $v_x(0) = 0$.

below a graph of $v_x(t)$. In this section, we will learn how to do the process in reverse, constructing graphs of $v_x(t)$ and $x(t)$ from a graph of $a_x(t)$.

There are essentially two ways to do this. The **slope method** simply re-verses the approach that we used in chapter N3. For example, imagine we are trying to construct a graph of $v_x(t)$ from a graph of $a_x(t)$. If we were instead constructing the graph of $a_x(t)$ from $v_x(t)$, we would look at the slope of the upper graph and plot its value on the lower graph. To do this in reverse, we should, for each of a set of specific instants of time, draw a little line segment on the upper graph whose slope reflects the *value* that we see at that instant on the lower graph. If you arrange these line segments so that they nearly touch end to end, as shown in figure N4.1a, then they will sketch out the de-sired upper curve.

The slope method

Unfortunately, there is a certain ambiguity in this process, because we can draw *several* upper graphs that satisfy this criterion, as shown. These different graphs all have the same *shape*, but are offset vertically from one another by different constant values. The only way to resolve this ambiguity is to *choose* the value that the upper graph should have at a certain time (generally at $t = 0$). In some cases, the problem statement may suggest an appropriate value (e.g. it might say that the object starts at *rest*), but otherwise we simply have to *pick* an arbitrary value. Once we have chosen or found a value for $v_x(t)$ at any given instant of time, the rest of the graph is completely determined.

An alternative approach (which I will call the **area method**) is to recog-nize that the fundamental theorem of integral calculus (see appendix NB on integral calculus) implies that the difference between the *value* on the upper graph at time t and its *value* at $t = 0$ corresponds to the *area* under the lower graph during this interval:

The area method

$$v_x(t) - v_x(0) = \text{area under curve of } a_x(t) \text{ between 0 and } t \quad \text{(N4.3}a\text{)}$$

$$x(t) - x(0) = \text{area under curve of } v_x(t) \text{ between 0 and } t \quad \text{(N4.3}b\text{)}$$

The phrase to remember is "*value* above equals *area* below" when we stack graphs in the conventional way (x above v_x above a_x).

This method is illustrated in figure N4.1b. Note that for the first half of the graph, the acceleration is constant and positive, so as t increases, the area below the graph between t and 0 increases steadily, and thus so does the velocity. However, after the acceleration has fallen to zero, no new area is added as t increases, so the velocity remains constant. Note that we *still* have to choose a value for the upper graph at some time (usually $t = 0$) to get started.

Example N4.1

Problem   Imagine a car (initially at rest at $x = 0$) powered by a rocket engine that pivots so that we can direct its thrust either forward or backward. The rocket engine ignites at $t = 0$, subsequently exerting a constant forward thrust on the car for a certain amount of time. Then the engine is turned around so that it exerts a rearward thrust on the car of the same magnitude for the same amount of time. Draw graphs of $a_x(t)$, $v_x(t)$, and $x(t)$ for this car. (Ignore drag or friction.)

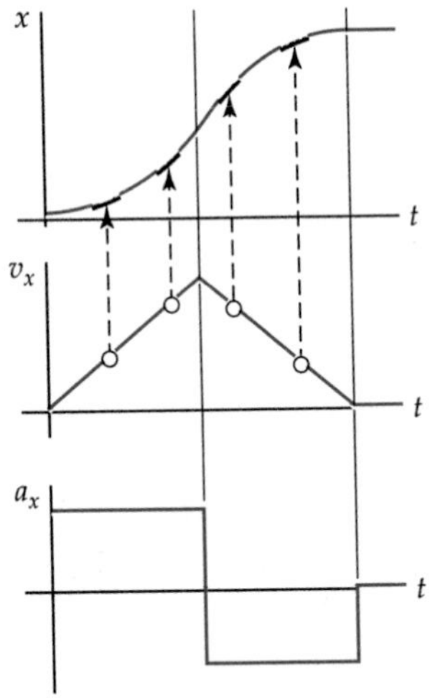

Model   Assuming that there is no significant drag or friction force opposing the thrust on the car, and assuming that the car moves along a level road (so that the vertical normal and gravitational forces on the car cancel out), then the net force on the car will be the same as that applied by the rocket engine's exhaust. If we take the $+x$ direction to be forward, the problem description implies that the car's x-acceleration is some positive constant a for a certain Δt, and then $-a$ for Δt.

Solution   This means the graph of $a_x(t)$ should look as shown in the bottom graph in figure N4.2.

Working upward, we can use either the slope or area method to show that the car's velocity first increases linearly and then decreases linearly. The problem description states that $v_x(0) = 0$, so a graph of the car's $v_x(t)$ must look as shown in the middle graph.

To create the position graph, we start with $x = 0$ at $t = 0$. I find the slope method somewhat easier to use to construct this graph. Note that as the velocity increases, the slope of the $x(t)$ graph increases; and as the velocity decreases, the slope of $x(t)$ decreases, as shown in the top graph.

Note that we have (qualitatively at least) completely described the car's motion, given the forces applied to it!

Figure N4.2

Graphs of $a_x(t)$, $v_x(t)$, and $x(t)$ for the rocket car.

You can get more practice drawing such graphs by making sure that you can generate the upper graph in figures N3.10 and N3.9b, given the lower graph.

Exercise N4X.1

Plot $z(t)$ for the basketball described in example N3.7.

N4.3   Integrals for One-Dimensional Motion

Using antiderivatives to do the same thing mathematically

Another whole approach to constructing the functions $v_x(t)$ and $x(t)$ from $a_x(t)$ is to use the techniques of integral calculus to determine these quantities mathematically. In this section, we will explore this method in the context of one-dimensional motion.

Since $dx/dt = v_x(t)$ and $dv_x/dt = a_x(t)$, the reverse kinematic chain implies that

$$v_x(t) = \int a_x(t)\,dt + C_1 \qquad \text{and} \qquad x(t) = \int v_x(t)\,dt + C_2 \qquad \text{(N4.4)}$$

where C_1 and C_2 are unknown **constants of integration** and $\int f(t)\,dt$ is the conventional notation for a particular antiderivative of $f(t)$, that is, any function $F(t)$ whose time derivative is $f(t)$. The constants C_1 and C_2 express mathematically what figure N4.1a expresses graphically: in the absence of additional information, *we can only determine a function's antiderivative up to an overall constant.* (See appendix NB for more discussion of this issue.)

We can, however, determine C_1 if we know the object's initial velocity $v_x(0)$: we need only substitute $t = 0$ into the expression for $v_x(t)$ and choose the value of C_1 that gives the correct initial velocity $v_x(0)$. We can similarly determine C_2 by choosing it so that we get the correct value of $x(0)$. We call $v_x(0)$ and $x(0)$ the object's **initial conditions**; we can completely determine an object's position and velocity from its acceleration *only* if we also know these quantities.

Example N4.2

Problem   Imagine a car waiting at a stoplight. After the stoplight turns green at time $t = 0$, the car accelerates with a constant acceleration of $\vec{a} = 2.0 \text{ m/s}^2$ forward along a straight stretch of road (which we will take to define our x axis). (a) Find a general expression for the car's x-velocity and x-position as a function of time t in terms of the car's x-acceleration a_x and its initial conditions $v_{0x} \equiv v_x(0)$ and $x_0 \equiv x(0)$ for as long a_x is constant. (b) Assuming that the car starts from rest and we define $x = 0$ to be the car's initial position, what are the car's x-velocity and x-position after 5.0 s?

Solution   (a) According to equation N4.4a,

$$v_x(t) = \int a_x(t)\,dt + C_1 = \int a_x\,dt + C_1 = a_x t + C_1 \qquad \text{(N4.5a)}$$

since $d(a_x t)/dt = a_x$. Plugging $t = 0$ into this result, we find that

$$v_x(0) = a_x \cdot 0 + C_1 \qquad \Rightarrow \qquad C_1 = v_x(0) \equiv v_{0x}$$

$$\Rightarrow \qquad v_x(t) = a_x t + v_{0x} \qquad \text{(N4.5b)}$$

To find the x-position as a function of time, we take the antiderivative again:

$$x(t) = \int v_x(t)\,dt + C_2 = \int (a_x t + v_{0x})\,dt + C_2 = \int a_x t\,dt + \int v_{0x}\,dt + C_2$$

$$= \tfrac{1}{2}a_x t^2 + v_{0x}t + C_2 \qquad \text{(N4.6a)}$$

since $d(\tfrac{1}{2}a_x t^2)/dt = at$ and $d(v_{0x}t)/dt = v_{0x}$. Plugging $t = 0$ into this result, we get

$$x(0) = \tfrac{1}{2}a_x \cdot 0^2 + v_{0x} \cdot 0 + C_2 \qquad \Rightarrow \qquad C_2 = x(0) \equiv x_0$$

$$\Rightarrow \qquad x(t) = \tfrac{1}{2}a_x t^2 + v_{0x}t + x_0 \qquad \text{(N4.6b)}$$

(b) We are told that $x_0 = 0$ and $v_{0x} = 0$. We are also told that the car's acceleration is directed forward (i.e., in the $+x$ direction), so in this case $a_x = +a$, where $a \equiv \text{mag}(\vec{a}) = 2.0 \text{ m/s}^2$. Therefore, at time $t = 5$ s,

$$v_x(t) = a_x t + v_{0x} = (+2.0 \text{ m/s}^2)(5.0 \text{ s}) + 0 = 10 \text{ m/s} \qquad \text{(N4.7a)}$$

$$x(t) = \tfrac{1}{2}a_x t + v_{0x}t + x_0 = \tfrac{1}{2}(+2.0 \text{ m/s}^2)(5.0 \text{ s})^2 + 0 + 0 = 25 \text{ m} \quad \text{(N4.7b)}$$

Useful equations for problems involving constant acceleration in one dimension

Equations N4.5b and N4.6b are useful whenever an object's x-acceleration is constant:

$$v_x(t) = a_x t + v_{0x} \qquad \text{and} \qquad x(t) = \tfrac{1}{2}a_x t^2 + v_{0x}t + x_0 \qquad \text{(N4.8)}$$

Purpose: These equations specify an object's x-velocity $v_x(t)$ and x-position $x(t)$ as functions of time t whenever its x-acceleration a_x is constant.
   Symbols: $v_{0x} \equiv v_x(0)$ is the object's initial x-velocity, and $x_0 \equiv x(0)$ is its initial x-position.
   Limitations: *Important!* This equation *only* applies if object's x-acceleration is constant (or we can reasonably model it as constant).

These equations are important for at least two reasons. First, this is a relatively simple case that illustrates the integral approach nicely. Second, these equations are useful because the assumption that $a_x = $ constant is a reasonably good *model* for a variety of situations. Physicists often use this model to make quick estimates even when it is clear that a_x is *not* particularly constant.

On the other hand, one of the most common student errors I have seen is thinking that these equations are more general then they actually are. These equations really apply only to the very *specific* case in which a_x is *constant*. While this is occasionally an excellent approximation to a realistic acceleration and more often an adequate first approximation, there are many more situations where equations N4.8 are not even close to being correct.

It is also worth noting that in example N4.2, the constants of integration C_1 and C_2 happen to be equal to the initial conditions $v_x(0)$ and $x(0)$. This is *not* generally true! In every problem where an object's acceleration is *not* constant, you *must* carefully evaluate these constants of integration by using the specified initial conditions.

Exercise N4X.2

A car whose initial x-position and x-velocity are $x(0) = 0$ and $v_x(0) = +12$ m/s experiences forces that give it an x-acceleration of $a_x(t) = -a$, with $a = 2.0 \text{ m/s}^2$. Find $v_x(t)$ and $x(t)$ and evaluate these quantities at $t = 5.0$ s.

Using definite integrals to avoid evaluating constants of integration explicitly

One can, however, *automatically, correctly,* and *implicitly* evaluate these constants of integration as follows. If we take the definitions $a_x \equiv dv_x/dt$ and $v_x \equiv dx/dt$ and "integrate both sides from t_A to t_B," the fundamental theorem of calculus implies that

$$\int_{t_A}^{t_B} \frac{dv_x}{dt}\,dt = \int_{t_A}^{t_B} a_x(t)\,dt \quad \Rightarrow \quad v_x(t_B) - v_x(t_A) = \int_{t_A}^{t_B} a_x(t)\,dt \qquad \text{(N4.9a)}$$

$$\int_{t_A}^{t_B} \frac{dx}{dt}\, dt = \int_{t_A}^{t_B} v_x(t)\, dt \quad \Rightarrow \quad x(t_B) - x(t_A) = \int_{t_A}^{t_B} v_x(t)\, dt \quad \text{(N4.9b)}$$

where $\int_{t_A}^{t_B} f(t)\, dt$ is the **definite integral** of $f(t)$, which we calculate by finding any function $F(t)$ whose derivative is $f(t)$ and evaluating the difference $F(t_B) - F(t_A)$ between that function's values at the limits of the integration. In particular, if we define $t_A = 0$ and t_B to be some arbitrary time t, then these equations become

$$v_x(t) - v_x(0) = \int_0^t a_x(t)\, dt \qquad \text{(N4.10a)}$$

$$x(t) - x(0) = \int_0^t v_x(t)\, dt \qquad \text{(N4.10b)}$$

Note how these equations *automatically* include the initial conditions. Moreover, since evaluating the definite integrals involves computing the *difference* of antiderivatives evaluated at two different times, the unknown constant of integration appearing in the antiderivative will cancel out of the difference, so we never even need to think about it! I think that once you get used to this method, you will find that it reduces the likelihood of error and is simpler in complicated problems.

Example N4.3

Problem   Imagine a similar situation as discussed in example N4.2: a car is sitting at a stoplight and then accelerates from rest when the light turns green. But in this case, let us assume that a drag force that increases with speed opposes the constant forward force applied to the car, so that the car's x-acceleration ends up being $a_x(t) = b/(t + T)^3$, where $b = 2000\text{ m·s}$ and $T = 10$ s (both are constants). What are the car's x-velocity and x-position as a function of time in this case? What are the car's x-velocity and x-position at $t = 5.0$ s?

Solution   Note that the opposing drag force makes $a_x(t)$ *decrease* as t increases: $a_x(t)$ decreases to one-eighth its original value by time $t = T$. Note also that at $t = 0$, $a_x = (2000\text{ m·s})/(10\text{ s})^3 = 2.0\text{ m/s}^2$, which is the same as the *constant* acceleration value in example N4.2. Since $v_x(0) = 0$ in this case, equation N4.10b implies that

$$v_x(t) = \int_0^t a_x(t)\, dt = \int_0^t \frac{b}{(t + T)^3}\, dt = b \int_0^t \frac{dt}{(t + T)^3}$$

$$= \frac{b}{-2}\left[\frac{1}{(t + T)^2} - \frac{1}{(0 + T)^2}\right] = \frac{b}{2T^2} - \frac{b}{2(t + T)^2} \qquad \text{(N4.11)}$$

since the derivative of $-\frac{1}{2}b(t + T)^{-2}$ is $+b(t + T)^{-3}$. Plugging this into equation N4.10a, again defining the car's initial position to be $x(0) \equiv 0$, and using the constant and sum rules for integration (see appendix NB), we get

$$x(t) = \int_0^t v_x(t)\, dt = \int_0^t \left[\frac{b}{2T^2} - \frac{b}{(t + T)^2}\right] dt = \frac{b}{2T^2}\int_0^t dt - \frac{b}{2}\int_0^t \frac{dt}{(t + T)^2}$$

$$= \frac{b}{2T^2}(t - 0) - \frac{b}{2}\left(\frac{-1}{t + T} - \frac{-1}{T}\right) = \frac{b}{2}\left(\frac{t}{T^2} + \frac{1}{t + T} - \frac{1}{T}\right) \qquad \text{(N4.12)}$$

since the derivative of $(t + T)^{-2}$ is $-(t + T)^{-1}$. Evaluating $v_x(t)$ and $x(t)$ at time $t = 5.0$ s, we get

$$v_x(5.0 \text{ s}) = \frac{2000 \text{ m·s}}{2(10 \text{ s})^2} - \frac{2000 \text{ m·s}}{2(15 \text{ s})^2} = 5.6 \text{ m/s} \qquad (\text{N4.13}a)$$

$$x(5.0 \text{ s}) = \frac{2000 \text{ m·s}}{2} \left[\frac{5.0 \text{ s}}{(10 \text{ s})^2} + \frac{1}{15 \text{ s}} - \frac{1}{10 \text{ s}} \right] = 17 \text{ m} \qquad (\text{N4.13}b)$$

Evaluation   Note that these values are smaller than the results in example N4.2, as we might expect.

Exercise N4X.3

Why might we expect the results to be smaller?

Exercise N4X.4

Check that the expressions given in equations N4.11 and N4.12 for $v_x(t)$ and $x(t)$ yield $v_x(0) = 0$ and $x(0) = 0$ when $t = 0$, consistent with our assumptions about the car's initial position and velocity.

Exercise N4X.5

Check that if you use equation N4.4 to evaluate $v_x(t)$ in example N4.3 and evaluate the constant of integration C_1 so that $v_x(0) = 0$, you get the same result as in equation N4.11.

The importance of initial conditions

In closing this section, let me emphasize again that just knowing $a_x(t)$ is *not* sufficient to determine an object's motion: we also need to specify the object's initial x-position $x_0 \equiv x(0)$ and x-velocity $v_{0x} \equiv v_x(0)$ to completely determine its motion. This is *always* the case when we are trying to determine an object's trajectory from its acceleration; so dealing with initial conditions is going to be a recurring theme in chapters N10 through N13.

N4.4   Free Fall in One Dimension

The definition of a freely falling object

An important application of equations N4.8 is the case of a *freely falling* object. If the only significant force acting on an object near the earth is its weight, we say that is **freely falling.** According to Newton's second law, a freely falling object's acceleration is

$$m\vec{a} = \vec{F}_{\text{net}} = \vec{F}_g = m\vec{g} \qquad \Rightarrow \qquad \vec{a} = \vec{g} \qquad (\text{N4.14})$$

where $\vec{g}$ is a *vector* whose magnitude is $g = 9.8 \text{ m/s}^2$ and whose direction is downward. Note that this equation implies that *all* objects fall with the same acceleration $\vec{g}$, independent of their mass! (The vector $\vec{g}$ is properly called the **gravitational field vector,** but people often call it the *acceleration of gravity* for this reason.) This is an interesting and nonobvious result that has important implications that we will explore in chapters N9 and N10.

N4.6   Constructing Trajectory Diagrams

Equations N4.19 and N4.21 provide a very general method for calculating trajectories, but actually computing the integrals in these equations can be *very* difficult in many situations of interest. In this section, we will see that we can reverse the procedure that we used to construct motion diagrams (which allowed us to determine an object's acceleration from its trajectory) to construct an object's trajectory from its acceleration under *any* circumstances.

Figure N4.3a reviews how we construct a motion diagram. We first plot the object's position at equally spaced instants of time and draw displacement arrows $\Delta \vec{r}_{12} = \vec{v}_{12}\, \Delta t$, $\Delta \vec{r}_{23} = \vec{v}_{23}\, \Delta t$, and so on between the dots. Since Δt is the same for all these arrows, these arrows depict both the direction and the relative magnitudes of the average velocities $\vec{v}_{12}, \vec{v}_{23}, \dots$ between the dots. To construct an arrow representing the object's acceleration at, say, point 2, we move the velocity arrow $\vec{v}_{23}\, \Delta t$ just after the point back so that its tail end coincides with the tail end of the velocity arrow $\vec{v}_{12}\, \Delta t$ just before the point, and we construct the vector difference $\Delta \vec{v}\, \Delta t \approx \vec{a}_2\, \Delta t^2$. As long as Δt is reasonably small compared to the time that it takes the object's acceleration to change significantly, this process yields reasonably accurate acceleration arrows $\vec{a}_2\, \Delta t^2, \vec{a}_3\, \Delta t^2$, and so on.

To construct a **trajectory diagram** of the object's motion, we simply do this process backward, as illustrated in figure N4.3b. We start by drawing the arrow $\vec{v}_0\, \Delta t$ (where $\vec{v}_0$ is the object's initial velocity) so that it is *centered* on the object's initial position $\vec{r}_0$. We *center* this arrow on the object's initial position because $\vec{v}_0$ is the object's *instantaneous* velocity at time $t = 0$, the same instant that the object has position $\vec{r}_0$. The arrow $\vec{v}_0\, \Delta t$ most closely approximates the object's *average* velocity during a time interval centered on $t = 0$. So if we define $t_1 = -\frac{1}{2}\Delta t$ and $t_2 = +\frac{1}{2}\Delta t$, then the initial velocity arrow $\vec{v}_0\, \Delta t$ will be very nearly equal to the average velocity $\vec{v}_{12}\, \Delta t$ between these instants of time. This in turn means that the endpoints of the $\vec{v}_0\, \Delta t$ arrow will approximately locate the object's positions 1 and 2 at times t_1 and t_2.

Note that doing the construction in this way implies that the object's first position dot (the dot for time t_1) is *not* the same as its initial position $\vec{r}_0$, but is rather its position one-half time step *before* $t = 0$. This may seem unnecessarily strange and complicated, but the approach described above more accurately models the object's true trajectory than the seemingly more straightforward method of locating the first dot at $\vec{r}_0$. This increased accuracy will be important to us in future applications.

Now, to find the object's position at time $t_3 = \frac{3}{2}\Delta t$, we carefully draw the object's acceleration arrow $\vec{a}_2\, \Delta t^2$ (which we have presumably calculated by

To draw a trajectory diagram, first *center* the object's initial velocity arrow on its initial position

Then use the known acceleration vector to construct future position points from the initial arrow's two endpoints

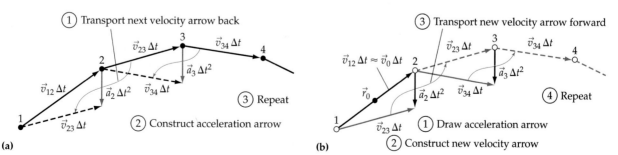

(a)   **(b)**

Figure N4.3

(a) How to construct acceleration arrows from a given trajectory. (b) How to construct a trajectory from given acceleration arrows. In both figures, black arrows represent known vectors, colored arrows represent constructed vectors, dashed arrows represent transported vectors, and white dots represent constructed positions.

Figure N4.4
A multiflash photograph of an actual falling ball. The ball's initial launch velocity was upward and to the right.

This process works in all cases (if Δt is small)

using Newton's second law) so its tail end coincides with point 2. We then construct the arrow $\vec{v}_{23} \Delta t$ so that the difference $\vec{v}_{23} \Delta t - \vec{v}_{12} \Delta t = \vec{a}_2 \Delta t^2$, that is, so that

$$\vec{v}_{23} \Delta t = \vec{a}_2 \Delta t^2 + \vec{v}_{12} \Delta t \qquad (N4.23)$$

We then transport the arrow $\vec{v}_{23} \Delta t$ so that its tail end coincides with point 2; its tip now indicates the position of the object at time t_3.

We can then repeat the process indefinitely, using the velocity arrow $\vec{v}_{23} \Delta t$ and the computed acceleration arrow $\vec{a}_3 \Delta t^2$ to find the object's position at time t_4, and so on. Figure N4.3b shows how this works. If you carefully compare figure N4.3a and N4.3b, you will see that the construction processes are simple inverses of each other. Again, note that since $t_1 = -\frac{1}{2}\Delta t$ instead of $t = 0$, the nth position dot you construct is the object's position at time $t_n = (n - \frac{3}{2})\Delta t$, not $n \Delta t$ or $(n-1)\Delta t$.

We can use this construction process *anytime* that we know the object's velocity and position at time $t = 0$ and the object's acceleration at all times. However, the construction method assumes that the object's average velocity during an interval is actually *equal* to the instantaneous velocity during that interval, and the same holds for the acceleration. In general, this is true only in the limit that Δt goes to zero, so the constructed trajectory is only an *approximation* if $\Delta t \neq 0$. Still, as long as Δt is fairly small compared to the time it takes the acceleration to change appreciably, the constructed trajectory will be reasonably accurate. Figure N4.3a was constructed assuming that the object's acceleration is constant and downward, as it would be for a freely falling object. Figure N4.4 shows that the trajectory of a falling object is indeed similar.

Example N4.5

Problem Imagine that we launch a marble with an initial velocity of 1.0 m/s at an angle of 30° above the horizontal, and that it subsequently falls freely. Use the graphical construction method to predict its future trajectory. (Suggestion: Set $\Delta t = 0.03$ s; this yields a diagram with a convenient size.)

Solution With $\Delta t = 0.03$ s, the marble's acceleration arrow at all points on the diagram (according to equation N4.14) should have a magnitude of $g\,\Delta t^2 = (9.8 \text{ m/s}^2)(0.03 \text{ s})^2 = 0.0088 \text{ m} = 0.88 \text{ cm}$ long and always points downward. The length of the marble's initial velocity vector arrow should be $\vec{v}_0 \Delta t = (1.0 \text{ m/s})(0.03 \text{ s}) = 0.03 \text{ m} = 3.0 \text{ cm}$. Figure N4.5 shows the resulting constructed trajectory. You may already know that a freely falling object generally follows a parabolic path: our constructed trajectory looks plausibly parabolic. Again, note that I have drawn the initial velocity arrow *centered* on the marble's initial position.

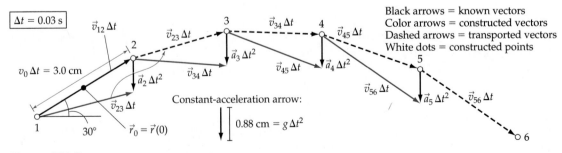

Figure N4.5
An actual-size trajectory diagram for the freely falling marble discussed in example N4.5.

Exercise N4X.10

If we had chosen $\Delta t = 0.09$ s in example N4.5, how long should we have drawn the acceleration and initial velocity arrows?

Exercise N4X.11

Construct the first few steps of the trajectory diagram for a marble whose initial velocity is 0.80 m/s and horizontal. Use $\Delta t = \frac{1}{30}$ s = 0.033 s.

Example N4.6

Problem A hockey puck initially slides due north on a level, frictionless plane of ice. Imagine that the puck has a tiny rocket engine that exerts a gentle but constant eastward force on the puck. Qualitatively, what will be the puck's trajectory after the rocket engine is started?

Solution Because the ice is level and frictionless, the gravitational force on the puck and the force exerted on it by its interaction with the ice will cancel, meaning that the net force on the puck will then be the force exerted by the rocket engine. Therefore Newton's second law implies that the puck's acceleration will be constant. Therefore, a top view of its trajectory must look qualitatively as shown in figure N4.6. This trajectory is consistent with what we might expect from a momentum transfer analysis. The puck should pick up speed in the eastward direction as its interaction with the engine's exhaust transfers eastward momentum to it, but its northward momentum should be unchanged. Thus the puck's trajectory should slowly bend toward the east, as shown.

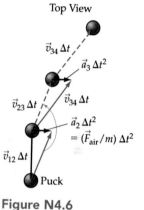

Top View

Figure N4.6
A northbound puck that is being pushed to the east.

N4.7   The Newton Program

Constructing trajectory diagrams by hand is not very accurate and quickly becomes tedious. However, because the process is simple, repetitive, and universally applicable, it provides an excellent foundation for a computer algorithm to compute trajectories.

    The computer program Newton (available for free from the *Six Ideas* website) uses precisely this algorithm to construct two-dimensional trajectories rapidly and accurately whenever you can specify the object's acceleration. When you start the program, you will see a window showing a graph window and a row of buttons on the left. If you press the Setup button, you will get the dialog box shown in figure N4.7. This dialog box prompts you to specify the duration of the time step Δt (at the upper left), the object's position components (upper middle) and velocity components (upper right) at time $t = 0$, and the object's acceleration (the four lines in the lower half of the dialog box). You can specify up to four terms that contribute to the object's total acceleration: a constant term (the first line), a term that is a constant times some power of the object's distance r from the origin (the second line), one that is a constant times some power of the object's speed v (the third line), and/or another term having the same form (the last line). You specify each contributing term by typing a value for the constant in the box on the left, typing in a power for r or v (if the term involves such a variable), and choosing a direction using the pop-up menu at the end of the line. The available

This process provides a good basis for a computer calculation

Setting up a Newton trajectory construction

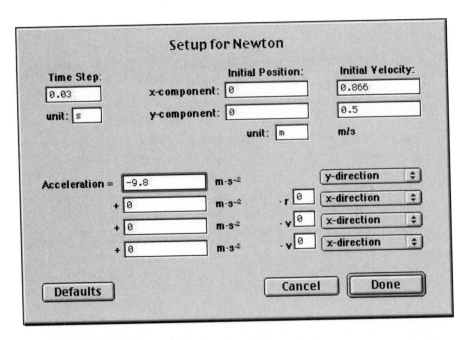

Figure N4.7

The Setup dialog box for the Newton trajectory-drawing program.

directions are the x (horizontal) direction, y (vertical) direction, r direction (directly away from the origin at the object's position), v direction (the direction of the object's current velocity), and the "left of $\vec{v}$" direction (the direction in the plane of the motion perpendicular to the object's current velocity and to the left of that vector if you look along it). The object's total acceleration can be any combination of such terms.

Figure N4.7 shows the setup for the particular case discussed in example N4.5. Note that $(1.0 \text{ m/s})(\cos 30°) = 0.866 \text{ m/s}$ and $(1.0 \text{ m/s})(\sin 30°) = 0.50 \text{ m/s}$. In this case, we need only one term for the acceleration—a constant term that says the acceleration is -9.8 m/s^2 in the $+y$ direction (which is equivalent to saying that the acceleration is 9.8 m/s^2 in the $-y$ direction). The constant factors for the other terms are set to zero so that only the first term contributes.

How to tell Newton to construct steps of a trajectory diagram

When you have specified a time step, an initial velocity, and as many contributing terms to the acceleration as you want, you can reenter the main program by pressing the Done button. If you check the "v-arrows" and "a-arrows" checkboxes, the graph will show your initial velocity arrow centered on the initial position, flanked by two points, which correspond to the first two position points on the diagram. The Step button then executes one step of the construction; the Do 5, Do 20, and Do 100 buttons execute the corresponding number of steps; the Go Slow button executes two steps per second; and the Reset button returns you to the first step. The graph will automatically scale to display all the steps you have executed so far, unless you check the "show last step" checkbox, which ensures that the graph only displays the last constructed step.

Figure N4.8 shows the graph for the situation discussed in example N4.5 after pressing the Step button twice. Although you can't see it in figure N4.8, the arrows are colored: the "known" velocity vector for the step (the upper vectors in each step in figure N4.8) is black, the acceleration arrow is blue, and the "constructed" velocity vector is green. The program automatically transports the constructed vector so that its tail coincides with the known vector's head, making it the known vector for the next step. Compare the graph shown here with figure N4.4, and I think you will see how the program works.

Other useful hints for using Newton

To get accurate results from this program, the time step should be small enough that the change in velocity is pretty small—much smaller than I have

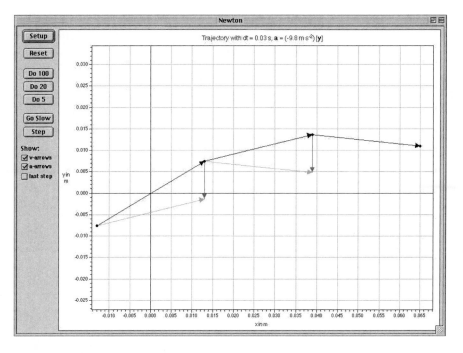

Figure N4.8
A Newton graph showing two steps of a trajectory construction.

chosen here. When Δt is appropriately small, it can be hard to see the acceleration and constructed arrows on the screen. However, you can zoom in on any part of the display by using the mouse to drag a rectangle around the area you want enlarged. The graph will resize to display the area of interest. Clicking anywhere in the graph restores it to its original size. You can use this or the "show last step" option to display the construction arrows if you want to see them.

You can print any graph by selecting Print from the File menu.

In chapters N10 through N13, we will find the Newton program to be a very powerful tool for computing trajectories in cases in which symbolic integration is difficult or impossible.

TWO-MINUTE PROBLEMS

N4T.1 An object's x-velocity $v_x(t)$ is shown in the boxed graph at the top left. Which of the other graphs in the set most correctly describes its x-position?

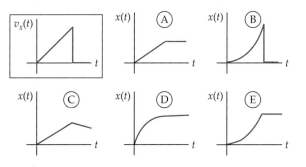

N4T.2 An object's x-acceleration $a_x(t)$ is shown in the boxed graph at the top left. Which of the other graphs in the set most correctly describes its x-velocity?

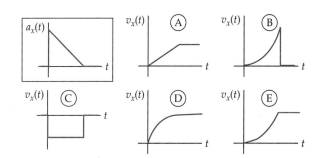

N4T.3 If a car has an x-acceleration of $a_x(t) = -bt + c$, and its initial x-velocity at time $t = 0$ is $v_x(0) = v_0$, which function below best describes $v_x(t)$?

A. $-b$
B. $-b + v_0$
C. $\frac{1}{2}bt^2 + ct + v_0$
D. $-\frac{1}{2}bt^2 + ct + v_0$ *(more)*

E. $-2bt^2 + v_0$
F. $-\frac{1}{2}bt^2 + v_0$

N4T.4 If a car's x-position at time $t = 0$ is $x(0) = 0$ and it has an x-velocity of $v_x(t) = b(t - T)^2$, where b and T are constants, which function below best describes $x(t)$?
A. $x(t) = 2b(t - T)$
B. $x(t) = 3b(t - T)^3$
C. $x(t) = \frac{1}{3}b(t - T)^3$
D. $x(t) = \frac{1}{2}b(t - T)$
E. $x(t) = \frac{1}{3}b[(t - T)^3 + T^3]$
F.   Other (specify)

N4T.5 Imagine that you are preparing an actual-size trajectory diagram of a freely falling object. The time interval between positions is 0.02 s. How long should you draw the acceleration arrows on your diagram?
A.   9.8 m
B.   0.20 m
C.   0.04 m
D.   3.9 cm
E.   0.39 cm
F.   Other (specify)

N4T.6 At time $t = 0$, a person is sliding due east on a flat, frictionless plane of ice. The net force on this person is due to a battery-powered fan the person holds that exerts a northward thrust force on the person. Assuming that drag is negligible, the eastward component of the person's velocity is unaffected by this force, true (T) or false (F)?

N4T.7 Consider the person described in problem N4T.6. The person's trajectory will look most like which of the following? (The dot shows the person's position at $t = 0$, and east is to the right and north to the top.)

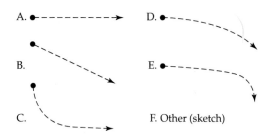

HOMEWORK PROBLEMS

Basic Skills

N4B.1 Construct a graph of the object's x-velocity as a function of time for an object whose x-acceleration is as given to the right. Assume that $v_x(0) = 0$.

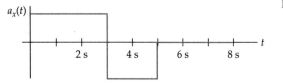

N4B.2 Construct a graph of x-position as a function of time for an object whose x-velocity is as given below. Assume that $x(0) = 0$.

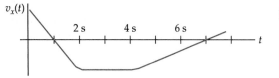

N4B.3 Integrate the following functions from 0 to t.
(a) $f(t) = bt$
(b) $f(t) = b(t - T)$ (b and T are constants)

N4B.4 Integrate the following functions from 0 to t.
(a) $f(t) = bt^3$
(b) $f(t) = b/(t + T)^{1/2}$ (b and T are constants)

N4B.5 Imagine that you are trying to construct an actual-size motion diagram of a falling object with an initial horizontal velocity of 2.0 m/s. Take the time interval between position points to be 0.05 s. How long should you draw the object's initial velocity arrow on the diagram? Its acceleration arrows?

N4B.6 Imagine that you are trying to construct an actual-size motion diagram of a falling object with an initial horizontal velocity of 4.0 m/s. Take the time interval between position points to be 0.15 s. How long should you draw the object's initial velocity arrow on the diagram? Its acceleration arrows?

N4B.7 A stone dropped from rest from the middle of a bridge hits the water below 2.5 s later. How far is the bridge above the water? (Ignore air resistance.)

N4B.8 An object freely falling from rest for 5.0 s will have what final speed? (Express your result in both meters per second and miles per hour, and ignore air resistance.)

Synthetic

N4S.1 Imagine that the net force on a car moving in one dimension is initially large and forward, but subsequently decreases linearly to zero and then remains zero thereafter. Draw graphs of $a_x(t)$, $v_x(t)$, and $x(t)$ for this car, assuming that it starts from rest at $x = 0$.

N4S.2 A car starts at $t = 0$ from rest at $x = 0$ and accelerates at a constant rate until it reaches a cruising speed of

15 m/s. After maintaining that speed for a while, the driver of the car, seeing that a bridge ahead is out, brakes suddenly, and the car comes to rest in a relatively short time. After remaining at rest for a few seconds, the driver then backs up at about 3 m/s. Draw qualitatively accurate graphs of the car's x-position, x-velocity, and x-acceleration as functions of time. Assume that the road is straight and the car initially travels in the positive x direction.

N4S.3 Imagine that the x-position of one car is given by the function $x_1(t) = bt^2$ (with $b = 2.5$ m/s^2), whereas the position of another car is given by the function $x_2(t) = ct$ (where $c = 18$ m/s). Note that both cars have a position equal to zero at $t = 0$. At time $t = 5.0$ s, which car is moving faster? Which one is farther ahead?

N4S.4 Imagine that a car's x-acceleration during a certain time period is given by the function $a_x(t) = -bt$. If the car's x-velocity at $t = 0$ is 32 m/s, and $b = 1.0$ m/s^3, how long will it take the car to come to rest?

N4S.5 The z-velocity of deep-sea probe descending into the ocean is computer-controlled to be $v_z(t) = c + bt^2$ (where c is a constant and $b = 0.040$ m/min^3) until the probe comes to rest. At $t = 0$, $z(0) = -22$ m (that is, 22 m below the ocean surface) and $v_z(0) = -120$ m/min.
(a) What are the units of c? What value must c have?
(b) Find the probe's z-acceleration as a function of time.
(c) At what time and position will the probe come to rest?

N4S.6 A spaceship is approaching Starbase Beta at an initial velocity of 130 km/s in the $+x$ direction. The ensign sets its computer to initiate a braking program starting at time $t = 0$. After this time, the computer controls the ship until the ship docks at time $t = T$ so that its x-acceleration is $a_x(t) = b(t - T)$, where $b = 0.26$ m/s^3. Note that $a_x < 0$ for $t < T$, but gets smaller as t approaches the docking time T.
(a) What must T be so that the ship's x-velocity is also zero at the time of docking?
(b) How far is the ship from the station at $t = 0$? (*Hint:* Use equation NB.10d in appendix NB.)

N4S.7 A mad scientist invents an antigravity device that shields the region of space above a large metal plate from the earth's gravitational field. Specifically, if the device is turned on at time $t = 0$, the effective magnitude of the gravitational acceleration in the region above the plate decreases exponentially with time according to mag$(\vec{a}) = ge^{-qt}$, where g is the usual gravitational field strength and q is a constant. When enough time has passed that qt becomes large, the effective gravitational field above the plate becomes very small compared to g. Assume that the value of q is 3.0 when expressed in the appropriate SI units.

(a) What are the SI units of q?
(b) The mad scientist keeps a 7-kg bowling ball on a shelf 2.0 m above the plate. When the mad scientist throws the gigantic wall switch to turn on the apparatus one day, the vibrations jostle the bowling ball loose, and it rolls off the shelf. The scientist watches in horror as the ball smashes into the plate 1.0 s later. If the ball leaves the shelf with essentially zero vertical velocity at time $t = 0$, what is its speed when its smashes into the delicate apparatus? (*Hints:* See equation NB.12a in appendix NB, and set $b = -q$. Note that $e^0 = 1$.)
(c) How high was the shelf above the plate?

N4S.8 Imagine that if you step hard a certain car's accelerator at $t = 0$, its forward acceleration is given by $a_x(t) = a_0 \sin \omega t$ for $0 \le \omega t \le \pi$ and becomes zero afterward, where $a_0 = 5.0$ m/s^2 and q is a constant with a value of $\pi/5$ when expressed in the appropriate SI units. Note that the car's acceleration is initially zero when $t = 0$ but increases to a maximum of a_0 as the engine reaches its maximally efficient speed, but then decreases as drag and friction forces begin to oppose the car's motion significantly. When the car reaches its cruising speed at a time t such that $\omega t = \pi$, the car's acceleration becomes zero. (This is not likely to be the actual acceleration function for any realistic car, but we can use it as a simple model.)
(a) What are the SI units of q?
(b) Assume that the car starts from rest at $x = 0$. What are the car's x-velocity and x-position at time $t = 5.0$ s? (*Hint:* The derivative of $\sin \omega t$ is $\omega \cos \omega t$, and the derivative of $\cos \omega t$ is $-\omega \sin \omega t$. What, therefore, are the antiderivatives of $\sin \omega t$ and $\cos \omega t$?)

N4S.9 Imagine that we throw a marble with an initial speed of 2.2 m/s at an angle of 60° above the horizontal.
(a) Draw a trajectory diagram for the marble, using $\Delta t = \frac{1}{30}$ s = 0.033 s for at least 10 time steps, and estimate from your diagram how long it takes the marble to reach the peak of its trajectory.
(b) Use Newton to check your work.

N4S.10 Imagine that an outfielder throws a baseball with an initial speed of 12 m/s in a direction 30° up from the horizontal.
(a) Use a trajectory diagram to determine the height of the peak of its trajectory. I suggest using a time step of $\Delta t = 0.2$ s and a scale of 1.0 m (in reality) = 2.0 cm (on diagram).
(b) Use Newton to check your work.

N4S.11 Imagine an air puck sliding frictionlessly on a level air table. The puck is connected to a string going through a hole in the table's center. A clever student pulls on the string so that it always exerts a force of constant magnitude on the puck, and as a result, the magnitude of the puck's acceleration is always

1.0 m/s². Assume that at time $t = 0$, the puck is 4.0 cm from the center of the table in the $-x$ direction, and is moving at a speed of 13 cm/s in the $+y$ direction.

(a) Construct a trajectory diagram that shows the puck's position at subsequent instants of time separated by $\Delta t = 0.1$ s until $t = 1.15$ s. [*Hint:* the main difference between Figure N4.5 and what you will draw is that the acceleration arrow that you should draw through each position dot during step 1 should point from the puck's position at that instant toward the table's center, since the string exerts its tension force in that direction.]

(b) Check your work using the Newton program.

N4S.12 (a) According to chapter N2, an object moving in a circle of radius r at constant speed v experiences an acceleration of constant magnitude $a = v^2/r$ directed toward the circle's center. But is the converse true? Does an object having a constant acceleration toward the origin necessarily move with a constant speed in a circle? Set the time step to be 0.05 s, the object's initial position to be $[x, y] = [1 \text{ m}, 0 \text{ m}]$, its initial velocity to be $[v_x, v_y] = [0 \text{ m/s}, 1 \text{ m/s}]$ (note: $\vec{v} \perp \vec{r}$), and the acceleration to be -1 m/s² in the r direction ($= 1$ m/s² in the $-r$ direction). Note that the acceleration has a magnitude equal to v^2/r for the given initial conditions. Does this generate a circular trajectory with constant speed and the right radius? Submit a graph justifying your conclusions.

(b) Reset the time step size to 0.05 s, but increase the velocity to 1.2 m/s in the y direction while leaving everything else the same. Now is the path circular? Does making the time step smaller change your result? Submit at least one graph in support of your claims.

(c) Instead of making the acceleration constant and directed toward a central point, make it

constant and always perpendicular to the object's velocity. Is the orbit circular now? For all speeds? If so, are the trajectories you get consistent with $a = v^2/r$? Submit at least one graph with a speed not equal to 1 m/s in support of your case.

Rich-Context

N4R.1 You are a police officer. Your squad car is at rest on the shoulder of an interstate highway when you notice a car fitting the description given in an all-points bulletin passing you at its top speed of 85 mi/h. You jump in your car, start the engine, and find a break in the traffic, a process which takes 25 s. You know from the squad car's manual that when it starts from rest with its accelerator pressed to the floor, the magnitude of its acceleration is $a = a_0 - bt^2$ (where $a_0 = 2.5$ m/s² and $b = 0.0028$ m/s⁴) until $a_0 = bt^2$, and then it remains zero thereafter. Can you catch the car before it reaches the next exit 5.3 mi away?

N4R.2 You are traveling in the fabled Mines of Moria when you come across an open water well in the stone floor of a tunnel. To find out how deep the well is, you drop a stone into the well and count the seconds until you hear the splash. If you *hear* the splash a time $t_h = 5.0$ s after you drop the stone, roughly what is the depth D of the well? Be sure you describe any approximations you have to make to do the problem. [*Hint:* The speed of sound is $v_s = 340$ m/s. Let the actual time of the splash be t_s. This is different from t_h: why? I strongly recommend that you solve for D symbolically before plugging in numbers (or you will likely become totally lost), but pay very careful attention to signs: in particular, make sure that D is positive. If you end up with two possible answers, select the answer that leads to a positive value for t_s. Why is a negative t_s absurd?)

ANSWERS TO EXERCISES

N4X.1 Graphs of $z(t)$ and $v_z(t)$ for the basketball look like this (note that I used the slope method to construct the upper graph):

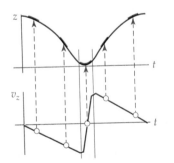

N4X.2 In this case, equations N4.8 imply that

$$v_x(t) = v_{0x} - at \qquad \text{and} \qquad x(t) = v_{0x}t - \tfrac{1}{2}at^2$$
(N4.24)

At $t = 5.0$ s, $v_x(5.0 \text{ s}) = 2.0$ m/s, $x(5.0 \text{ m}) = +35$ m.

N4X.3 The car's x-acceleration at $t = 0$ is the same for both cases, but in the second example the acceleration decreases with time and thus is generally less than in the first case. Therefore it makes sense that it should reach a smaller final x-velocity and not go as far in the same time interval.

N4X.4   At time $t = 0$, we have

$$v_x(0) = \frac{b}{2T^2} - \frac{b}{2(0+T)^2} = \frac{b}{2T^2} - \frac{b}{2T^2} = 0$$
(N4.25a)

$$x(0) = \frac{b}{2}\left(\frac{0}{T^2} + \frac{1}{0+T} - \frac{1}{T}\right) = \frac{b}{2}\left(\frac{1}{T} - \frac{1}{T}\right) = 0$$
(N4.25b)

N4X.5   According to equation N4.4,

$$v_x(t) = \int a_x(t)\, dt + C_1 = \int \frac{b}{(t+T)^3}\, dt + C_1$$

$$= \frac{-b}{2(t+T)^2} + C_1$$
(N4.26)

Setting $t = 0$, we find that

$$v_x(0) = \frac{-b}{2(0+T)^2} + C_1 = \frac{-b}{2T^2} + C_1$$
(N4.27)

Since $v_x(0) = 0$ in the case at hand, $C_1 = +\frac{1}{2}b/T^2$ and

$$v_x(t) = \frac{-b}{2(t+T)^2} + \frac{b}{2T^2}$$
(N4.28)

which is consistent with equation N4.11.

N4X.6   We can find when the ball will hit the ground by setting $z(t) = 0$ and solving for t:

$$0 = z(t) = -\frac{1}{2}gt^2 + z_0 \quad \Rightarrow \quad \frac{1}{2}gt^2 = z_0$$

$$\Rightarrow \quad t^2 = \frac{2z_0}{g}$$

$$t = \sqrt{\frac{2z_0}{g}} = \sqrt{\frac{2(25\text{ m})}{9.8\text{ m/s}^2}} = 2.26\text{ s}$$
(N4.29)

N4X.7   Plugging the numbers again into equations N4.15 yields $v_z(2.2\text{ s}) = -9.6$ m/s (meaning that the ball is going downward) and $z(2.2\text{ s}) = +27.7$ m (meaning that it is still well above the ground).

N4X.8   The downward component of a falling object's velocity is $-v_z(t)$ [since $v_z(t)$ is the upward component]. The rate of change of this downward component is then

$$\frac{d(-v_z)}{dt} = \frac{d}{dt}(-v_{0z} + gt) = 0 + g = g$$
(N4.30)

So g expresses the constant rate at which the downward component of a falling object's velocity increases. Since

$$g = \left(9.8\,\frac{\text{m}}{\text{s}^2}\right)\left(\frac{2.24\text{ mi/h}}{1\text{ m/s}}\right) = 22\,\frac{\text{mi/h}}{\text{s}}$$
(N4.31)

a falling object's downward velocity component increases at a rate of 22 mi/h per second.

N4X.9   For a freely falling object, $\vec{a} = \vec{g} = [0, 0, -g]$. So equation N4.20 becomes

$$\begin{bmatrix} v_x(t) \\ v_y(t) \\ v_z(t) \end{bmatrix} = \begin{bmatrix} \int_0^t a_x(t)\, dt \\ \int_0^t a_y(t)\, dt \\ \int_0^t a_z(t)\, dt \end{bmatrix} + \begin{bmatrix} v_{0x} \\ v_{0y} \\ v_{0z} \end{bmatrix}$$

$$= \begin{bmatrix} \int_0^t 0\, dt \\ \int_0^t 0\, dt \\ \int_0^t (-g\, dt) \end{bmatrix} + \begin{bmatrix} v_{0x} \\ v_{0y} \\ v_{0z} \end{bmatrix}$$

$$\Rightarrow \quad \begin{bmatrix} v_x(t) \\ v_y(t) \\ v_z(t) \end{bmatrix} = \begin{bmatrix} 0 \\ 0 \\ -gt \end{bmatrix} + \begin{bmatrix} v_{0x} \\ v_{0y} \\ v_{0z} \end{bmatrix}$$

$$= \begin{bmatrix} v_{0x} \\ v_{0y} \\ -gt + v_{0z} \end{bmatrix}$$
(N4.32)

Equation N4.21 then becomes

$$\begin{bmatrix} x(t) \\ y(t) \\ z(t) \end{bmatrix} = \begin{bmatrix} \int_0^t v_x(t)\, dt \\ \int_0^t v_y(t)\, dt \\ \int_0^t v_z(t)\, dt \end{bmatrix} + \begin{bmatrix} x_0 \\ y_0 \\ z_0 \end{bmatrix}$$

$$= \begin{bmatrix} \int_0^t v_{0x}\, dt \\ \int_0^t v_{0y}\, dt \\ \int_0^t (-gt + v_{0z})\, dt \end{bmatrix} + \begin{bmatrix} x_0 \\ y_0 \\ z_0 \end{bmatrix}$$

$$= \begin{bmatrix} v_{0x}t - 0 \\ v_{0y}t - 0 \\ -\frac{1}{2}gt^2 + v_{0z}t - 0 \end{bmatrix} + \begin{bmatrix} x_0 \\ y_0 \\ z_0 \end{bmatrix}$$

$$= \begin{bmatrix} v_{0x}t + x_0 \\ v_{0y}t + y_0 \\ -\frac{1}{2}gt^2 + v_{0z}t + z_0 \end{bmatrix}$$
(N4.33)

N4X.10   The acceleration arrow should be 7.9 cm long, and the velocity arrow should be 9 cm long.

N4X.11   The trajectory diagram should look like this:

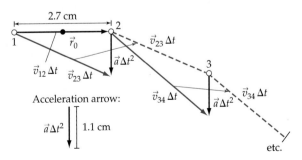

N5 Statics

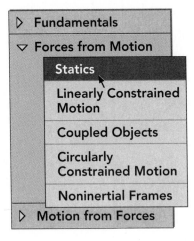

Chapter Overview

Section N5.1: Forces from Motion: An Overview

This chapter begins the "Forces from Motion" subdivision of the unit. In this subdivision, we will consider progressively more complicated cases in which we use Newton's second law and our knowledge about an object's motion to infer things about the forces acting on the object. This chapter kicks off the subdivision by considering the simplest kind of motion: *no* motion.

Section N5.2: Introduction to Statics

We call any problem involving an extended object at rest a **statics problem.** Since the acceleration of the center of mass of an object at rest is zero, Newton's second law implies that *the net external force on the object is zero:*

$$\vec{F}_1 + \vec{F}_2 + \cdots = m\vec{a} = 0 \tag{N5.1}$$

The simplest statics problem involves *only* this principle.

    This section carefully discusses the seemingly simple case of determining the force that a hanging mass m exerts on the hook from which it is suspended. While we all know that the force is mg downward, carefully arguing this displays how even simple cases can entail a multistep argument involving both Newton's second and third laws. This section also notes that we generally call interactions mediated by strings *tension* interactions even when the actual physical connection between the string and the object involves a compression interaction.

Section N5.3: Statics Problems Involving Torque

Some statics problems involve cases in which forces act at different places on an extended object at rest. The angular momentum of an object at rest is not changing; *the net torque on the object* (around any origin O you might choose) *must also be zero:*

$$\vec{\tau}_1 + \vec{\tau}_2 + \cdots = \vec{r}_1 \times \vec{F}_1 + \vec{r}_2 \times \vec{F}_2 + \cdots = \frac{d\vec{L}}{dt} = 0 \tag{N5.2}$$

where $\vec{r}_1$ is the position relative to the origin of the place where $\vec{F}_1$ acts, and so on. An extended object at rest must satisfy equations N5.1 and N5.2 simultaneously. As the examples illustrate, these equations taken together provide a very powerful tool for calculating forces that intuitively look impossible to determine.

Section N5.4: Solving Force-from-Motion Problems

This section describes a problem-solving framework for *any* problem in which we use Newton's second law and information about an object's motion to determine forces. The *translation* step of such a problem should do the following:

1. Display the object in its context.
2. Define necessary *symbols*.
3. Provide a list of known symbols.
4. Define an appropriate coordinate system.

The *model* step of such a problem must answer these questions:

1. What is the object's acceleration $\vec{a}$?
2. What are *all* the forces acting on the object?
3. How can we write these force and/or acceleration vectors as column vectors?
4. What equations do we need in addition to Newton's second law?
5. What approximations and/or assumptions must we make?

To address these needs efficiently:

1. Draw an acceleration arrow (or write $\vec{a} = 0$) on your *translation* diagram.
2. Draw a free-body diagram of the object and write Newton's second law as $m\vec{a} = \vec{F}_1 + \vec{F}_2 + \cdots$, using the force symbols you define in that diagram.
3. Rewrite this equation in column-vector form with forces in the same order.
4. Circle unknowns and connect unknowns to additional equations until you have enough information to solve the problem.
5. Use cartoon balloons to explain any assumptions and approximations.

Following these steps will produce an adequate conceptual model for the problem. The *solution* and *evaluation* steps are as before.

Section N5.5: Solving Statics Problems

If a problem also requires that we balance torques on an object, we should modify the steps as follows:

1'. Write $\vec{a} = 0$ *and* $\vec{\tau}_{\text{net}} = 0$ on your *translation* diagram.
2'. Indicate the origin O on your free-body diagram; and in addition to displaying forces, draw and label *position* vectors from O to where each force acts. Also write a *torque* equation in the form $0 = \vec{r}_1 \times \vec{F}_1 + \vec{r}_2 \times \vec{F}_2 + \cdots$.
3'. Rewrite Newton's second law *and* the torque equation in column-vector form.
4'. Work with unknowns in both equations simultaneously.

The section includes example problem solutions displaying this approach.

N5.1   Forces from Motion: An Overview

An overview of the chapters in this subdivision

This chapter opens the "Forces from Motion" subdivision of the unit. In this subdivision, we will explore a variety of situations in which we use an object's known motion and Newton's second law to determine one or more of the forces acting on the object.

Knowing how to infer forces from motion is useful in many practical physical contexts. For example, knowing the force that a rope must exert in a given circumstance can help us ensure that we choose a rope that will not break. In other cases, the presence or absence of contact forces can tell us whether an object is touching another object. Knowing how strongly friction forces oppose an object's motion can tell us how hard we have to push on the object to get it moving or to keep it moving. These are just a sample of the many practical applications of knowing how to determine forces from motion.

The simplest kind of motion is *no* motion, so in *this* chapter we focus on what we can learn about the forces acting on an object that either we see does not move or we want to ensure does not move. In subsequent chapters, we consider progressively more complicated motions: linear motion in chapters N6 and N7 and circular motion in chapter N8. Chapter N9 closes the subdivision by considering cases in which observing an object's motion from an improper reference frame can tempt us to infer the existence of forces that are really not there (learning to recognize and handle such cases correctly is essential if one is to apply Newton's second law successfully in many realistic situations). Taken together, these chapters provide a solid introduction to this important use of Newton's second law.

N5.2   Introduction to Statics

Definition of a statics problem

With this big picture in mind, let's begin considering what we can learn from the fact that an object is at rest. Imagine an extended object that has various external forces ($\vec{F}_1, \vec{F}_2, \ldots$) acting on it. We call any problem involving an extended object at rest a **statics problem.**

If the object is *completely* at rest, it means that its center of mass is at rest and thus is not accelerating. Newton's second law then implies that *the net external force on the object is zero:*

$$\vec{F}_1 + \vec{F}_2 + \cdots = m\vec{a} = 0 \qquad (N5.1)$$

The simplest statics problems involve only this principle.

Example N5.1

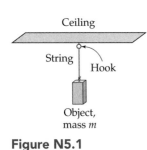

Figure N5.1
An object hanging from the ceiling.

Problem   An object of mass m hangs in the earth's gravitational field from a string attached to a hook embedded in the ceiling. What is the magnitude of the downward force that the hanging mass exerts on the hook?

Translation   Figure N5.1 shows a picture of the situation.

Model and Solution   We know intuitively that the answer is mg, which is the magnitude of the gravitational force on the object. But this gravitational force acts on the hanging object, *not* on the hook directly! So how is the force on the hook connected to this gravitational force? This may seem like a trivial question, but thinking about this simple situation *carefully* illustrates issues and techniques that we will need to deal with more complicated problems.

The hanging object is not moving, so it is not accelerating. Therefore, by Newton's second law, the downward gravitational force on it must be canceled by some *upward* force of the same magnitude for the net force to be zero. Since the string is the only thing the object touches that could exert such a force, the string must exert an upward tension force of magnitude mg on the mass.

Now, by Newton's *third* law, this means that the mass exerts a *downward* force of magnitude mg on the string. Since the string is also at rest, its acceleration is zero, implying (by Newton's *second* law again) that the net external force on the *string* is zero. Therefore, to cancel the downward force on the string, there must be an upward force of magnitude mg acting on it, which must be supplied by its interaction with the hook. By Newton's *third* law, then, if the hook must exert an upward force of magnitude mg on the string, the string must exert a downward tension force of magnitude mg on the hook.

Evaluation   This is what we expected.

Now that we know that we *can* analyze this situation fully, we will not do it again: you may safely assume in the future (as you may have already done in the past) that a hanging mass *at rest* exerts a force equal to its weight on whatever it is attached to. Note, however, that this simple statement is actually a summary of a rather involved line of reasoning!

One more item of minutia before we leave this case. In this example, I called the force exerted on the hook by its contact interaction with the string a *tension* force. But interaction between the hook and the loop of string tied around it is technically a *compression* interaction, as shown in figure N5.2. (Think: if the atoms of the loop and hook could freely intermingle, the loop would simply slip *through* the hook as a wire would through pudding!) However, we *usually* consider a wire or string *not* as an object in its own right but rather simply as a mediator of an interaction between two other objects (the hanging weight and the hook in this case). Since this interaction opposes those objects' separation, it makes sense to think of it as a tension interaction. Even when we *do* consider the wire or string as an object in its own right (as we will in chapter N7), we usually ignore the exact nature of the connection and simply describe the interaction as being a tension interaction.

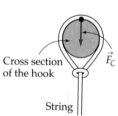

Figure N5.2
The top of the loop of the string connecting the hook to the hanging mass technically exerts a downward *compression* force on the hook. But we usually ignore the detailed nature of the connection between an object and a string and simply call this a *tension* force anyway.

N5.3   Statics Problems Involving Torque

A more interesting set of statics problems involves situations in which multiple forces act at various *different positions* on an extended object at rest. Newton's second law still implies that the net external force on that object must be zero. But forces acting at different positions on an extended object also exert *torques* on that object. If the object is completely at rest, it is not rotating, so its angular momentum is not changing. This implies that *the net torque on the object* (around any origin you might choose) *must also be zero*:

The net torque on an object must also be zero if it is at rest

$$\vec{\tau}_1 + \vec{\tau}_2 + \cdots = \vec{r}_1 \times \vec{F}_1 + \vec{r}_2 \times \vec{F}_2 + \cdots = \frac{d\vec{L}}{dt} = 0 \qquad \text{(N5.2)}$$

Note that $\vec{r}_1$ is the position vector of the point where $\vec{F}_1$ acts on the object (relative to the origin), $\vec{r}_2$ is the same for $\vec{F}_2$, and so on.

Equations N5.1 and N5.2 *both* apply *simultaneously* to any extended object fully at rest.

Example N5.2

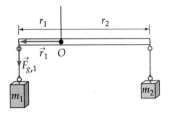

Figure N5.3
Where should we attach the wire to suspend this rod?

Problem   Two objects with masses m_1 and m_2 hang from the ends of a comparatively lightweight rod of length L. (*a*) At what position (relative to m_1) would you want to suspend the rod so that the rod could hang horizontally at rest? (*b*) What upward force must we exert on the rod at that point to suspend the whole system at rest?

Translation   Figure N5.3 illustrates the situation. Imagine that we suspend the rod from a point O that is a distance r_1 from where the object with mass m_1 is suspended. The distance to the point from which the object with mass m_2 is suspended is then $r_2 = L - r_1$.

(a) Model   The rod will only remain at rest if the net torque on the rod around the suspension point O is zero. The supporting tension force $\vec{F}_3$ acts at O by definition; it exerts zero torque around O. However, you can see that the tension force $\vec{F}_1$ exerted by mass m_1 will exert a counterclockwise torque on the rod (that is, $\vec{\tau}$ points directly out of the plane of the picture) while $\vec{F}_2$ from the other weight exerts a clockwise torque (i.e., directly into the plane of the picture). Since these torques point in opposite directions, they will add to zero if they have the same magnitude. If the rod is horizontal, the force exerted on the rod by each hanging mass is perpendicular to the rod, and thus perpendicular to the position vector $\vec{r}_1$ or $\vec{r}_2$ from O to the place where the force is applied. This means that the sine of the angle θ between $\vec{r}$ and $\vec{F}$ in each case is $\sin 90° = 1$, so the magnitude of each torque is simply the product of the magnitude of the force and the distance between O and the point where the force is applied. Since these torques must have equal magnitudes, we must have

$$r_1 F_1 = r_2 F_2 \qquad (N5.3)$$

But the magnitudes of the applied forces in this case are simply the magnitudes of the objects' weights ($F_1 = m_1 g$ and $F_2 = m_2 g$), so equation N5.3 becomes

$$r_1 m_1 g = r_2 m_2 g \qquad (N5.4)$$

The problem says that we are given m_1, m_2, and L (and we know g). We need to find either r_1 or r_2 to locate the hanging point. Equation N5.4 and the relationship $r_2 = L - r_1$ give us the two equations we need to solve for r_1 and r_2.

Solution   Plugging $r_2 = L - r_1$ into equation N5.4 and dividing through by g, we get

$$r_1 m_1 = (L - r_1) m_2 \quad \Rightarrow \quad r_1 (m_1 + m_2) = L m_2$$

$$\Rightarrow \quad r_1 = \frac{m_2 L}{m_1 + m_2} \qquad (N5.5)$$

Evaluation   If you look back at equation C4.6, you will see that this is the same as the distance between m_1 and the system's center of mass, implying that the suspension point O should be at the same horizontal position as the system's center of mass! (Perhaps this is no great surprise to you.)

(b) Model and Solution   The external forces on the rod are the upward tension force $\vec{F}_3$ exerted by the supporting wire and the downward tension forces exerted by the hanging weights. Requiring that the total external force

This mobile's suspension point has the same horizontal position as its center of mass.

be zero means that the total upward force should balance the total downward force: $F_3 = F_1 + F_2 = (m_1 + m_2)g$.

Evaluation   Again, this should be no surprise.

In example N5.1, equations N5.1 and N5.2 were not linked, so we were able to work out the implications of these equations separately. In many problems, however, we have to combine both equations to solve the problem. Doing this allows us to calculate the magnitudes of contact forces that we might think at first glance are impossible to determine. Example N5.3 illustrates such a case.

Problems linking Newton's second law and the law of zero torque

Example N5.3

Problem   A uniform plank of mass m and length L sits on two supports, one a distance $2L/5$ and the other a distance $L/5$ from the plank's center. What are the magnitudes of the normal forces that each exerts on the plank (in terms of mg)?

Translation   The situation is shown in figure N5.4.

Model   Three forces act on the plank: the two supporting forces and the plank's weight (which can be considered to act on the plank's center of mass). If we take the origin O to be the plank's center of mass, then the torque that the weight force exerts around the origin is zero (since the distance between the point where the weight force is applied and the origin is zero by definition). The interaction with the left support exerts a torque into the

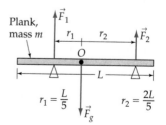

Figure N5.4
How to support a plank.

plane of the drawing, while that with the right support exerts a torque out of the plane of the drawing. These torques will cancel if their magnitudes are equal. Since the normal forces exerted by these interactions act perpendicular to the plank, the magnitude of the torque due to each will be simply rF, where F is the magnitude of the force that the interaction exerts and r is the distance between the origin and the point where the force acts. Since the magnitudes of the torques exerted by the supports must be equal, we have

$$r_1 F_1 = r_2 F_2 \qquad\qquad (N5.6)$$

We know that $r_1 = L/5$ and $r_2 = 2L/5$, but this is not enough information to find F_1 and F_2 separately. How can we proceed?

Well, since the center of mass of the plank is not moving, the three forces acting on the plank must add to zero as well: $\vec{F}_1 + \vec{F}_2 + \vec{F}_g = 0$. Since $\vec{F}_1$ and $\vec{F}_2$ both act upward and $\vec{F}_g$ acts downward, this will happen only if

$$F_1 + F_2 = F_g = mg \qquad\qquad (N5.7)$$

This provides the second equation we need to solve for the two unknowns.

Solution   If we solve equation N5.7 for F_2 and plug the result into equation N5.6, we can solve for F_1:

$$r_1 F_1 = r_2(mg - F_1) \quad\Rightarrow\quad F_1(r_1 + r_2) = r_2 mg$$

$$F_1 = \frac{r_2}{r_1 + r_2} mg = \frac{2\,\cancel{L}/5}{3\,\cancel{L}/5} mg = \frac{2}{3} mg \qquad\qquad (N5.8)$$

Plugging this back into equation N5.7, we find that $F_2 = \frac{1}{3} mg$.

Exercise N5X.1

We do not *have* to choose the origin O to be the plank's center of mass. Redo the plank problem, taking point O to be the position of the left-hand support. (You should get the same answer.)

Exercise N5X.2

A plank 4.0 m long and having a negligible mass is supported by its ends. A person with a weight of 480 N stands 1.0 m from one end. What is the magnitude of the upward force exerted by each support?

N5.4   Solving Force-from-Motion Problems

A general framework for solving force-from-motion problems

This section describes a powerful problem-solving framework for *any* problem in which we use Newton's second law and information about an object's motion to determine something about the forces acting on that object. In section N5.5, I will describe how we can extend and adapt this general framework to handle the special case of statics problems.

What a good translation step should include

The *translation* step of *any* problem involving Newton's second law (including statics problems) should do the following:

1. Display the object of interest in the context of its surroundings.
2. Define necessary *symbols* for masses, distances, angles, and so on.

3. Provide a list of symbols having known values to the right of the drawing.
4. Define an appropriate coordinate system.

Since Newton's second law is a vector equation, setting up a coordinate system is *essential* so that we can unambiguously define vector components in column vectors describing forces acting on the object and the object's acceleration.

The core questions we must resolve in the *model* step of *any* force-from-motion problem solution are the following:

1. What is the object's acceleration $\vec{a}$?
2. What are *all* the forces acting on the object?
3. How can we write these force and/or acceleration vectors as column vectors in terms of symbols defined in the translation step?
4. What equations do we need in addition to Newton's second law?
5. What approximations and/or assumptions must we make?

Core questions to be resolved in the model

The model section of a problem solution is not complete if it does not address each and every one of these concerns. To address these needs as efficiently as possible, do the following:

1. Draw an acceleration arrow (or write $\vec{a} = 0$) on your *translation* diagram.
2. Draw a free-body diagram of the object (isolated from its surroundings) that displays and labels *all* the forces acting on it. Then write Newton's second law (perhaps below the diagram) in the form $m\vec{a} = \vec{F}_1 + \vec{F}_2 + \cdots$ except replace the symbols $\vec{F}_1, \vec{F}_2, \ldots$ with the symbols you have just defined.
3. Rewrite this equation in column-vector form, expressing the components in terms of symbols you have defined earlier. Write the column vectors for the forces in the same order as you did in step 3, so that it is clear which column vector corresponds to which force.
4. This becomes your master equation. Circle any unknown symbols, and link these unknowns to additional equations and/or cancel common factors until you have enough equations to solve for all the unknowns.
5. Add cartoon balloons to any additional equations or value assignments to explain the assumptions and/or approximations you are making.

How to construct an efficient conceptual model

In the second step, think about *everything* that touches the object, and then draw arrows for all forces that might arise from those contact interactions. Then add any forces that might arise from long-range interactions. Note that you should *never* draw arrows representing velocities or accelerations on a free-body diagram: the sole purpose of a free-body diagram is to display the forces that act on the object and where they act. Note also that $m\vec{a}$ is *not* a force acting on the object: it instead describes how the object *responds* to the vector sum of all forces acting on the object! Therefore, the quantity $m\vec{a}$ should *never* appear on a free-body diagram.

In the third step, I find it *most* useful to express vector components in terms of symbols whose values I know to be *positive*. For example, if I knew that a contact force on a certain object acts in the $-z$ direction, I would write

Important comments and suggestions about doing these steps

$$\begin{bmatrix} 0 \\ 0 \\ -F_C \end{bmatrix} \quad \text{instead of} \quad \begin{bmatrix} 0 \\ 0 \\ F_{Cz} \end{bmatrix} \tag{N5.9}$$

where $F_C \equiv \text{mag}(\vec{F}_C) \geq 0$. The point is that this explicitly *displays* the sign of the last component instead of hiding it inside the F_{Cz} symbol, and I find that displaying the sign in this way greatly reduces the probability that I will make a sign error. If I don't *know* whether $\vec{F}_C$ points in the $+z$ or $-z$ direction (or want to allow for both possibilities), then I have to use F_{Cz}; but otherwise I try to use positive symbols exclusively.

If it helps you keep track of things in the fourth step, you can put a check mark by the symbols you know in the master equation and supporting equations as a way of helping you zero in on the unknown quantities. Be sure to circle each unknown only once, so that you can more easily count how many unknowns you have.

The rule of thumb about cartoon balloons is that you should supply a balloon for *any* equation you add to supplement the master equation (unless it is purely a definition) or to any value assignment you make that is not explicitly specified in the problem statement. For example, if the problem statement tells you to ignore air friction, then you can write $\vec{F}_D = 0$ in your list of knowns in the translation step; but if you have to make this assumption unbidden, then you should include $\vec{F}_D$ in your statements of Newton's second law, but add an explanatory balloon to the master equation where you list the components of this column vector as being zero.

Following these five steps will produce the conceptual model for these problems. If you are asked for a prose model, simply explain the issues in steps 1, 4, and 5 in words, but still include a free-body diagram for step 2 and write down the master equation in column-vector form for step 3.

Solution and evaluation steps

The *solution* and *evaluation* steps have pretty much the same structure as in previous problems. You should solve symbolically for any unknowns, *then* plug in numbers (keeping track of units), and check that your result has the correct units and a plausible sign and magnitude.

Following these steps carefully not only provides an efficient way to present your reasoning, but also gives you a way to start a problem even when you are unsure about how to find a solution. Simply going through the steps will often help you work through any confusion you might have.

N5.5   Solving Statics Problems

Modifications of the conceptual model steps for static problems

The conceptual-model outline given in section N5.4 is adequate as it stands for any statics problem that does *not* involve torque. If a problem also requires that we balance torques on an object, we should modify the steps of that outline as follows:

1'. Write $\vec{a} = 0$ *and* $\vec{\tau}_{net} = 0$ on your *translation* diagram.
2'. On your free-body diagram, indicate the origin O around which you are calculating torques; and in addition to drawing force vectors, draw and label *position* vectors from the origin to each point where a force is applied. In addition to the equation expressing Newton's second law, write a torque equation in the form $0 = \vec{r}_1 \times \vec{F}_1 + \vec{r}_2 \times \vec{F}_2 + \cdots$, again supplying your own symbols.
3'. In addition to writing Newton's second law in column-vector form, write the torque equation in column-vector form.
4'. Both of these column-vector equations are your master equations: they need to be satisfied simultaneously in your solution.

In other regards, your solution for a statics problem will be the same as for any other force-from-motion problem involving Newton's second law.

Comments on these modifications

In step 2, remember that you can define the origin O to be anywhere that you like. It is especially helpful to define O to be at the location where an unknown force is applied, because that unknown force will not then appear in the torque equation.

In step 3, the torque column vector for, say, $\vec{r}_1 \times \vec{F}_1$ will usually have only one nonzero component, which you can easily express in terms of $r_1 F_1 \sin\theta$, where θ is the angle between $\vec{r}_1$ and $\vec{F}_1$. You can determine the component's sign by using the right-hand rule. (I try to avoid using negative angles for the reasons discussed in section N5.4.) In rare cases, you may need to evaluate the component form of the cross product.

The examples on the following pages display how one can use these problem-solving outlines to solve fairly complicated problems pretty easily. Example N5.4 does not involve torque, whereas problem N5.5 does.

Example N5.4

Problem   A 65-kg rock climber hangs from two ropes while crossing a gap between two rocks. One end of each rope is tied to the climber's harness; the other ends are anchored 5.5 m to the left and 2.1 m vertically up, and 7.2 m to the right and 1.5 m vertically up from the climber's harness, respectively. If the climber's full weight is to be supported by the ropes, what tension force must each rope exert? If the ropes can supply 2500 N of tension force without breaking, is the climber safe?

Translation

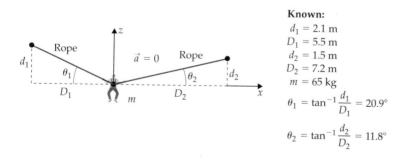

Known:
$$d_1 = 2.1 \text{ m}$$
$$D_1 = 5.5 \text{ m}$$
$$d_2 = 1.5 \text{ m}$$
$$D_2 = 7.2 \text{ m}$$
$$m = 65 \text{ kg}$$

$$\theta_1 = \tan^{-1}\frac{d_1}{D_1} = 20.9°$$

$$\theta_2 = \tan^{-1}\frac{d_2}{D_2} = 11.8°$$

Conceptual Model

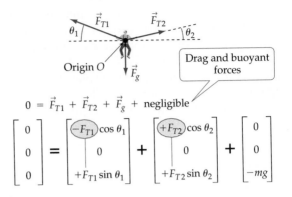

$$0 = \vec{F}_{T1} + \vec{F}_{T2} + \vec{F}_g + \text{negligible}$$

$$\begin{bmatrix} 0 \\ 0 \\ 0 \end{bmatrix} = \begin{bmatrix} -F_{T1}\cos\theta_1 \\ 0 \\ +F_{T1}\sin\theta_1 \end{bmatrix} + \begin{bmatrix} +F_{T2}\cos\theta_2 \\ 0 \\ +F_{T2}\sin\theta_2 \end{bmatrix} + \begin{bmatrix} 0 \\ 0 \\ -mg \end{bmatrix}$$

Prose Model   Torque is not an issue in this problem: we are interested in only the motion of the climber's center of mass here. A climber "supported" by the ropes will be at rest, so $\vec{a} = 0$. The climber only touches the ropes and the air. The buoyant force exerted by the air on a person is negligible, and (unless a very stiff wind is blowing) any drag force exerted by the air will be negligible. Newton's second law therefore implies the equation above. The tension magnitudes F_{T1} and F_{T2} are unknown, but the top and bottom rows of this equation provide two meaningful equations we can solve for these quantities.

Solution   The top row of this equation implies that

$$F_{T1}\cos\theta_1 = F_{T2}\cos\theta_2 \quad \Rightarrow \quad F_{T2} = F_{T1}\frac{\cos\theta_1}{\cos\theta_2} \tag{1}$$

Plugging this into the bottom row, we get

$$mg = F_{T1}\sin\theta_1 + F_{T1}\frac{\cos\theta_1}{\cos\theta_2}\sin\theta_2 = F_{T1}(\sin\theta_1 + \cos\theta_1\tan\theta_2) \tag{2}$$

Solving this for F_{T1} (and noting that $\tan\theta_2 = d_2/D_2$), we get

$$F_{T1} = \frac{mg}{\sin\theta_1 + (\cos\theta_1)(d_2/D_2)}$$

$$= \frac{(65\,\text{kg})(9.8\,\text{m/s}^2)}{\sin 20.9° + (\cos 20.9°)(1.5\,\text{m}/7.2\,\text{m})}\left(\frac{1\,\text{N}}{1\,\text{kg·m/s}^2}\right) = 1160\,\text{N} \tag{3}$$

Plugging this into equation 2, we get $F_{T2} = (1160\,\text{N})(\cos 20.9°)/(\cos 11.8°) = 1100\,\text{N}$.

Evaluation   These both come out with the right units and sign (plus signs are appropriate for magnitudes), and seem reasonable. So the ropes *should* hold!

Example N5.5

Problem   A 580-kg drawbridge 6.6 m long spans a moat around a castle. One end of the drawbridge is connected by a hinge to the castle wall. The drawbridge is also held up by two chains, each of which is connected to a point on the drawbridge 4.4 m from the hinge. The bridge is raised by reeling in the chains, which enter the castle wall 4.4 m above the hinge. When the bridge is raised just above its support on the far side of the moat (but the bridge is still essentially horizontal), what is the magnitude of the tension force that each chain must exert on the drawbridge? Also find the components of the force that the *hinge* must exert on the drawbridge.

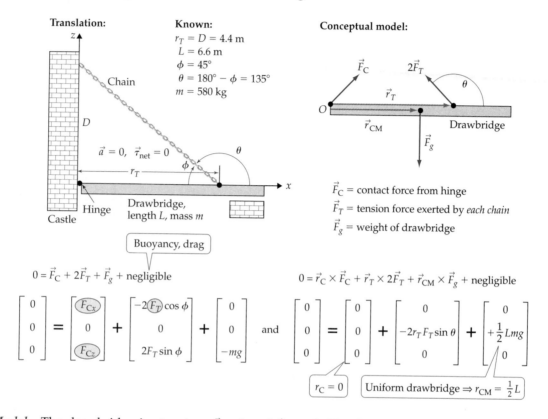

Prose Model   The drawbridge is at rest, so $\vec{a} = 0$ and $\vec{\tau}_{net} = 0$. The drawbridge touches only the hinge, the air, and the two chains. We will treat contact forces with the air as negligible, and we assume that each chain exerts the same tension force $\vec{F}_T$ on the drawbridge. Assuming that the drawbridge is uniform, its center of mass (which is the effective position where its weight acts) is located halfway along its length. We know nothing about either the magnitude or the direction of the contact force $\vec{F}_C$ the hinge exerts on the drawbridge, so it is convenient to define the origin O at the hinge so that this force drops out of the torque equation. Newton's second law and the requirement that the net torque be zero then imply the equations above. Note that the right-hand rule implies that the torque exerted by $\vec{F}_T$ points out of the plane of the drawing (i.e., the $-y$ direction) while the gravitational torque goes into that plane (the $+y$ direction).

Solution   The middle line of the torque equation implies that $2r_T F_T \sin\theta = \frac{1}{2}Lmg$, so

$$F_T = \frac{Lmg}{4r_T \sin\theta} = \frac{(6.6\text{ m})(560\text{ kg})(9.8\text{ N/kg})}{4(4.4\text{ m})\sin 135°} = 3000\text{ N} \qquad (1)$$

Plugging this into the first and third rows of Newton's second law implies that

$$F_{Cx} = +2F_T \sin\phi = 2(3000\text{ N})(\sin 45°) = 4200\text{ N} \qquad (2a)$$

$$F_{Cz} = mg - 2F_T \sin\phi = (580\text{ kg})(9.8\text{ N/kg}) - 2(3000\text{ N})(\sin 45°) = 1400\text{ N} \qquad (2b)$$

The units all come out right, and $\vec{F}_C$ is in the first quadrant (as we guessed in the drawing: this is the only way that the vectors can add to zero). The magnitudes of F_{Cx} and F_{Cz} are comparable to that of $\vec{F}_T$, so everything looks pretty good.

TWO-MINUTE PROBLEMS

N5T.1   A weight hangs from a string but is pulled to one side by a horizontal string, as shown. The tension force exerted by the angled string is
A.   Less than the hanging object's weight.
B.   Equal to the hanging object's weight.
C.   Greater than the hanging object's weight.

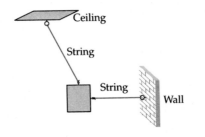

N5T.2   A person would like to pull a car out of a ditch. This person ties one end of a chain to the car's bumper and wraps the other end around a tree so that the chain is taut. The person then pulls on the chain perpendicular to its length, as shown in the picture. The magnitude of the force that the chain exerts on the car in this situation is
A.   Much smaller than the force the person exerts on the chain.
B.   About equal to the force the person exerts on the chain.
C.   Much bigger than the force the person exerts on the chain.

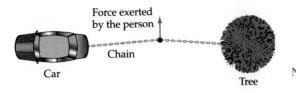

N5T.3   The lid of a grand piano is propped open as shown. Which arrow most closely approximates the direction of the force that the hinge exerts on the lid?

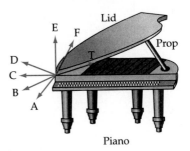

N5T.4   Imagine that a helicopter's rotor spins clockwise. The helicopter engine must continually exert a torque on the rotor to keep it spinning against the drag that the air exerts on the rotor. Note that a helicopter is usually designed so that its center of mass is directly under the rotor. In order for the helicopter to hover motionless in the air, a small rotor at the helicopter's tail is necessary. As viewed by someone looking at the tail from the helicopter's front, the small rotor must blow air
A.   To the left.
B.   To the right.
C.   Vertically upward.
D.   Vertically downward.
E.   In some combination of these directions.

N5T.5   A board of mass m lies on the ground. What is the magnitude of the force that you would have to exert to lift *one end* of the board barely off the ground (assuming that the other end still touches the ground)?
A. $2mg$
B. mg
C.   The answer depends on the length of the board.
D. $\frac{1}{2}mg$
E.   Other (explain).

N5T.6   Imagine that you continue to lift the board described in problem N5T.5. Assume that the force you exert is always perpendicular to the board, and that one end of the board always remains on the ground. What happens to the magnitude of the force you exert on the end as the angle between the board and the ground increases? It
A.   Increases
B.   Decreases
C.   Remains the same

HOMEWORK PROBLEMS

Basic Skills

N5B.1   Consider the situation shown in the drawing associated with problem N5T.1. If the diagonal string makes an angle of θ from the vertical, what is the magnitude of the tension force it exerts on the hanging mass, as a fraction or multiple of the magnitude mg of the hanging mass's weight? Please explain your reasoning.

N5B.2   Imagine a glider with mass m on a frictionless air track that is inclined at an angle θ with respect to the horizontal. The glider is tied to the upper end of the track with a string that is parallel to the track. Find an expression for the tension force that this string exerts on the glider in terms of m, g, and θ. Please explain your reasoning.

N5B.3   Imagine that we place a 10-g weight at the 10-cm mark on a uniform meter stick, and we find that the meter stick now balances at the 45-cm mark. What is the mass of the meter stick? (*Hint:* Put the origin at the point O.)

N5B.4   Imagine that a ladder leans against the house at an angle of 20° with respect to the vertical. A person with a weight of 650 N stands on the ladder at a point 1.0 m from its bottom. What are the magnitude and direction of the torque that this person exerts on the ladder around the point where the ladder touches the ground? (Express the direction as seen by the person standing on the ladder.)

N5B.5   A skateboard rider with mass m stands on the middle of a skateboard of length L. Let's place the origin O at the skateboard's rear wheels. Let the upward forces exerted by the road on the skateboard's front and rear wheels be $\vec{F}_{CF}$ and $\vec{F}_{CR}$, respectively.
   (a) In terms of these forces, what are the magnitude and direction of the torque exerted on the skateboard by the front wheels?
   (b) What are the magnitude and direction of torque exerted by the rear wheels?
   (c) What are the magnitude and direction of the torque exerted by the skateboard rider? (Express your directions relative to the forward direction of the skateboard.)

N5B.6   A gymnast stands on a balance beam of length L that is supported above the gym floor by two supports at its ends. Let's place the origin O at the point where the gymnast's foot touches the beam. Let's say that the gymnast is $\frac{1}{5}L$ from the right end, the mass of the gymnast is m, and the mass of the beam is M. Let the contact forces exerted on the beam by the right and left supports be $\vec{F}_{CR}$ and $\vec{F}_{CL}$, respectively.

(a) In terms of these quantities, what are the magnitude and direction of the torque exerted by the right support?
(b) What are the magnitude and direction of the torque exerted by the left support?
(c) What are the magnitude and direction of the torque exerted by the beam's weight?
(Express your directions relative to the picture shown below.)

(See problem N5B.6.)

N5B.7   Pole-vaulters hold their poles in front of them as they run at the beginning of their vault. Explain qualitatively why the upward force that a vaulter exerts on the pole with his or her front hand must exceed the pole's weight.

(See problem N5B.7.)

Synthetic

The starred problems are well suited for practicing the problem-solving framework discussed in this chapter.

*N5S.1   In the situation discussed in problem N5T.2, find the force that the person can exert on the car if the distance between the car and the tree is 5.0 m and the length of the chain between the car and the tree is 5.2 m.

*N5S.2 A plank 3.0 m long and having a mass of 20 kg is supported at its ends. Imagine that a person with a mass of 50 kg stands 1.0 m from one end. What are the magnitudes of the forces exerted by the supports?

*N5S.3 Imagine that you have a plank 2.0 m long. You put a small piece of wood under the plank 60 cm from its far end, making the plank into a lever. A friend with a mass of 65 kg stands on the far (short) end of the plank. How much downward force do you have to exert on your end to lift your friend?

*N5S.4 Imagine that a pole with a mass of 8 kg and a length of 1.8 m is connected to a wall so that the pole sticks out horizontally from the wall. One end of the pole is connected directly to the wall, while the other end is connected to a higher point on the wall by a chain that makes a 45° angle with respect to the pole. If a 65-kg person hangs from the center of the pole, what is the tension on the chain?

*N5S.5 A certain board is 4.0 m long and rests horizontally and somewhat above the ground on two cylinders of wood, each supporting the board 0.5 m from the corresponding end of the board. (The axis of each cylinder is perpendicular to the length of the board.) The board has a mass of 21 kg. If a person with a mass of 68 kg steps on one end of the board, will it support that person?

*N5S.6 A board has one end wedged under a rock having a mass of 380 kg, and is supported by another rock that touches the bottom side of the board at a point 85 cm from the end under the rock. The board is 4.5 m long, has a mass of about 22 kg, and projects essentially horizontally out over a river. Is it safe for an adult with a mass of 62 kg to stand at the unsupported end of the board? If not, how far out on the board can one safely go?

*N5S.7 A 5-m ladder of negligible mass leans against the side of a house, with the bottom of the ladder 2.5 m from the wall. The contact interaction between the ladder and the ground is able to exert a horizontal static friction force of no more than $0.4mg$, where mg is the weight of the person on the ladder. Assume that the contact interaction between the wall and the top of the ladder is essentially frictionless. How high can a person safely climb on this ladder without running the risk of the ladder's slipping out from under him or her? (*Hint:* the answer is independent of the person's mass.)

*N5S.8 A ladder of length L and mass m leans against the side of a house, making an angle of θ with the vertical. Assume that the ladder is free to slide at the point where it touches the side of the house (there is no significant friction). Find an expression for the normal force that the side of the house exerts on that end of the ladder in terms of m, g, L, and θ.

How high can this person safely climb? (See problem N5S.7.)

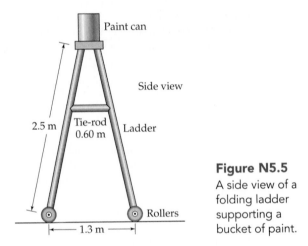

Figure N5.5
A side view of a folding ladder supporting a bucket of paint.

*N5S.9 Figure N5.5 shows a 20-kg bucket of paint on top of a folding ladder that is much lighter. The feet of the ladder have rollers to make it easy to move the ladder around the hard floor. What forces must the tie-rods exert on the ladder to keep the ladder from collapsing?

Rich-Context

N5R.1 A person's forearm consists of a bone (which we can model by a rigid rod) that is free to rotate around the elbow. When the forearm is held level with the elbow bent at a 90° angle, the bone is supported by the biceps muscles, which are attached by a tendon to the bone a few centimeters away from the axis represented by the elbow. By making measurements on your own forearm and appropriate approximations and estimations (be sure that you describe them), estimate the tension force on the biceps tendon when you hold an object with a mass of 10 kg in your hand (with your forearm level and elbow bent at 90°).

(See problem N5R.1.)

greater when a person is hanging near one side of the gorge or in the middle? Assume that the rope doesn't stretch much when under tension.

(See problem N5R.2.)

*N5R.2 You are part of a research team in the deepest jungles of South America. At one point, the team has to cross a 20-m-wide gorge by stringing a rope across it and then crossing hand over hand while hanging from the rope. After someone has managed to hook the rope on a rock on the other side of the gorge, a relatively new member of the team ties the near end of the rope to a tree in such a way that the rope is very tight. A more experienced member makes the neophyte untie the rope and retie it so that there is some slack. Why? If the rope can exert a tension force of 3500 N without breaking and the heaviest member of the group has a mass of 110 kg, how much slack do you need? Is the tension on the rope

ANSWERS TO EXERCISES

N5X.1 The torque exerted by the left-hand support is now zero, but gravity exerts a clockwise torque on the plank around the new O. Assuming that the gravitational force effectively acts as if it is applied to the object's center of mass, the distance from our new O to the center of mass is $r_{CM} \equiv \frac{1}{5}L$. Force $\vec{F}_2$, which is applied at a distance of $r_{2n} = \frac{3}{5}L$, exerts a counterclockwise torque on the plank. In each case the forces are exerted perpendicular to the plank, so the torque magnitudes are simply $r_{CM}F_g$ for the gravitational force and $r_{2n}F_2$ for the other force. These torques can cancel only if their magnitudes are equal, so we must have $r_{CM}mg = r_{2n}F_2$

$$\Rightarrow \quad F_2 = \frac{r_{CM}}{r_{2n}}mg = \frac{L/5}{3L/5}mg = \frac{1}{3}mg \quad (N5.10)$$

as we found before. Plugging this into equation N5.7 yields the same result as in equation N5.8.

N5X.2 Let's put the origin at the nearest support to the person. Since the person is at rest, the person's weight must be equal in magnitude to the contact

force that the plank exerts on the person, which by Newton's third law must have the same magnitude as the force that the person exerts on the plank. So the person exerts a downward force equal in magnitude to mg at a distance $r_g = 1.0$ m from the origin, and the far support exerts an upward contact force $\vec{F}_{CF}$ at a point a distance $L = 4.0$ m from the origin. Assuming that the plank is level, both forces are exerted perpendicular to the plank, so the magnitude of the torque exerted by each is the simple product of the force magnitude and the distance from the origin. For the torques to balance, their magnitudes must be equal, implying that $r_{CM}mg = LF_{CF}$

$$\Rightarrow \quad F_{CF} = \frac{r_{CM}}{L}mg = \frac{480\,\text{N}}{4} = 120\,\text{N} \quad (N5.11)$$

Newton's second law implies that the upward forces exerted by the support on the plank must be equal in magnitude to the downward force of $mg = 480$ N, so the near support must exert an upward force of magnitude $F_{CN} = 360$ N.

N6

Linearly Constrained Motion

Chapter Overview

Introduction

This chapter continues our exploration of what we can learn about forces by observing motion. In this chapter we will consider problems involving objects that move in straight lines with either constant or nonconstant speeds (these are **linearly constrained motion** problems).

Section N6.1:   Free-Particle Diagrams

In the remainder of unit N, we will consider torque only very rarely. We will instead focus on the *center-of-mass* motion of objects, modeling them simply as particles located at their centers of mass. A **free-particle diagram** portrays the object as a particle responding to external forces represented by arrows, whose directions are displayed relative to appropriately chosen coordinate axes. Such a diagram greatly facilitates determining force components, which we need to know to apply Newton's second law $\vec{F}_{net} = m\vec{a}$ in column-vector form. A free-particle diagram should include

1. A sketch of the object with a central dot representing its center of mass.
2. A labeled arrow (with its tail attached to the dot) for each external force acting on that object.
3. Reference frame axes.

We will distinguish force symbols that would otherwise be the same by using a superscript to indicate the external thing exerting the force on the object.

Section N6.2:   Motion at a Constant Velocity

If an object moves in a straight line at a constant velocity, its acceleration is zero, implying that the net force on the object must be zero, just as if the object were at rest. It is convenient in such problems to orient the reference frame so that as many forces as possible lie along coordinate axes.

Section N6.3:   Static and Kinetic Friction Forces

A contact interaction between two solid objects can exert a force *parallel* to the surfaces in contact. If this force prevents the surfaces from sliding relative to each other, we call it a **static friction force.** This force arises because atoms in the surfaces become essentially "cold-welded" to one another. A static friction force automatically adjusts its magnitude to whatever value keeps the surfaces from sliding as long as it is less than a certain limit $\vec{F}_{SF,max}$ that depends on the characteristics of the surfaces and how strongly they are pressed together. Empirically, we find that

$$\text{mag}(\vec{F}_{SF}) \leq \text{mag}(\vec{F}_{SF,max}) \approx \mu_s \, \text{mag}(\vec{F}_N) \qquad \text{(N6.5)}$$

Purpose: This equation describes the magnitude of the static friction force that a contact interaction between solid objects exerts on those objects.

Symbols: $\vec{F}_{SF}$ is the force of static friction; $\vec{F}_N$ is the normal force exerted by the same contact interaction; and μ_s is the **coefficient of static friction,** a unitless constant that depends on the characteristics of the surfaces.
Limitations: This is a simplified model that is only approximately true.
Notes: This equation only specifies an *upper limit* for $\text{mag}(\vec{F}_{SF})$.

If the surfaces are sliding relative to one another, then we call the part of the contact force that acts parallel to the surfaces a *kinetic friction* force. Empirically,

$$\text{mag}(\vec{F}_{KF}) \approx \mu_k \text{mag}(\vec{F}_N) \qquad (\text{N6.6})$$

Purpose: This equation describes the magnitude of the kinetic friction force exerted by a sliding contact interaction between solid objects.
Symbols: $\vec{F}_{KF}$ is the kinetic friction force; $\vec{F}_N$ is the normal force exerted by the same contact interaction; and μ_k is the **coefficient of kinetic friction,** a unitless constant that depends on the characteristics of the surfaces.
Limitations: This is a simplified model that is only approximately true.
Notes: Note also that (almost always) $\mu_k < \mu_s$ for a given pair of surfaces.

Although these equations look similar, the static friction equation only specifies an *upper limit* to the force while the other specifies the force's nearly steady value.

Section N6.4: Drag Forces

When a sufficiently large object moves through a fluid, its contact interaction with the fluid exerts an opposing drag force whose magnitude is roughly

$$\text{mag}(\vec{F}_D) \approx \tfrac{1}{2} C \rho A v^2 \qquad (\text{N6.11})$$

Purpose: This equation describes the magnitude of the drag force $\vec{F}_D$ exerted by a fluid on an object moving through it with speed v.
Symbols: ρ is the density of the fluid, A is the object's cross-sectional area, and C is a dimensionless **drag coefficient** that depends on the object's shape and surface characteristics.
Limitations: This expression works only if A and v are sufficiently large, ρ is sufficiently small, and the fluid is not very viscous. It works for most sports projectiles in air, but $F_D \propto v$ for slow-moving objects in water.
Note: The coefficient C is 0.5 for a sphere and is typically less than 2.

Section N6.5: Linearly Accelerated Motion

Solving problems in which the object accelerates while moving along a line is much easier if you *orient your reference system so that one axis is parallel to* $\vec{a}$. Some such problems are hybrid problems in which we use information about motion (or lack thereof) in one component direction to determine characteristics of the object's motion in another direction.

Section N6.6: A Constrained-Motion Framework

You can use the general framework discussed in section N5.4 for solving **constrained-motion problems,** except that (1) you should draw a free-*particle* diagram instead of a free-body diagram and (2) you should draw an acceleration arrow $\vec{a}$ on your translation diagram and make *sure* that it is right.

N6.1   Free-Particle Diagrams

Newton's second law

In this chapter and in much of the rest of the text, we will be exploring the consequences of applying Newton's second law to the *center-of-mass* motion of objects large and small. Newton's second law $\vec{F}_{net,ext} = m\vec{a}_{CM}$ tells us that the *net external force* acting on an object causes its center of mass to accelerate in inverse proportion to its total mass m. In the interests of simpler notation, from now on I will simply write Newton's second law as

$$\vec{F}_{net} = m\vec{a} \qquad (N6.1)$$

and we will automatically assume that when we are talking about an extended object, $\vec{F}_{net}$ refers to the net *external* force on that object and $\vec{a}$ is the acceleration of its *center of mass*.

We will work with the component form of the law

It is crucial to note that this is a *vector* equation, and thus is equivalent to three completely independent component equations:

$$\begin{bmatrix} F_{net,x} \\ F_{net,y} \\ F_{net,z} \end{bmatrix} = m \begin{bmatrix} a_x \\ a_y \\ a_z \end{bmatrix} \quad \text{or} \quad \begin{array}{l} F_{net,x} = ma_x \\ F_{net,y} = ma_y \\ F_{net,z} = ma_z \end{array} \quad \begin{array}{r} (N6.2a) \\ (N6.2b) \\ (N6.2c) \end{array}$$

where $F_{net,x}$ is the sum of the x components of all the (external) forces acting on the object, $F_{net,y}$ is the sum of the y components of the same, and so on.

A free-particle diagram helps us determine force components

To determine the components of the net external force on an object, we need to know the components of *each* of the forces acting on that object. Starting in this chapter, we will use a variant of the free-body diagram to help us do this. A **free-particle diagram** combines aspects of the free-body and net-force diagrams introduced in chapters N1 and N3 in a way that is especially well suited to this task. A free-particle diagram consists of the following items:

1. A sketch of the object, with a dot representing its center of mass.
2. A labeled arrow representing *each* (external) force acting on the object *with its tail attached to the center of mass.* (If multiple forces act in the same direction, draw them in sequence to display the total force acting in that direction.)
3. A set of reference frame axes.

The differences between free-body and free-particle diagrams

A free-particle diagram differs from a free-body diagram in several crucial ways. First, a free-particle diagram makes *explicit* the idea that we are modeling the object as if it were a particle located at its center of mass. Second, drawing the force arrows so that they are attached to the center of mass (rather than the point where the forces are applied) makes it easier to read the components of these forces visually. It is also usually easier to see what the net force on an object will be from a free-particle diagram than from a free-body diagram, making drawing a separate net-force diagram less crucial. Finally, we did not worry about drawing reference frame axes in chapters N3 and N4 because they were not needed for a qualitative discussion of forces; but we will need them from now on. Figure N6.1 illustrates these differences.

We will not be concerned with torque in what follows

In chapter N5, we were concerned with the necessary conditions for an object to be at rest and also *not rotating*. Free-*body* diagrams, which show exactly where the external forces act on an object, are useful in such cases because they make it easier to compute torques. From now on, however, we will be primarily interested in the motion of an object's *center of mass*. Information about where forces act on the object is thus not relevant (and can be

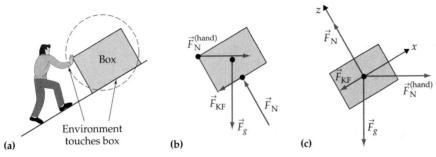

Figure N6.1

(a) A person pushes a large box up an incline. Seeing where the environment touches the box helps us determine and classify the forces acting on it. (b) A free-body diagram of the box. (c) A free-particle diagram of the box. Note that the tails of all vectors are placed at the box's center of mass, even though that is not actually where the force is applied.

distracting), and so is removed from a free-particle diagram. Because of this, a free-particle diagram is more abstract than a free-body diagram. You may find it easiest (particularly at first) to draw a free-*body* diagram before you deal with the abstractions involved in drawing a free-*particle* diagram: if so, I encourage you to do so.

We generally label the forces in a free-particle diagram using the force classification scheme in chapter N1. Sometimes forces arising from distinct interactions end up with the same symbol, which could be confusing. In such cases, let us follow the convention of attaching to the force symbol a super-script (in parentheses) that specifies the other object involved in the interaction exerting the force. For example, in figure N6.1 there is a normal force acting on the box due to its contact interaction with the incline and another normal force arising from its contact interaction with the hand of the person pushing it. We distinguish the two by attaching a superscript "(hand)" to the symbol for the latter force.

N6.2   Motion at a Constant Velocity

When an object moves at a constant velocity, its acceleration is zero. Newton's second law then implies that the net force acting on that object must be *zero*, just as if the object were at rest. Indeed, we can consider *rest* to be simply a special case ($\vec{v} = 0 =$ constant) of constant velocity.

Constant-$\vec{v}$ problems are like statics problems

Example N6.1

Problem   Imagine that you push a 50-kg cart with frictionless wheels at a constant speed up a ramp that makes an angle of 30° with the horizontal. If you push parallel to the ramp, how hard do you have to push?

Translation   Figure N6.2 shows a picture of the situation and a free-particle diagram of the cart. The two normal forces are perpendicular to each other, so if we tilt our reference frame axes so that the x axis is parallel to the incline, then each normal force has only one nonzero component.

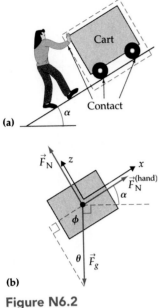

Figure N6.2
(a) A person pushing a cart up a ramp. (b) A free-particle diagram for the cart in this situation.

Model Since the cart's velocity is constant, we must have $0 = \vec{F}_{net} = \vec{F}_g + \vec{F}_N + \vec{F}_N^{(hand)}$, which in component form reads

$$\begin{bmatrix} 0 \\ 0 \\ 0 \end{bmatrix} = \begin{bmatrix} -mg\sin\theta \\ 0 \\ -mg\cos\theta \end{bmatrix} + \begin{bmatrix} 0 \\ 0 \\ +F_N \end{bmatrix} + \begin{bmatrix} +F_N^{(hand)} \\ 0 \\ 0 \end{bmatrix}$$

$$\Rightarrow \quad \begin{aligned} 0 &= -mg\sin\theta + F_N^{(hand)} \\ 0 &= 0 + 0 \\ 0 &= -mg\cos\theta + F_N \end{aligned} \qquad (N6.3)$$

where $F_N^{(hand)} \equiv \text{mag}(\vec{F}_N^{(hand)})$, $mg = \text{mag}(\vec{F}_g)$, and $F_N \equiv \text{mag}(\vec{F}_N)$. Note that since $\phi + 90° + \alpha = 180° = 90° + \phi + \theta$, θ must be the same as the angle of the ramp: $\theta = \alpha = 30°$. The top and bottom rows give us enough information to solve for our unknowns F_N and $F_N^{(hand)}$.

Solution Solving the top row (x component) of equation N6.3 for $F_N^{(hand)}$, we get

$$F_N^{(hand)} = mg\sin\theta = (50\text{ kg})(9.8\text{ N/kg})(\sin 30°) = 245\text{ N} \qquad (N6.4)$$

Evaluation This has the right units and is positive, as a magnitude should be, and seems reasonable. Although we were not asked for F_N, note that in the bottom row (z component) of equation N6.3 $F_N = mg\cos\theta < mg$: the magnitude of the normal force that the contact interaction with the incline exerts on the cart is thus *smaller* than the cart's weight here!

The key to solving force-from-motion problems

The most important thing to note in example N6.1 is how I expressed Newton's second law in column-vector form, expressing the components of each force vector in terms of that vector's *magnitude* (keeping careful track of signs), and finally reduced the vector equation to three separate *component* equations. This process is really the *key* to solving quantitative force-from-motion problems.

How to orient frame axes when $\vec{a} = 0$

Note also that when we solve a problem in which the object's acceleration is zero, it is usually most convenient to orient our reference frame so that as many force components as possible are zero. I did this in example N6.1 by tilting the reference frame so that the x axis is parallel to the incline.

N6.3   Static and Kinetic Friction Forces

Many realistic situations involving objects at rest or objects moving at constant speeds involve *friction* forces. The purpose of this section is to explore the quantitative nature of static and kinetic friction forces exerted by contact interactions between two solid objects.

Static friction

Imagine the following situation. A large and heavy box sits on the floor of your room. Imagine that you push on the box in a direction parallel to the floor, as shown in figure N6.3a. What happens? You know from experience that if you don't push hard enough, you can exert a steady horizontal force on the box and yet it doesn't move.

How can we reconcile this behavior with Newton's second law? Since the box remains at rest, it is not accelerating, and therefore there must be *zero* net force on the box. Since you are pushing horizontally forward on the box, there must be another force acting backward on the box that exactly balances whatever force you exert, as shown in figure N6.3b. But what exerts this force, and why does it seem to automatically adjust its value to match however hard we push on the box?

The following model helps us understand this force. When you place a box on the floor, the box's atoms and the floor's atoms come into contact, microscopic hills in the box's surface become interlocked with microscopic valleys in the floor and vice versa, and some of the box's atoms actually become "cold-welded" to floor atoms. When you then push the box parallel to the floor, the interactions between the atoms of the box and the atoms of the floor under the box push the floor atoms forward, and thus (by Newton's third law) also push the box atoms backward. We call the total force exerted on the box due to this interaction between cold-welded atoms a *static friction* force.

This force adjusts its magnitude to keep the box at rest. If you push the box harder forward, the interaction intensifies, pushing the floor atoms farther forward and the box atoms more strongly backward. If you ease up, the interaction becomes less intense, allowing atoms on the surface of both the floor and the box to relax back closer to their original positions. Thus, however hard you push, the static friction force adjusts to exactly cancel the sum of horizontal components of the other forces acting on the box, keeping the box at rest.

However, this works only up to a point. If you push hard enough, the box will suddenly start to slide. Once it starts sliding, you can keep the box moving at a constant velocity by exerting a steady (and usually smaller) force.

Again, how do we reconcile this behavior with Newton's second law? Here we are exerting a steady horizontal force on the box, and yet it is only moving with a constant velocity, not accelerating (there may be only a brief moment of acceleration as the box began to move). This means that there *still* must be some horizontal force acting to cancel the effect of your push. This force is the *kinetic friction force between* the rubbing surfaces.

Here is a simplified model for understanding this force. When you finally get the box moving, the microscopic hills in the box's surface have to push past microscopic hills in the floor's surface. When two hills come in contact, they momentarily weld together and deform as the surfaces continue to move before ultimately the hills break apart. As the hills on the box and floor surfaces meet and deform, their interaction exerts a force on the box's hills that opposes the deformation. The net result of the tiny forces exerted by untold numbers of interacting hills is a (roughly) steady force opposing the box's motion.

It turns out that while the magnitude of this friction force can depend weakly on the speed of the box, it is generally *less* than the magnitude of force required to start the box moving at first (the welds that form on the fly are not as strong as ones that can form when surfaces are at rest relative to each other). This means that you have to exert a lot of force to get the box going, but once it is moving, not as much force is required to *keep* it moving.

This phenomenon is illustrated by the graph shown in figure N6.4. This graph illustrates what happens when you slowly and steadily increase a manually applied horizontal force $\vec{F}$ to a motionless box. At first, the magnitude of the static friction force due to the box's interaction with the floor increases exactly in step with that of applied horizontal force, keeping the box at rest. But as the magnitude of your applied force continues to increase, there comes a point where the bonds between floor atoms and box atoms cannot stretch any farther without breaking: at this point the atoms are exerting the maximum possible static friction force that they can (we call the magnitude of this force $F_{SF,max}$). If you push still harder, these bonds break, and the horizontal force exerted by the contact interaction suddenly drops to the lower kinetic friction value. There is now a difference in magnitude between the horizontal applied force and the opposed friction force that allows

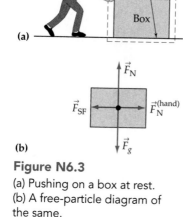

Figure N6.3
(a) Pushing on a box at rest.
(b) A free-particle diagram of the same.

Kinetic friction

F_{KF} is usually less than the maximum value of F_{SF}

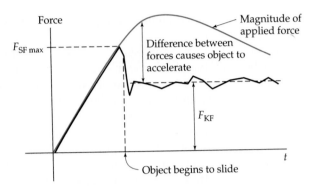

Figure N6.4
A graph illustrating the behavior of the static and kinetic friction forces. If we manually apply a horizontal force to an object sitting on a surface, the magnitude of the static friction will at first increase in step with that of the applied force until a certain maximum is reached, after which the object suddenly begins to slide and the friction force drops to the smaller kinetic friction value for sliding surfaces. (Usually a person pushing will ease up after the object starts to move, which is why the applied force shown turns over.)

the box to accelerate from rest. The box will continue to accelerate until either the kinetic friction force has increased (with increasing speed) to match your applied force or (more likely) you ease up automatically on the box so as to match your push to its new lower friction force.

This experimentally observed behavior has some interesting implications. *Antilock brakes* on cars prevent the brakes from locking the cars' wheels, so that the tires stay locked to the road instead of skidding. Since the maximum static friction force that the road can exert on the tires is greater than the kinetic friction force that it can exert if the tires are skidding, antilock brakes help the driver to brake the car more rapidly (as well as keep the car under better control).

Empirically, the maximum possible magnitude of the static friction force that a contact interaction can exert seems to be roughly *proportional* to the magnitude of the normal force exerted by the same contact interaction:

$F_{SF,max}$ is proportional to F_N (for same interaction)

$$\text{mag}(\vec{F}_{SF}) \leq \text{mag}(\vec{F}_{SF,max}) \approx \mu_s \, \text{mag}(\vec{F}_N) \qquad (N6.5)$$

Purpose: This equation describes the magnitude of the static friction force that a contact interaction between solid objects exerts on those objects.
Symbols: $\vec{F}_{SF}$ is the force of static friction; $\vec{F}_N$ is the normal force exerted by the same contact interaction; and μ_s is the **coefficient of static friction**, a unitless constant that depends on the characteristics of the surfaces.
Limitations: This is a simplified empirical model that is only approximately true.
Notes: This equation only specifies an *upper limit* for mag($\vec{F}_{SF}$), not its actual value.

It makes sense that mag($\vec{F}_{SF}$) should depend on mag($\vec{F}_N$), because the normal force reflects how strongly the surfaces are pressed together and thus how extensively the surface atoms are likely to interpenetrate and bond together. Note that μ_s is unitless (since $\vec{F}_{SF}$ and $\vec{F}_N$ have the same units). Its value depends on the nature of the interacting surfaces.

Since the normal force on an object sitting on a surface typically (but not always!) cancels the object's weight, the applied force needed to get an object moving is often proportional to the object's weight. This is the newtonian explanation for the intuitive idea that one has to exert a force strong enough to "overcome an object's inertia" before it starts to move. This Aristotelian notion thus works well enough in many cases, but it fails to explain why even a tiny force suffices to accelerate an object in space or in an otherwise frictionless environment. If a freight train car could be mounted on sufficiently frictionless wheels, even a toddler could cause it to accelerate!

A *similar* relation usefully models the relatively steady magnitude of the kinetic friction forces exerted by surfaces moving relative to one another:

F_{KF} is proportional to F_N (for the same interaction)

$$\text{mag}(\vec{F}_{KF}) \approx \mu_k \, \text{mag}(\vec{F}_N) \qquad\qquad \text{(N6.6)}$$

Purpose: This equation describes the magnitude of the kinetic friction force exerted by a sliding contact interaction between solid objects.

Symbols: $\vec{F}_{KF}$ is the kinetic friction force; $\vec{F}_N$ is the normal force exerted by the same contact interaction; and μ_k is the **coefficient of kinetic friction,** a unitless constant that depends on the characteristics of the surfaces (and sometimes weakly on speed).

Limitations: This is a simplified empirical model that is only approximately true.

Table N6.1 lists measured values of μ_s and μ_k for various kinds of surfaces. Note that (almost always) $\mu_k < \mu_s$ for a given pair of surfaces.

Table N6.1   Some coefficients of friction **(These values are approximate.)**

	μ_s	μ_k
Climbing boots on rock	1.0	0.8
Tires on dry concrete	1.0	0.7
Rubber shoes on wood	0.9	0.7
Steel on steel (dry)	0.7	0.5
Tires on wet concrete	0.7	0.5
Tires on asphalt	0.6	0.4
Leather shoes on carpet	0.6	0.5
Rope or metal on wood	0.5	0.3
Wood on wood	0.4	0.2
Tires on icy concrete	0.3	0.2
Leather shoes on wood	0.3	0.2
Waxed wood on wet snow	0.14	0.1
Steel on steel (lubricated)	0.1	0.05
Shoes on ice	0.1	0.05
Ice on ice	0.1	0.03
Teflon on Teflon	0.04	0.04
Best human joints	0.01	0.003

Although they look very similar, there is an important difference between equations N6.5 and N6.6. Equation N6.6 expresses the *actual value* of the kinetic friction force (which has a fairly fixed value), while equation N6.5 expresses an *upper limit* on the adjustable static friction force.

Example N6.2

Problem   Imagine that a box sits on a plank of wood. If we gradually lift one end of the plank, we find that the box suddenly starts to slide down the plank when the plank's angle with the horizontal reaches 28°. What is the coefficient of static friction between the box and the plank?

Translation   Figure N6.5 illustrates the situation in which the plank is inclined at an angle $\theta < 28°$ and the box is still at rest.

Model   Since the box is at rest, it is not accelerating, so the net force acting on it must be zero: $\vec{F}_{net} = 0$. If we orient the reference frame as shown in figure N6.5b, then writing $0 = \vec{F}_{net} = \vec{F}_g + \vec{F}_N + \vec{F}_{SF}$ in column-vector form yields

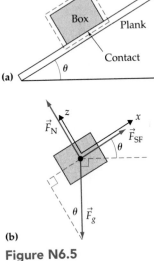

$$\begin{bmatrix} 0 \\ 0 \\ 0 \end{bmatrix} = \begin{bmatrix} -mg \sin \theta \\ 0 \\ -mg \cos \theta \end{bmatrix} + \begin{bmatrix} 0 \\ 0 \\ +F_N \end{bmatrix} + \begin{bmatrix} +F_{SF} \\ 0 \\ 0 \end{bmatrix} \qquad (N6.7)$$

where $F_{SF} \equiv \text{mag}(\vec{F}_{SF})$, $F_N \equiv \text{mag}(\vec{F}_N)$, and $mg = \text{mag}(\vec{F}_g)$. We also know that

$$F_{SF} \le \mu_s F_N \qquad (N6.8)$$

This gives us three useful equations in the four unknowns m, F_{SF}, F_N, and μ_s, but it turns out that the mass will cancel out when we solve for μ_s.

Solution   If we solve the top and bottom rows (the x and z components) of this vector equation for the force magnitudes F_{SF} and F_N, we get

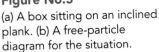

$$F_{SF} = mg \sin \theta \qquad \text{and} \qquad F_N = mg \cos \theta \qquad (N6.9)$$

Figure N6.5
(a) A box sitting on an inclined plank. (b) A free-particle diagram for the situation.

If we use these equations to eliminate F_{SF} and F_N from equation N6.8, we get

$$mg \sin \theta \le \mu_s mg \cos \theta \qquad \Rightarrow \qquad \mu_s \ge \frac{\cancel{mg} \sin \theta}{\cancel{mg} \cos \theta} = \tan \theta \quad (N6.10a)$$

This says that in order for the box to remain at rest, $\tan \theta$ must be smaller than or equal to the unknown value of μ_s. Since this condition breaks down if θ exceeds $\theta_{max} = 28°$,

$$\mu_s = \tan \theta_{max} = \tan(28°) = 0.53 \qquad (N6.10b)$$

Evaluation   The value of μ_s is unitless, as we would expect, and reasonable.

Exercise N6X.1

When the plank in example N6.2 is inclined at 20°, what is the ratio of the magnitude of the actual static friction force on the box to the maximum possible value $\mu_s F_N$ for a box sitting on a plank at this angle?

Exercise N6X.2

Imagine that you have to exert a horizontal force of 49 N on a 10-kg box to get it to begin to slide along a level floor. What is the coefficient of static friction between the box and the floor?

N6.4   Drag Forces

While the kinetic friction force between two solid objects is fairly independent of relative speed, the drag force experienced by an object as it moves through a fluid depends strongly on speed. Empirically, the magnitude of the drag force on most objects moving through air is given by the formula

The drag on a large object moving quickly through air

$$\text{mag}(\vec{F}_D) \approx \tfrac{1}{2} C \rho A v^2 \qquad\qquad \text{(N6.11)}$$

Purpose: This equation describes the magnitude of the drag force $\vec{F}_D$ exerted by a fluid on an object moving through it with speed v.
   Symbols: ρ is the density of the fluid, A is the object's cross-sectional area, and C is a dimensionless **drag coefficient** that depends on the object's shape and surface characteristics.
   Limitations: This expression works only if A and v are sufficiently large, ρ is sufficiently small, and the fluid is not very viscous. This expression works for most sports projectiles in air, but $F_D \propto v$ for slow-moving objects in water.

The drag coefficient C is about 0.5 for a sphere and is smaller for smooth and streamlined shapes, but it can be as large as 2 for irregular shapes.

Exercise N6X.3

What are the SI units for C?

Exercise N6X.4

A certain streamlined car has a frontal area of about 3.0 m² and a drag coefficient of 0.3. What is the approximate magnitude of the drag force exerted on the car when it travels at a speed of 25 m/s (55 mi/h)?

Equation N6.11 accurately models the drag force on most sports projectiles and other large objects moving through air. The drag force on objects that are very small, move very slowly, and/or move through a more viscous fluid (such as water) turns out to be proportional to v and thus is better

The drag on objects that are very small, move very slowly, and/or are in a viscous medium (such as water)

described by

$$F_D = \mathrm{mag}(\vec{F}_D) = bv \tag{N6.12}$$

where b is some constant with abbreviated units of $\mathrm{N \cdot s / m = kg/s}$.

Exercise N6X.5

If the forward force on a motorboat has to be 550 N to drive it through the water at a constant speed of 5 m/s, what is the constant b for that particular motorboat (assuming tht equation N6.12 applies in this case)?

N6.5   Linearly Accelerated Motion

Frame alignment is key to solving problems with $\vec{a} \neq 0$

Consider now an object moving along a straight line with nonzero acceleration. Newton's second law $\vec{F}_{net} = m\vec{a}$ implies that the net force on the object must point in the same direction as its observed acceleration. This means that analyzing the motion of an object becomes *much* simpler if we *orient the reference frame axes so that one axis coincides with the direction of the observed acceleration*. Then the two components of the net force perpendicular to this axis are zero (just as in a constant-velocity problem), and analysis of the object's *motion* becomes a one-dimensional problem.

Some constrained-motion problems involving accelerating objects are *hybrid* problems in that we use information about the motion to determine things about the forces acting on the object, which in turn we use to determine things about the linear motion of the object (such as its acceleration). Example N6.3 illustrates such a problem.

Example N6.3

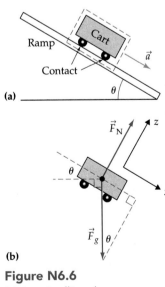

Figure N6.6

(a) A cart rolling down an incline. (b) A free-particle diagram of this situation.

Problem   Find the acceleration of a cart rolling on small frictionless wheels down a ramp inclined at an angle θ with respect to the horizontal. How does the magnitude of the normal force exerted on the cart by the contact interaction with the ramp compare to the magnitude of the object's weight?

Translation   Figure N6.6 illustrates this situation.

Model   If there are no significant friction forces acting on the object, then the only forces acting on the cart are the normal force (which points perpendicular to the incline) and the cart's weight (which points vertically downward). Since the object is confined to moving on the surface of the incline, its acceleration has to point parallel to the incline. Let us orient our reference frame so that the x axis points down the incline and the z axis is perpendicular to the incline. If we express Newton's second law $m\vec{a} = \vec{F}_{net} = \vec{F}_g + \vec{F}_N$ in column-vector form, we get

$$\begin{bmatrix} ma_x \\ 0 \\ 0 \end{bmatrix} = \begin{bmatrix} +mg\sin\theta \\ 0 \\ -mg\cos\theta \end{bmatrix} + \begin{bmatrix} 0 \\ 0 \\ +F_N \end{bmatrix} \tag{N6.13}$$

where $mg = \mathrm{mag}(\vec{F}_g)$ and $F_N \equiv \mathrm{mag}(\vec{F}_N)$. This gives us two useful equations in the three unknowns m, a_x, and F_N, so it does not look as if we have enough information to solve.

Solution   However, if we solve the first and third rows of this equation for a_x and F_N, respectively, we get

$$a_x = +g \sin \theta \quad \text{and} \quad F_N = mg \cos \theta = F_g \cos \theta \quad \text{(N6.14)}$$

We see that an object moving frictionlessly down an incline will have a constant acceleration whose magnitude is $a = g \sin \theta$, and the magnitude of the normal force is $\cos \theta$ times that of the object's weight. Since $\cos \theta < 1$, this means that $F_N < F_g$. (This last statement is true whether the object is accelerating or not.) So we are able to answer both questions definitively without knowing the object's mass m.

Evaluation   Note that the units of both expressions make sense, and that a_x is positive, which also makes sense considering how we defined the reference frame.

Exercise N6X.6

How long would it take such a cart to travel from rest down a 3-m ramp inclined at an angle of 15°?

Exercise N6X.7

Imagine that we analyzed the situation here using a reference frame where the x axis was horizontal. Why would this be more difficult?

N6.6   A Constrained-Motion Framework

Constrained-motion problems are a subcategory of force-from-motion problems in which the object in question is constrained by its environment to move in a certain way (e.g., with a constant velocity along a straight line, constant acceleration along a straight line, or constant speed in a circle), and we use this information to determine the forces acting on the object. Constrained-motion problems will be our focus through chapter N8; our focus in this particular chapter is on **linearly constrained motion** problems, in which the object is constrained to move in a line.

In section N5.4, we discussed a general approach for solving force-from-motion problems. This general framework works very well for constrained-motion problems if we make two minor adjustments.

(1) Because we will almost always be interested in center-of-mass motion in constrained-motion problems (and not in balancing torques), drawing a free-*particle* diagram is usually more useful than a free-*body* diagram in the conceptual model step. Draw a free-body diagram *only* in the rare cases where torques are of interest and/or if you find a free-body diagram helpful as a first step toward drawing a free-particle diagram.

(2) It is *especially* important in constrained-motion problems to understand and think about the constraint that the environment places on the object's *acceleration*. (This is partly because you want to be *sure* to align one of your coordinate axes with the direction of that acceleration.) Because of this, it is *essential* that you draw an acceleration arrow on the translation step diagram (if $\vec{a} \neq 0$) and check that the arrow you draw is consistent with your free-particle diagram (use a net-force diagram if that helps).

Adjustments to the problem-solving framework

Use a free-particle diagram instead of a free-body diagram

Draw an acceleration arrow and make sure it's right

Examples N6.4 and N6.5 show the complete framework and how it is used.

Example N6.4

Problem   A 14-kg box sits in the back of a 2800-kg pickup truck waiting at a stoplight. When the stoplight turns green, the driver of the truck drives forward with an acceleration whose magnitude is 5.0 m/s². If the coefficient of static friction between the box and the truck bed is 0.40, will the box be able to accelerate with the truck, or will it slide backward relative to the truck bed?

Translation

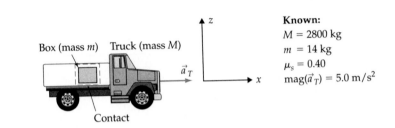

Known:
$M = 2800$ kg
$m = 14$ kg
$\mu_s = 0.40$
$\text{mag}(\vec{a}_T) = 5.0$ m/s²

Conceptual Model

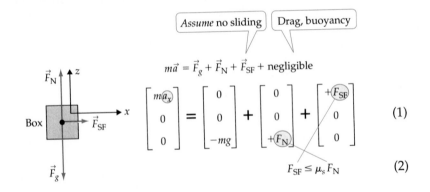

$$m\vec{a} = \vec{F}_g + \vec{F}_N + \vec{F}_{SF} + \text{negligible}$$

$$\begin{bmatrix} ma_x \\ 0 \\ 0 \end{bmatrix} = \begin{bmatrix} 0 \\ 0 \\ -mg \end{bmatrix} + \begin{bmatrix} 0 \\ 0 \\ +F_N \end{bmatrix} + \begin{bmatrix} +F_{SF} \\ 0 \\ 0 \end{bmatrix} \quad (1)$$

$$F_{SF} \le \mu_s F_N \quad (2)$$

Prose Model   *Assuming* the box accelerates along with the truck, the box's acceleration $\vec{a}$ points in the $+x$ direction, as shown in the diagram. The box only touches the truck bed (which exerts an upward normal force and a *forward* static friction force that pulls the box along with the truck) and the air (which we will assume exerts negligible forces on the box). The box also interacts gravitationally with the earth. Note that the forces in the free-particle diagram add up to a net force in the $+x$ direction, which is what is necessary to give the box an acceleration in this direction. Newton's second law then implies equation 1. We have *three* unknowns here, because we actually want to find the maximum possible *box* acceleration magnitude a and compare it with the *truck's* actual acceleration magnitude a_T to see whether the box will slide. The third equation we need to solve for our three unknowns is equation 2, which links the maximum static friction force magnitude with that of the normal force.

Solution   The top and bottom lines of equation 1 imply, respectively,

$$ma_x = F_{SF} \quad (3a)$$

and

$$F_N = mg \tag{3b}$$

Plugging equation 3b into equation 2 and equation 2 into equation 1, we get

$$ma_x = F_{SF} \leq \mu_s F_N = \mu_s mg$$

$$\Rightarrow \quad a_x \leq \mu_s g = 0.4(9.8 \text{ m/s}^2) = 3.9 \text{ m/s}^2 \tag{4}$$

Since the truck's x-acceleration $a_{Tx} = 5.0 \text{ m/s}^2$ exceeds this, the box will indeed slip backward as the truck accelerates forward.

Evaluation   Note that this result is independent of the mass of the box! The result for a_x also does have the correct sign (positive) and the correct units. The result is also believable: an acceleration of 5.0 m/s² is pretty large ($\approx \frac{1}{2}g$) and so might plausibly result in slipping.

Example N6.5

Problem   A 55-kg person in an elevator traveling upward is standing on a spring scale that reads 420 N. What are the magnitude and direction of the elevator's acceleration? [*Note:* An ordinary scale does not directly register your weight, since weight is a force that acts directly on you and cannot be intercepted by the scale. Rather the scale registers the magnitude of the upward normal force that its spring has to exert to support you. This same contact interaction puts pressure on your feet that gives you a sense of what your "perceived weight" is. When you are at rest, this normal force is equal to your actual weight; but if you are accelerating vertically, the net vertical force on you will *not* be zero, and thus the normal force (and thus your perceived weight) will not be equal to your actual weight (which does not change significantly no matter how the elevator moves). This is why your perceived weight changes as you ride an elevator.]

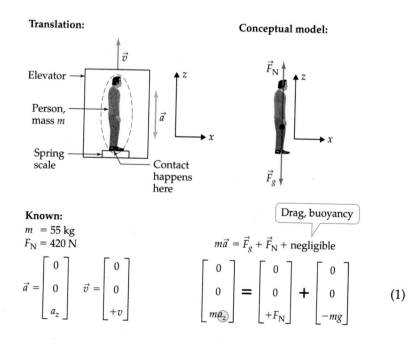

Translation:

Elevator

Person, mass m

Spring scale

Contact happens here

Conceptual model:

Drag, buoyancy

$m\vec{a} = \vec{F}_g + \vec{F}_N + \text{negligible}$

Known:

$m = 55$ kg
$F_N = 420$ N

$$\vec{a} = \begin{bmatrix} 0 \\ 0 \\ a_z \end{bmatrix} \quad \vec{v} = \begin{bmatrix} 0 \\ 0 \\ +v \end{bmatrix} \qquad \begin{bmatrix} 0 \\ 0 \\ ma_z \end{bmatrix} = \begin{bmatrix} 0 \\ 0 \\ +F_N \end{bmatrix} + \begin{bmatrix} 0 \\ 0 \\ -mg \end{bmatrix} \tag{1}$$

Prose Model   The person touches only the scale and the air. The scale exerts an upward normal force on the person, and we will ignore any forces resulting from contact with the air. The person also interacts gravitationally with the earth. Since the person is constrained to move vertically with the elevator, the person's acceleration (if there is any) must be vertical. This is consistent with the free-particle diagram: there are no horizontal forces, and the relative magnitudes of the vertical normal and gravitational forces will determine the person's (and thus by implication the elevator's) acceleration. Newton's second law in this case therefore reads as shown in equation 1. Only the bottom line (z component) of this equation tells us anything useful, but it is sufficient for us to solve for our single unknown variable a_z.

Solution   Solving that last line for a_z, we get

$$a_z = \frac{F_N}{m} - g = \frac{(420\ \text{N})}{(55\ \text{kg})}\left(\frac{1\ \text{kg·m/s}^2}{1\ \text{N}}\right) - 9.8\ \text{m/s}^2 = -2.2\ \text{m/s}^2 \qquad (2)$$

Evaluation   Since this is negative, the elevator must be accelerating *downward* at a rate of 2.2 m/s² (perhaps as it slows when it approaches an upper-floor destination). This makes the person apparently weigh less, since the normal force of 420 N is lower than the person's actual unchanging weight ($mg \approx 540$ N). This coincides with our experience: when an elevator slows down while going up, we feel momentarily lighter. The acceleration also has the right units and has a reasonable magnitude ($< g$). (Note that the full three-dimensional machinery of equation 1 is probably overkill here, but it is *essential* in more complicated problems.)

TWO-MINUTE PROBLEMS

N6T.1   The magnitude of the normal force on a box sitting on an incline is equal to that of its weight, true (T) or false (F)?

N6T.2   A certain crate sits on a rough floor. You find that you have to apply a horizontal force of 200 N to get the crate moving. If you put some massive objects in the crate so that its mass is doubled, how much force does it take to get the crate moving now?
A.   Still 200 N
B.   400 N
C.   800 N
D.   It depends (specify)

N6T.3   Two boxes of the same mass sit on a rough floor. These boxes are made of the same kind of cardboard and are identical except that one is twice as large as the other. If it takes 200 N to start moving the smaller box, how much force does it take to start moving the larger one?
A.   Still 200 N
B.   400 N

C.   800 N
D.   It depends (specify)

N6T.4   The coefficient of static friction between Teflon and scrambled eggs is about 0.1. What is the smallest tilt angle from the horizontal that will cause the eggs to slide across the surface of a tilted Teflon-coated pan?
A.   0.002°
B.   5.7°
C.   15°
D.   33°
E.   Other (specify)

N6T.5   If you want to stop a car as quickly as possible on an icy road, you should
A.   Jam on the brakes as hard as you can.
B.   Push on the brakes as hard as you can without locking the car's wheels and thus making the car skid.
C.   Pump the brakes.
D.   Do something else (specify).

N6T.6 Putting wider tires on your car will clearly give you more traction, T or F?

N6T.7 Assume that the coefficient of static friction between your car's tires and a certain road surface is about 0.75. Your car can climb a 45° slope, T or F?

N6T.8 Imagine that an external force of 100 N must be applied to keep a bicycle and rider moving at a constant speed of 12 mi/h against opposing air drag. To double the bike's speed to 24 mi/h, we must increase the magnitude of the force exerted on the bike to

A. 141 N
B. 200 N
C. 400 N
D. It depends on the bike's shape and area
E. Other (specify)

N6T.9 Imagine that a certain engine can cause the road to exert a certain maximum forward force F_{SF} on a certain car. If we change the car's design to reduce its drag coefficient by a factor of 2, by what factor will the car's maximum speed increase (other things being equal)?

A. No increase
B. 1.41
C. 2
D. 4
E. Depends (specify)
F. Other (specify)

N6T.10 A truck is traveling down a steady slope such that for each 1 m the truck goes forward along the slope it goes down 0.04 m (we call this a 4% grade). Imagine that the truck's brakes fail. What is the approximate increase in the truck's speed after 30 s, assuming the engine is not used and there is little drag or other friction? [$g = 22$ (mi/h)/s.]

A. 0.9 mi/h
B. 11 mi/h
C. 26 mi/h
D. 53 mi/h
E. 660 mi/h
F. Other (specify)

HOMEWORK PROBLEMS

Basic Skills

N6B.1 Imagine that you have to exert 200 N of horizontal force on a 30-kg crate to get it moving on a level floor. What is the value of μ_s for the surfaces involved here?

N6B.2 Imagine that the coefficient of static friction μ_s between a 25-kg box and the floor is 0.55. How hard would you have to push on the box to get it moving?

N6B.3 Imagine that you have to exert a horizontal force of magnitude 80 N to push a 20-kg box at a constant speed of 3 m/s. What is the coefficient of kinetic friction μ_k between the box and the floor in this case?

N6B.4 Imagine that the coefficient of kinetic friction between a certain 15-kg box and the floor is 0.35. How hard would you have to push on it to move it at a constant speed of 2 m/s across the floor?

N6B.5 Assuming that air drag is the main opposing force here, that the area which the bike and rider present to the wind is about 0.75 m², and that C for such an irregular shape is about 1, what is the approximate forward force required to give the bicycle a constant speed of 10 m/s?

N6B.6 The drag force on a car moving at 65 mi/h is how many times bigger than the drag force on a car moving at 45 mi/h?

N6B.7 Imagine that a cart rolls without friction down a slope that makes an angle of 6° with respect to the horizontal. If the cart rolls for 12 s, how far does it go? (*Hint:* you can use the result of example N6.3.)

N6B.8 Imagine that a glider slides without friction down a tilted air track. If it takes 3.0 s to slide the 1.5-m length of the track, at what angle was the track inclined? (*Hint:* you can use the result of example N6.3.)

Synthetic

The starred problems are particularly well suited for practicing the use of the problem-solving framework.

N6S.1 The coefficient of static friction between a certain car's tires and an asphalt road is about 0.60. Only the rear tires are powered. The magnitude of the car's acceleration can be at most what value? Explain your response carefully.

N6S.2 Why do the tires of a car grip the road better on level ground than when the car is going up or down an incline? Explain carefully. (You may find that a couple of force diagrams will help you make your case.)

*N6S.3 A 2.0-kg box slides down a 25° incline at a constant velocity of 3.0 m/s. What are the magnitude and direction of the kinetic friction force acting on this box?

*N6S.4   Imagine that the coefficient of static friction be-
tween the tires of a certain 2250-kg all-terrain vehicle and a typical gravel roadbed is 0.45. What is
the maximum possible incline that the vehicle can
climb?

How steep a slope can this vehicle climb?
(See problem N6S.4.)

*N6S.5   A 12-kg box sits at rest on a 15° incline. If the coeffi-
cient of static friction is 0.3 between the box and the
incline, what additional force would you have to
exert directly down the incline to get the box to
start sliding?

*N6S.6   A certain car has a drag coefficient of 0.32. If you look
at it from the front, the car's cross section looks
roughly like a rectangle that is 1.5 m high by 2.1 m
across. What is the *minimum* horsepower of its engine if its top cruising speed is 45 m/s ($\approx$ 100 mi/h)?

*N6S.7   A 200-kg motorboat is cruising in the $+x$ direction
at a speed of v_0. The motor suddenly dies at a time
we will call $t = 0$, and the boat's x-velocity thereafter is observed to be described by the equation
$v_x(t) = v_0 e^{-qt}$, where $q = 0.5 \text{ s}^{-1}$. Does the observed motion of this boat seem consistent with the
idea that the drag force on the boat is given by
$F_D = bv$? If so, determine the value of b for this
boat. (*Hint:* The time derivative of e^{-qt} is $-qe^{-qt}$.)

*N6S.8   A crane hauls a crate (mass 250 kg) upward at a
constant acceleration of 2.2 m/s². What is the magnitude of the tension force exerted on the crate by
the crane's cable?

*N6S.9   A 65-kg person is standing on a bathroom scale in
an elevator moving downward. If the scale reads
720 N, what are the magnitude and direction of the
elevator's acceleration?

*N6S.10   A railroad flatcar is loaded with crates having a
coefficient of static friction of 0.50 with respect
to the car's floor. If a train is moving at 22 m/s
($\approx$ 48 mi/h), within how short a distance can the
train be stopped without letting the crates slide?

*N6S.11   A pickup truck carries cans of paint in the back bed.
The coefficient of static friction between the cans
and the bed of the truck is 0.54. There is no back
gate to the truck. How long should the driver take
to accelerate to a speed of 55 mi/h to avoid losing
the paint cans out of the rear of the truck?

*N6S.12   An 1100-kg car with a frontal cross-sectional area
of 3.5 m² and a drag coefficient of 0.42 rolls down a
slope that makes a constant angle of 8° with respect
to the horizontal. After accelerating a while, the car
will eventually reach a maximum constant speed.
What is this speed? (Assume that the wheels rotate
frictionlessly.)

*N6S.13   An 1800-kg car with four-wheel drive travels up a
12° incline at a speed of 15 m/s. What can you infer
about the coefficient of static friction between the
tires of this car and the road? Which pieces of information provided are relevant, and which are not?

*N6S.14   Antilock brakes keep a car's tires from skidding
on a road surface. A certain 1500-kg car equipped
with such brakes and initially traveling at 27 m/s
($\approx$ 59 mi/h) is able to come to rest within a time
interval of 6.0 s.
(a) What is the minimum coefficient of static friction between the tires and the road in this case?
(b) How far does the car travel before it stops?

Rich-Context

N6R.1   A bicyclist (whose mass is 54 kg and whose bike has
a mass of 11 kg) coasting down an essentially end-
less 5° slope is observed to reach a maximum speed
of 42 mi/h. At these kinds of speeds, air drag dom-
inates over all other kinds of friction. *Estimate* the
drag coefficient C for the bike and rider.

N6R.2   You are driving a 12,000-kg truck at a constant
speed of 65 mi/h down a 6% slope (i.e., an incline
that goes down 0.06 m for every 1 m that one goes
along the incline). You suddenly see that a bridge
is out 425 ft ahead, and you jam on the brakes. The

coefficient of static friction between your tires and the wet asphalt road is 0.45, the coefficient of kinetic friction is 0.30, and the cross-sectional area of your truck is 6.6 m^2. Can you stop in time? (*Hints:* The magnitude of the drag force depends on speed, so it will not be constant as you slow down. Is the drag force significant? Make plausible estimates for quantities you are not given.)

Advanced

N6A.1   We can use the following model to understand equation N6.11. Imagine that we assume that an object with cross-sectional area A moving through air of density ρ has to accelerate all the air molecules that it touches from rest up to its speed. Show that if this were so, the drag force on the object would be

$$\mathrm{mag}(\vec{F}_D) = \rho A v^2 \qquad (N6.15)$$

(The actual drag force will be somewhat smaller, since some of the air will slip around the object without being fully accelerated to its speed.)

N6A.2   (This problem requires using some calculus involving exponentials and logarithms.) Imagine that an object of mass m is initially moving in the $+x$ direction through a viscous fluid at a speed v_0. If the only force acting on this object is a viscous drag force whose magnitude is given by $F_D = bv$, prove that the object's x-velocity is given by

$$v_x(t) = v_0 e^{-qt} \qquad (N6.16)$$

and determine how the constant q depends on b and m. *Hint:* This is actually a motion-from-force problem. Divide both sides of the x component of Newton's second law by v_x, and take the indefinite integral of both sides.

ANSWERS TO EXERCISES

N6X.1   According to equation N6.8, $F_{SF} = mg \sin\theta$. Equations N6.8 and N6.9, on the other hand, also indicate that $F_{SF,max} = \mu_s F_N = \mu_s mg \cos\theta$. Therefore,

$$\frac{F_{SF}}{F_{SF,max}} = \frac{\cancel{mg}\sin\theta}{\mu_s \cancel{mg}\cos\theta} = \frac{\tan\theta}{\mu_s} = 0.69 \qquad (N6.17)$$

N6X.2   Since $F_{SF,max} = 49$ N and $F_N = F_g = mg = 98$ N in this situation, $\mu_s = F_{SF,max}/F_N = 0.50$.

N6X.3   The SI units of $\rho A v^2$ are $(kg/\cancel{m^3})(\cancel{m^2})(m^{\cancel{2}}/s^2) = $ kg·m/s^2. Since these are already the units of force, C must be unitless.

N6X.4 $F_D = \frac{1}{2}(0.3)(1.2 \text{ kg/m}^3)(3.0 \text{ m}^2)(25 \text{ m/s})^2 = 340$ N.

N6X.5   The motorboat is not accelerating when it is moving at a constant speed, so the forward force acting on it must cancel the drag force. Solving $F_{forward} = F_D = bv$ for b, we get $b = F_{forward}/v = 110$ kg/s.

N6X.6   From chapter N4 we know that

$$x(t) - x_0 = \frac{1}{2}a_x t^2 + v_{0x}t \qquad (N6.18)$$

Since $v_{0x} = 0$ here, $x(t) - x_0 = D = 3.0$ m, and $a_x = g\sin\theta$,

$$D = \frac{1}{2}g\sin\theta t^2 \quad \Rightarrow \quad t = \sqrt{\frac{2D}{g\sin\theta}} = 1.54 \text{ s} \qquad (N6.19)$$

N6X.7   In this case, Newton's second law would read

$$\begin{bmatrix} ma_x \\ 0 \\ ma_z \end{bmatrix} = \begin{bmatrix} F_{N,x} \\ F_{N,y} \\ F_{N,z} \end{bmatrix} + \begin{bmatrix} F_{g,x} \\ F_{g,y} \\ F_{g,z} \end{bmatrix}$$

$$= \begin{bmatrix} F_N\sin\theta \\ 0 \\ F_N\cos\theta \end{bmatrix} + \begin{bmatrix} 0 \\ 0 \\ -mg \end{bmatrix} \qquad (N6.20)$$

But neither a_x or a_z is zero in this case, so it becomes much more difficult to solve this problem. It can be done: we can use what we know about the acceleration's direction to show that $a_x = a\cos\theta$ and $a_z = -a\sin\theta$, plug these into the equations above, and solve the two coupled equations for the two unknowns a and F_N. However, this is much more complicated than the process illustrated in the example.

N7 Coupled Objects

Chapter Overview

Introduction

This chapter continues our exploration of what we can learn about forces by observing motion. In this chapter we will consider pairs or sets of objects that are coupled together by internal interactions that constrain the objects to move with the same acceleration. We will see that Newton's third law can help us determine the magnitudes of these internal forces.

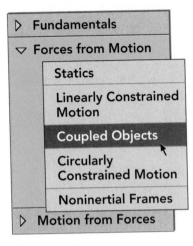
Section N7.1: Force Notation for Coupled Objects

Coupled objects are objects that are constrained by some connection so that their motions are linked in a well-defined manner. When we draw free-particle diagrams for the objects, it helps greatly if we use a notation for forces that describes not only the type of force we are talking about, and thus the type of interaction involved, but also the two objects involved in that interaction. In this chapter, we will use notation of the form $\vec{F}_N^{A(B)}$, where the *subscript* defines the type of force (a normal contact force in this example) and the *superscripts* define the objects involved in the interaction (in this case, the force is acting on object A and arises from its interaction with object B).

According to section N3.4, forces are *third-law partners* if

1. Each force acts on a different object.
2. The pair of forces represents the two ends of the *same* interaction.

This means that in our notation, any third-law partners will have reversed superscripts and the same subscript, for example, $\vec{F}_N^{A(B)}$ and $\vec{F}_N^{B(A)}$. Such partners are easy to spot on the free-body diagrams of coupled objects participating in a contact interaction: the partners are the forces applied to the surfaces in contact.

Section N7.2: Pushing Blocks

This section discusses in detail an example involving a block that is being pushed along a surface by another block. Since Newton's third law links the contact forces that the blocks exert on each other, we can use it in conjunction with Newton's second law to determine the magnitudes of the contact forces. We find that the pair of blocks behaves exactly as if it were a single object responding to the external forces exerted on the system.

Section N7.3: Strings, Real and Ideal

This section explores an example involving two blocks connected by a string that are being pulled upward by a known external force. The fully correct way to analyze this system involves applying Newton's second and third laws to the *three* objects involved (the string being the third object). This approach yields five equations that can be solved for five unknowns (four internal tension forces and the system's common acceleration). As before, we find that the entire set accelerates as if it were a particle responding to the external forces on the system.

Moreover, we find that the difference between the magnitudes of the forces exerted by each end of the internal string goes to zero as the string's mass goes to zero. The ideal string model assumes that the string is completely massless, inextensible, and flexible. In this ideal limit, the tension forces exerted by the string's ends have equal magnitudes (we call this magnitude *the* **tension on the string**), and the two objects linked by the string accelerate at exactly the same rate. This model makes it much easier to solve problems involving objects linked by a cable, chain, rope, or string.

Section N7.4: Pulleys

An **ideal pulley** (i.e., a frictionless, massless pulley) changes the direction of a string without affecting the tension forces it exerts. Real pulleys, of course, have nonzero mass and nonzero friction. Since rotating a real pulley thus requires a nonzero torque, the tension forces exerted on the pulley by the parts of the string entering and leaving the pulley must be different. However, this change in the string tension can be very small for a good pulley.

It is often helpful in pulley problems to use a separate coordinate system for each object with its x axis aligned with the object's direction of motion.

Section N7.5: Using the Framework

When solving coupled-object problems, you can use the general force-from-motion framework for solving coupled-object problems with the following adaptations:

1. Describe how the objects' motions are linked. On the main translation diagram in a pulley problem, draw a *separate* acceleration arrow for each object.
2. Draw a *separate* free-particle diagram for each object, using separate coordinate axes in pulley problems.
3. Circle and link any third-law partners in the free-particle diagrams.
4. Apply Newton's second law in column-vector form to each object, and use Newton's third law to link the magnitudes of any third-law partners.

In some cases, drawing free-body diagrams of the objects as well as free-particle diagrams can help you locate third-law partners.

Example N7.4 illustrates the use of this adapted framework.

N7.1   Force Notation for Coupled Objects

What are coupled objects?

Our general task in chapters N5 through N9 is to expand our ability to apply Newton's laws to determine the forces acting on objects whose motion is constrained in various ways. In this chapter, we will focus on situations involving *coupled objects*. A pair of **coupled objects** consists of two objects that are constrained by some kind of connection so that their motions (which we will assume here to be linear) are linked in some well-defined manner. We can use this constraint to analyze the forces acting on the individual objects in the pair and describe their motion.

When we are called on to describe and analyze the forces acting on more than one object, we begin to run into serious problems with notation. For example, imagine that a person pushes a box B up an incline, which in turn pushes box A in front of it up that incline (figure N7.1a). How can we distinguish the symbol for the normal force exerted on A (by A's contact interaction with the incline) from the symbol for the normal force exerted on B (by B's interaction with the incline) from the symbol for the normal force exerted on B (by B's contact interaction with the pusher's hands) from the symbol for the normal force exerted on A (by A's contact interaction with the box B behind it)? We would get very confused if we assigned all these different forces the same symbol $\vec{F}_N$!

A notation convention for force symbols

Let us adopt the following notation convention for cases like this. We start with the basic symbol for the type of force involved, using the standard symbols described in section N1.5. We add to this a pair of superscripts, the first indicating the object on which this force is exerted and the second (in parentheses) indicating the other object involved in the interaction that gives rise to this force. For example, we give the normal force exerted on box A by its contact interaction with box B (which is pushing A up the incline) the symbol $\vec{F}_N^{A(B)}$. The normal force exerted on box B by the person P we might call $\vec{F}_N^{B(P)}$, and so on. While it is not necessarily the prettiest notation imaginable, this notation (which is summarized in figure N7.2) is simple, relatively easy both to typeset and to write by hand, and easy to interpret (as long as we have well-defined single letters for every object involved). Figure N7.1b shows free-particle diagrams for the two boxes with all forces labeled using this notation.

Obviously, this kind of notation is too cumbersome to use all the time. When we are analyzing the forces acting on a single object, this notation is only rarely helpful. In this chapter, however, it is more than helpful; this notation (or something equivalent) is essential for keeping things straight.

Figure N7.1

(a) A sketch of a person pushing two boxes up an incline and free-body diagrams for each of the boxes. (b) Free-particle diagrams for each of the boxes. The forces in both the free-body and free-particle diagrams are labeled according to the notation convention established in this section. Note that we typically use the symbol E for the earth in the labels for gravitational forces.

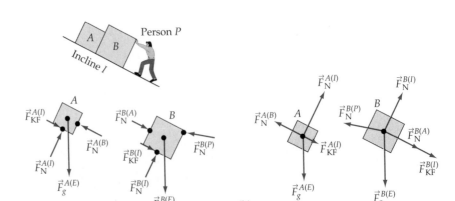

One advantage of this notation is that it makes it pretty easy to recognize third-law partners in a given situation. Newton's third law states that

Recognizing third-law partners

> When two objects interact, the force the interaction exerts on each is equal in magnitude and opposite in direction to the force it exerts on the other.

This means (as discussed in chapter N3) that the two forces linked by this law (third-law partners) have the following characteristics:

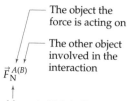

Figure N7.2
A convention for force symbols used in coupled-object problems.

1. The forces always act on *different* objects (when A and B interact, the interaction exerts one force on A and one on B).
2. They always reflect the *same* interaction.

This in turn implies that the symbols for third-law partners will have *subscripts* that are the *same* (since both forces must reflect the same interaction) and *superscripts* that are *reversed* (since the partner to the force exerted on A due to its interaction with B is the force on B due to its interaction with A). For example, if I exert a normal force $\vec{F}_N^{W(H)}$ on the wall W by pushing on it with my hand H, the contact interaction exerts a force $\vec{F}_N^{H(W)}$ back on my hand.

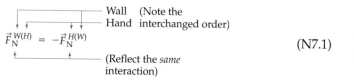

$$\vec{F}_N^{W(H)} = -\vec{F}_N^{H(W)} \tag{N7.1}$$

The symbols for third-law partners

by Newton's third law. The symbols for third-law partners will always have the characteristics noted in equation N7.1.

In the situation shown in figure N7.1, the only third-law partners among the forces shown are the normal forces $\vec{F}_N^{A(B)}$ and $\vec{F}_N^{B(A)}$ that the boxes exert on each other due to their contact interaction. Note that these symbols display the characteristics shown in equation N7.1. Figure N7.1a also illustrates that it is often particularly easy to spot third-law partners on the free-*body* (as opposed to free-*particle*) diagrams of a pair of objects participating in a *contact* interaction: one only has to look for the pair of forces that act on each object on the surface that is in contact with the other object. Check to see that the contact forces $\vec{F}_N^{A(B)}$ and $\vec{F}_N^{B(A)}$ in figure N7.1a fit this description.

Spotting third-law partners on free-body diagrams

Example N7.1

Problem   Consider a book sitting at rest on a table. Draw free-body diagrams (not free-particle diagrams) for both the book and the table, and determine which pairs of forces on these diagrams (if any) are third-law partners.

Solution   Figure N7.3 shows the free-body diagrams for these objects. The only forces acting on the book are a downward gravitational force $\vec{F}_g^{B(E)}$ and an upward normal force $\vec{F}_N^{B(T)}$ due to its contact interaction with the table. The forces acting on the table are a downward gravitational force $\vec{F}_g^{T(E)}$, a set of upward normal forces $\vec{F}_N^{T(F)}$ exerted by the contact interaction between the floor F on the table legs, and the downward normal force $\vec{F}_N^{T(B)}$ exerted by the contact interaction between the book and the table.

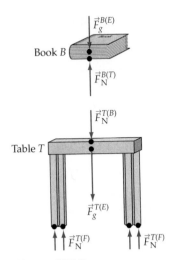

Figure N7.3
Free-body diagrams for the book and the table. Note that $E =$ earth and $F =$ floor.

The only third-law partners among these forces are the two forces that arise from the contact interaction between the book and the table. Note that this pair has all the characteristics described in this section: their symbols are consistent with the pattern shown in equation N7.1, and these forces appear on the free-body diagram acting on the surfaces where the book and table touch.

In contrast, note that the gravitational force on the book $\vec{F}_g^{B(E)}$ and the normal force $\vec{F}_N^{B(T)}$ are *not* third-law partners, even though these forces are opposite and have the same magnitude. These forces are equal and opposite because of Newton's *second* law: since the book is at rest, its acceleration is zero; so the net force on it is zero, and so $\vec{F}_g^{B(E)}$ and $\vec{F}_N^{B(T)}$ must be equal in magnitude and opposite in direction so that they cancel.

Note that even though the book's weight $\vec{F}_g^{B(E)}$ does not act directly on the table, the sum of the normal forces $\vec{F}_{N,tot}^{T(F)}$ acting on the table legs is larger when the book sits on the table than it would be if the book were not there, because Newton's second law requires that this force cancel both the table's weight $\vec{F}_g^{T(E)}$ and the downward contact force $\vec{F}_N^{T(B)}$, which is equal in magnitude to the book's weight (as discussed in the previous paragraph).

Exercise N7X.1

Consider someone in an elevator moving at a constant speed. Draw free-body diagrams (not free-particle diagrams) for both the person and the elevator (ignoring friction); label the forces, using the convention described in this section; and determine which pairs of forces on these diagrams (if any) are third-law partners. Explain why there is more tension on the elevator's cable when the person is in the elevator than when it is empty, even though the gravitational force (weight) of the person does not act directly on the elevator.

N7.2   Pushing Blocks

When two objects are connected in such a way that they become constrained to move together in a well-defined way, the constraint on their motion itself provides information that we can use to learn something about the forces acting on those objects. Example N7.2 illustrates how we can use Newton's second and third laws together to extract information from this constraint.

Example N7.2

Problem   Consider two blocks pushed along a frictionless surface by someone who exerts a constant force with his or her hand (see figure N7.4a). Imagine that block A has a mass of 4.0 kg, block B has a mass of 2.0 kg, and the hand exerts a force of 3.0 N (about 0.65 lb). What is the magnitude of the acceleration of the blocks? What is the magnitude of the force exerted by block B on block A?

Translation   Figure N7.4a provides a sketch of the situation.

Model   The first step in the conceptual model is to specify a coordinate system and draw free-particle diagrams of the two objects in question: these items are shown in figure N7.4c. Drawing free-*body* diagrams of the objects (see figure N7.4b) before drawing the free-particle diagrams more clearly displays the relationships involved in this case.

The contact interaction between blocks A and B exerts opposing normal forces on each block, which are labeled $\vec{F}_N^{A(B)}$ (read "the normal force on A due to its interaction with B") and $\vec{F}_N^{B(A)}$ (read "the normal force on B due to its interaction with A"). Newton's third law asserts that these forces have equal magnitudes, whether the blocks are accelerating or not.

In figure N7.4, $\vec{F}_N^{A(H)}$ represents the force exerted on A by the interaction with the person's hand, $\vec{F}_N^{A(S)}$ and $\vec{F}_N^{B(S)}$ represent the normal forces exerted on A and B by their contact interaction with the surface on which they slide, and $\vec{F}_g^{A(E)}$ and $\vec{F}_g^{B(E)}$ represent the forces exerted on each by their gravitational interaction with the earth (E = earth). Note that figure N7.4a explicitly defines the symbols S and H to make these single-letter abbreviations clearer.

The next step in developing the model for this problem is to list the known quantities. We know that $\text{mag}(\vec{F}_N^{A(H)}) = 3.0$ N, that $m_A = 4.0$ kg, and that $m_B = 2.0$ kg. Knowing these masses would allow us to compute the magnitudes of the weight forces $\vec{F}_g^{A(E)}$ and $\vec{F}_g^{B(E)}$, should we so desire. The implicit constraints in the problem are (1) that the blocks move only in the x direction, which means that $\vec{a} = [a_x, 0, 0]$, and (2) that the two boxes have the *same* horizontal acceleration a_x (since they are moving as a unit). On the other hand, we do not know either the value of a_x or the magnitudes of the forces $\vec{F}_N^{A(B)}$ and $\vec{F}_N^{B(A)}$ (the first of which we are asked to find) or the magnitudes of $\vec{F}_N^{A(S)}$ and $\vec{F}_N^{B(S)}$ (which are probably of no concern), a total of five unknowns.

The next model step in most constrained-motion problems is to apply Newton's second law in column-vector form. Newton's second law for block A reads

$$m_A \vec{a} = \vec{F}_g^{A(E)} + \vec{F}_N^{A(S)} + \vec{F}_N^{A(B)} + \vec{F}_N^{A(H)}$$

$$\Rightarrow \begin{bmatrix} m_A a_x \\ 0 \\ 0 \end{bmatrix} = \begin{bmatrix} 0 \\ 0 \\ -m_A g \end{bmatrix} + \begin{bmatrix} 0 \\ 0 \\ +F_N^{A(S)} \end{bmatrix} + \begin{bmatrix} -F_N^{A(B)} \\ 0 \\ 0 \end{bmatrix} + \begin{bmatrix} +F_N^{A(H)} \\ 0 \\ 0 \end{bmatrix}$$
(N7.2a)

(Force symbols without arrows or component subscripts refer to *magnitudes*, as usual.) The x and z components of this vector equation imply, respectively, that

$$m_A a_x = F_N^{A(H)} - F_N^{A(B)} \tag{N7.2b}$$

and

$$F_N^{A(S)} = m_A g \tag{N7.2c}$$

Similarly, Newton's second law for block B implies that $m_B \vec{a} = \vec{F}_g^{B(E)} + \vec{F}_N^{B(S)} + \vec{F}_N^{B(A)}$, or

$$\begin{bmatrix} m_B a_x \\ 0 \\ 0 \end{bmatrix} = \begin{bmatrix} 0 \\ 0 \\ -m_B g \end{bmatrix} + \begin{bmatrix} 0 \\ 0 \\ +F_N^{B(S)} \end{bmatrix} + \begin{bmatrix} +F_N^{B(A)} \\ 0 \\ 0 \end{bmatrix}$$

$$m_B a_x = F_N^{B(A)} \tag{N7.3a}$$

$$\Rightarrow \qquad 0 = 0 \tag{N7.3b}$$

$$F_N^{B(S)} = m_B g \tag{N7.3c}$$

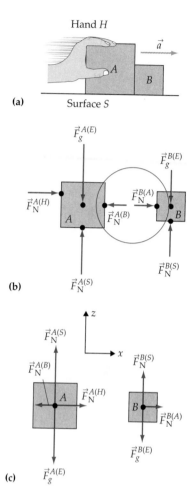

(a)   Surface S

(b)

(c)

Figure N7.4
(a) Two blocks being pushed along a frictionless surface. (b) Free-body diagrams of the blocks, with the single third-law pair circled. (c) Free-particle diagrams of the blocks.

The z component equations for both blocks tell us that the vertical normal force on each block cancels its weight: there are no surprises here. Newton's third law also tells us that

$$F_N^{A(B)} = F_N^{B(A)} \tag{N7.4}$$

Equations N7.2b, N7.2c, N7.3a, N7.3c, and N7.4 provide five equations for our five unknowns.

Solution   Plugging equation N7.4 into equation N7.3a yields

$$m_B a_x = +F_N^{A(B)} \tag{N7.5}$$

If we add this equation to equation N7.2b, the unknown force $F_N^{A(B)}$ cancels out, leaving us with

$$(m_A + m_B)a_x = F_N^{A(H)} \tag{N7.6a}$$

Note that this tells us that the *system* consisting of the two objects accelerates as if it were a particle of mass $M = m_A + m_B$ subject to the net external force acting on the system ($\vec{F}_N^{A(H)}$ in this case). We already knew that this *should* be the case from chapter C4: the important thing here is to notice how Newton's third law ensures that it *does* happen. Solving this equation for a_x yields

The block's common acceleration

$$a_x = \frac{F_N^{A(H)}}{m_A + m_B} = \frac{3.0\,\text{N}}{4.0\,\text{kg} + 2.0\,\text{kg}}\left(\frac{1\,\text{kg·m/s}^2}{1\,\text{N}}\right) = 0.50\,\text{m/s}^2 \tag{N7.6b}$$

Plugging this result into equation N7.5 and using $F_N^{A(B)} = F_N^{B(A)}$, we find that

The magnitude of the forces due to the contact interaction between the blocks

$$F_N^{A(B)} = F_N^{B(A)} = m_B a_x = (2.0\,\text{kg})(0.50\,\text{m/s}^2)\left(\frac{1\,\text{N}}{1\,\text{kg·m/s}^2}\right) = 1.0\,\text{N} \tag{N7.6c}$$

Evaluation   Note that the net x-force on object A is $F_N^{A(H)} - F_N^{A(B)} = 3.0\,\text{N} - 1.0\,\text{N} = 2.0\,\text{N}$, which is just the force required to give this 4.0-kg object an x-acceleration of $a_x = 0.5\,\text{m/s}^2$. Note also that the net force on B is $1.0\,\text{N}$, which is just what is required to give it an x-acceleration of $0.5\,\text{m/s}^2$. Everything is self-consistent!

Exercise N7X.2

What if the hand pushes with a force of 9.0 N on block A? What is the magnitude of the contact force between the blocks in this case?

The point of example N7.2 is that when objects are coupled so that they are constrained to have the same acceleration, the forces associated with the interaction between the objects will adjust themselves to whatever common magnitude gives each block the *same* horizontal acceleration. (Such forces are invariably contact forces, which *can* adjust themselves in this way.)

N7.3   Strings, Real and Ideal

Overview

Objects do not have to be in direct contact to exert forces on each other. Two objects connected by string or cord can exert forces on each other through the string that are similar to the forces that they exert when in direct contact.

Each end of a string should be thought of as exerting a tension force on the object to which it is connected. While these two forces are generally *not* equal in direction, they are often at least approximately equal in magnitude (for reasons that we will discuss shortly). Because of this, we sometimes refer to the common magnitude of these forces as being *the* **tension on** (or **of**) **the string.** It is not *always* the case that the forces exerted by the ends of a string have the same magnitude, and in such cases "the string's tension" is not well defined; but you should understand that when this phrase is used, it refers to the common magnitude of the forces exerted by each end of the string.

A specific example will help us understand these issues more clearly.

Example N7.3

Problem   Consider two blocks connected by a string (call this the *internal* string) and being pulled vertically upward by another string (the *external* string) attached to block A. Imagine that block A has a mass of 3.0 kg and block B has a mass of 2.0 kg, and imagine that the external string exerts a tension force of magnitude 54 N. How do the tension forces exerted by the ends of the internal string compare if the mass of the internal string is 0.010 kg?

Translation and Model   Again, the usual first step is to draw a sketch (see figure N7.5a) and a set of free-body diagrams (figure N7.5b). I have defined symbols for the internal string S and the external string X, set up a reference frame (we only need a z axis for this problem), and indicated knowns and unknowns (the latter with question marks). The implicit constraints on the motion are that the objects only move vertically, and that all three objects have the same z-acceleration a_z.

To solve the problem, we apply Newton's second law in component form to each of the three objects involved (block A, block B, and the internal string S). In this case, only the z component of Newton's second law is interesting (the others all read $0 = 0$). So, rather than write a bunch of zeros, we will cut to the chase: the z component of Newton's second law for each of the three objects is, respectively,

$$m_A a_z = F_{\text{net},z}^A = F_{g,z}^{A(E)} + F_{T,z}^{A(X)} + F_{T,z}^{A(S)} = -m_A g + F_T^{A(X)} - F_T^{A(S)} \quad \text{(N7.7a)}$$

$$m_S a_z = F_{\text{net},z}^S = F_{g,z}^{S(E)} + F_{T,z}^{S(A)} + F_{T,z}^{S(B)} = -m_S g + F_T^{S(A)} - F_T^{S(B)} \quad \text{(N7.7b)}$$

$$m_B a_z = F_{\text{net},z}^B = F_{g,z}^{B(E)} + F_{T,z}^{B(S)} = -m_B g + F_T^{B(S)} \quad \text{(N7.7c)}$$

where (as usual) the force symbols appearing on the right without arrows or component subscripts refer to magnitudes. In addition, Newton's third law tells us that forces circled in color in figure N7.5b have equal magnitudes:

$$F_T^{A(S)} = F_T^{S(A)} \quad \text{(N7.7d)}$$

$$F_T^{S(B)} = F_T^{B(S)} \quad \text{(N7.7e)}$$

Equations N7.7 represent five equations in the four unknown force magnitudes $F_T^{S(A)}$, $F_T^{A(S)}$, $F_T^{S(B)}$, and $F_T^{B(S)}$ and the one unknown acceleration a_z. We can therefore solve the problem.

Solution   First, we can solve equation N7.7c for $F_T^{B(S)}$. According to equation N7.7e, that force magnitude is equal to $F_T^{S(B)}$, so we can plug all this into equation N7.7b to eliminate $F_T^{S(B)}$ and rearrange things a bit to get

$$F_T^{S(A)} = m_S a_z + m_S g + m_B a_z + m_B g \quad \text{(N7.8)}$$

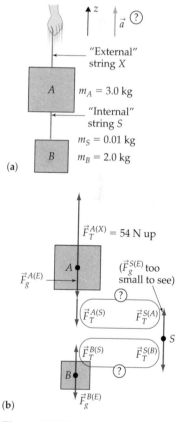

Figure N7.5

(a) A person pulls vertically on two blocks connected by a string. (b) Free-particle diagrams for the two blocks and the string. (Note that this diagram also specifies the knowns and unknowns in this problem.)

Exercise N7X.3

Verify equation N7.8.

Equation N7.7d tells us that $F_T^{S(A)} = F_T^{A(S)}$, so we can plug equation N7.8 into equation N7.7a, eliminating $F_T^{A(S)}$. After a bit more rearrangement, we get

$$(m_A + m_S + m_B)a_z = +F_T^{A(X)} - (m_A + m_S + m_B)g \qquad \text{(N7.9)}$$

Exercise N7X.4

Verify equation N7.9.

This equation says that the system consisting of the two blocks and the internal string accelerates as if it were a single object of mass $M = m_A + m_S + m_B$ acted on by two external forces: (1) the tension force exerted by the external string and (2) the system's total weight. All references to the internal forces acting in the system have disappeared in this equation! This is yet another illustration of the general principle that a system of interacting objects responds to external forces as if it were a single object. Notice again how the third law is required to ensure that this is true.

We can easily solve equation N7.9 for the unknown acceleration a_z:

$$a_z = \frac{F_T^{A(X)}}{M} - g = \frac{54\,\text{N}}{5.01\,\text{kg}}\left(\frac{1\,\text{kg·m/s}^2}{1\,\text{N}}\right) - 9.8\,\frac{\text{m}}{\text{s}^2} = 0.98\,\text{m/s}^2 \quad \text{(N7.10)}$$

Once a_z is known, we can plug its value back into equations N7.7b and N7.7c to find the magnitudes of the tension forces exerted by the string. A simple rearrangement of equation N7.7c tells us that

$$F_T^{B(S)} = m_B(g + a_z) \qquad \text{(N7.11)}$$

Note that this implies that the string exerts a force on block B that is *larger* than the weight of block B. (It is necessary to have a nonzero *net* force on B to accelerate it upward.) You can verify that $F_T^{B(S)} = 21.6\,\text{N}$.

Exercise N7X.5

Check this result.

Equation N7.7b in combination with Newton's third law (equations N7.7d and N7.7e) implies that the difference in the magnitudes of the tension forces exerted by each end of the string is

$$F_T^{A(S)} - F_T^{B(S)} = m_S(g + a_z) \qquad \text{(N7.12)}$$

Exercise N7X.6

Verify equation N7.12.

This difference in forces is necessary to provide a nonzero net force to accelerate the string. The difference is 0.11 N when $m_S = 0.010$ kg.

Note in example N7.3 that the difference in the tension forces between the two ends of the string is 0.11 N, which is about 0.5% of the magnitude of the tension forces themselves. Therefore, with a string this light, the magnitudes of the tension forces are almost imperceptibly different.

Even 0.010 kg = 10 g is a pretty large mass for a short string. A realistic string having a length of 20 cm or so might actually have a mass of 2 g or less.

The difference between the tension magnitudes at the string's ends decreases as its mass decreases

---

Exercise N7X.7

If the string's mass were 1 g, show that the system's acceleration would be 0.998 m/s^2, the tension force exerted on block B would still essentially be 21.6 N, and the difference in tension across the string would be less than 0.05% of this.

Exercise N7X.7 makes it clear that as the string's mass approaches zero, we can begin to think of 21.6 N as being *the* tension of the string: the difference between the tensions at its ends becomes *very* small.

If the string were completely massless, then equation N7.12 tells us that the tension forces exerted by the string's ends would be exactly the same, and the tension on the string would be a precisely defined number. While no string is truly massless, strings connecting objects usually have masses so much smaller than the objects that they connect that the idea that the string is massless represents an excellent approximation to the real situation.

If we adopt the fiction that the string is *massless*, then analyzing a situation like the one shown in figure N7.5 becomes much simpler. We can ignore the free-body diagram of the string altogether and simply assume that the forces exerted by the string's ends have the common magnitude F_T^S. Instead of the *five* equations N7.7a through N7.7e, the problem is completely described by *two* equations in the unknowns a_z and F_T^S that express the z component of Newton's second law as it applies the two blocks:

Treating strings as massless makes problems easy!

$$m_A a_z = F_{net,z}^A = F_{g,z}^{A(E)} + F_{T,z}^{A(X)} + F_{T,z}^{A(S)} = -m_A g + F_T^{A(X)} - F_T^S \quad (N7.13a)$$

$$m_B a_z = F_{net,z}^B = F_{g,z}^{B(E)} + F_{T,z}^{B(S)} = -m_B g + F_T^S \quad (N7.13b)$$

Simply *adding* these equations eliminates the unknown tension F_T^S, leaving

$$(m_A + m_B)a_z = F_{T,z}^{A(X)} - (m_A + m_B)g \quad (N7.14)$$

which can be quickly solved for a_z:

$$a_z = \frac{F_T^{A(X)}}{m_A + m_B} - g = \frac{54 \text{ N}}{5.0 \text{ kg}}\left(\frac{1 \text{ kg·m/s}^2}{1 \text{ N}}\right) - 9.8 \frac{\text{m}}{\text{s}^2} = 1.0 \text{ m/s}^2 \quad (N7.15)$$

Plugging this result into equation N7.13b, we find that

$$F_T^S = m_B(g + a_z) = (2.0 \text{ kg})(10.8 \text{ m/s}^2) = 21.6 \text{ kg·m/s}^2 = 21.6 \text{ N} \quad (N7.16)$$

Note how using this massless string approximation allows us to analyze the situation shown in figure N7.5 in a few lines, whereas it took more than a page of work before. Moreover, we get essentially the same results!

In physics, the technical term **ideal string** refers to a hypothetical massless, inextensible, and flexible cord binding two objects together. An ideal string is really a *model* that we can apply to any real string, chain, wire, or other cord connecting two objects as long as (1) the cord's mass is very much less than the masses of the objects that it connects, (2) the cord is reasonably inextensible and flexible, and (3) we state that we are using the approximation.

Ideal string model

The tension on a (presumably ideal) **string** is defined to be the tension force exerted on the object at *either* end of the string as a result of its interaction with the string.

N7.4   Pulleys

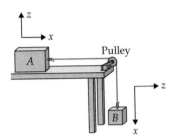

Figure N7.6
A situation involving two objects connected by a string going over a pulley.

Consider the situation shown in figure N7.6. The pulley in this situation redirects the string so that the tension forces exerted by its ends need no longer be opposite in direction: the string pulls *rightward* on object A and *upward* on object B. An **ideal pulley** simply redirects the string without creating any kind of difference in tension between the string on one side of the pulley and the string on the other. Like an ideal string, an ideal pulley is a mental construct that we use as an approximate model for a real pulley.

A real pulley will be approximately ideal if (1) it has a very small mass (particularly out near the rim) and (2) it has nearly frictionless bearings. If the pulley has a significant mass, then making significant changes in the pulley's rotation rate requires applying a significant torque to the pulley, which in turn requires that there be a significant difference in the tension of the string on one side and the string on the other. To see this, imagine tugging on a rope that goes over a very massive pulley. Perhaps you can imagine that you will have to pull very hard on the rope to accelerate the massive pulley wheel, even if the other side of the rope is almost slack. Similarly, a difference in tension would be needed to rotate the pulley if its bearings had friction: this difference in tension supplies a net force and thus a net torque to the rim of the pulley that keeps it turning against the opposing torque due to friction.

Pulleys these days are sufficiently light and frictionless that the ideal pulley model is actually a fairly reasonable approximation in many situations. Like the ideal string model, the ideal pulley model makes problems involving pulleys *much* easier to do.

In working problems involving pulleys, you will often find it advantageous to use a separate reference frame for each object, as shown in figure N7.6. This is perfectly acceptable as long as you (1) keep straight which reference frame applies to which object and (2) are careful to correctly describe in *each* reference frame any quantities that link the objects. For example, in the situation shown in figure N7.6, the reference frames have been chosen so that the acceleration of each object points entirely along the x axis, and a_x has the same positive value for both objects (since the string has a fixed length and thus any motion of object A along its x axis is exactly duplicated by object B along *its* x axis.)

N7.5   Using the Framework

We can easily adapt the constrained-motion framework to solve coupled-object problems. These problems are almost exactly like the linearly constrained motion problems discussed in chapter N6. The main *differences* are as follows. First, we need to do the following:

1. Draw a *separate* free-particle diagram for each object in the system (and provide single-letter labels for all objects involved).
2. Circle and link any *third-law partners* in these diagrams.
3. Describe in the *conceptual model* section any *joint* constraints on the objects' motion (e.g., we might say, "Because the objects are connected, a_x is the same for both").
4. Apply Newton's second law in column-vector form to *each* object.

In addition (depending on the details of the problem), it *may* be necessary to do the following:

5. Draw individual reference frames for each object.
6. Draw separate acceleration arrows for each object.
7. Use Newton's third law to connect the magnitudes of third-law partners.

You may also find it *helpful* to draw free-body diagrams of the objects involved in the problem before drawing free-particle diagrams because free-body diagrams can help you locate third-law partners.

Example N7.4 illustrates how to use the adapted constrained-motion framework to analyze a pulley problem based on the situation shown in figure N7.6.

Example N7.4

Problem In the situation shown in figure N7.6, assume that block A has a mass of 0.75 kg, block B has a mass of 0.25 kg, and the table is frictionless. What is the tension on the string connecting the blocks?

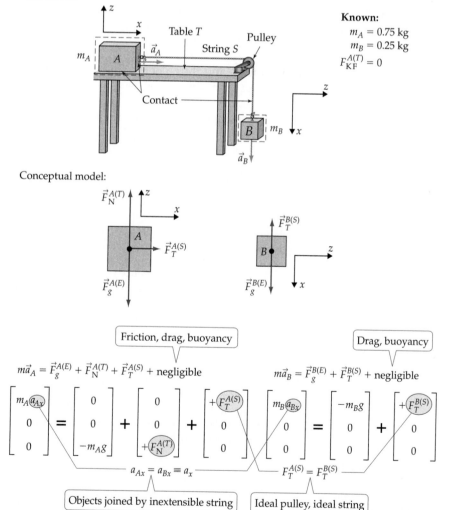

Translation:

Table T Pulley

String S

m_A A $\vec{a}_A$

Contact

B m_B

$\vec{a}_B$

Known:
$m_A = 0.75$ kg
$m_B = 0.25$ kg
$F_{KF}^{A(T)} = 0$

Conceptual model:

$\vec{F}_N^{A(T)}$ A $\vec{F}_T^{A(S)}$

$\vec{F}_g^{A(E)}$

$\vec{F}_T^{B(S)}$ B

$\vec{F}_g^{B(E)}$

Friction, drag, buoyancy

$m\vec{a}_A = \vec{F}_g^{A(E)} + \vec{F}_N^{A(T)} + \vec{F}_T^{A(S)} + \text{negligible}$

$$\begin{bmatrix} m_A a_{Ax} \\ 0 \\ 0 \end{bmatrix} = \begin{bmatrix} 0 \\ 0 \\ -m_A g \end{bmatrix} + \begin{bmatrix} 0 \\ 0 \\ +F_N^{A(T)} \end{bmatrix} + \begin{bmatrix} +F_T^{A(S)} \\ 0 \\ 0 \end{bmatrix}$$

Drag, buoyancy

$m\vec{a}_B = \vec{F}_g^{B(E)} + \vec{F}_T^{B(S)} + \text{negligible}$

$$\begin{bmatrix} m_B a_{Bx} \\ 0 \\ 0 \end{bmatrix} = \begin{bmatrix} -m_B g \\ 0 \\ 0 \end{bmatrix} + \begin{bmatrix} +F_T^{B(S)} \\ 0 \\ 0 \end{bmatrix}$$

$a_{Ax} = a_{Bx} \equiv a_x$

Objects joined by inextensible string

$F_T^{A(S)} = F_T^{B(S)}$

Ideal pulley, ideal string

Prose Model Each object is constrained to move in its $+x$ direction (according to its own set of frame axes, as defined above), and the string also constrains the magnitudes of the objects' accelerations to have the same magnitude, so in this case $a_{Ax} = a_{Bx}$. The forces shown in the free-particle diagrams look consistent with accelerations in the $+x$ direction (as long as $F_g^{B(E)} > F_T^{B(S)}$). Block A touches the string and the table, which exert a normal force and a tension force on A, respectively (note we are excluding any kinetic friction force). Block B touches the string. Both blocks also touch the air, but we will ignore drag and buoyancy forces resulting from these interactions. Newton's second law thus implies the two column-vector equations above. We have only three meaningful rows in these equations but five unknowns (a_{Ax}, a_{Bx}, $F_N^{A(T)}$, $F_T^{A(S)}$, and $F_T^{B(S)}$). However, $a_{Ax} = a_{Bx}$ and if we assume the ideal string and pulley approximations, then $F_T^{A(S)} = F_T^{B(S)}$. This gives us enough information to solve.

Solution   The bottom and top rows of Newton's second law for block A and the top row of the same for block B tell us $m_A g = F_N^{A(T)}$ (which is not relevant),

$$m_A a_x = F_T^{A(S)} \tag{1a}$$

and

$$m_B a_x = m_B g - F_T^{B(S)} \tag{1b}$$

Plugging $F_T^{A(S)} = F_T^{B(S)}$ into equation 1a yields $m_A a_x = F_T^{B(S)}$, and adding this to equation 1b yields $(m_A + m_B)a_x = m_B g$. Solving this for a_x and plugging the result into equation 1a, we get

$$F_T^{A(S)} = m_A \left(\frac{m_B g}{m_A + m_B} \right) = \frac{(0.25 \text{ kg})(0.75 \text{ kg})(9.8 \text{ N/kg})}{0.25 \text{ kg} + 0.75 \text{ kg}} = 1.84 \text{ N} \tag{2}$$

Evaluation   This has the right sign for a magnitude and the right units, and seems plausible, particularly as $m_B g = 2.45 \text{ N} > 1.84 \text{ N} = F_T^{A(S)} = F_T^{B(S)}$, as is required to produce a positive x-acceleration in block B.

TWO-MINUTE PROBLEMS

N7T.1   A jet airplane flies at a constant velocity through the air. Its jet engines exert a constant force forward on the plane that exactly balances the force of air friction exerted backward on the plane. These forces are equal in magnitude and opposite in direction. Do we know this because of Newton's second law or Newton's third law?
- A.   Newton's second law
- B.   Newton's third law
- C.   Both laws
- D.   Neither (explain)

N7T.2   Which of the following pairs of forces are third-law partners? Answer T if the two forces described are third-law partners, F if they are not.
- a.   A thrust force from its propeller pulls a plane forward; a drag force pushes it backward.
- b.   A car exerts a forward force on a trailer; the trailer tugs backward on the car.

- c.   A motorboat propeller pushes backward on the water; the water pushes forward on the propeller.
- d.   Gravity pulls down on a person sitting in a chair; the chair pushes back up on the person.

N7T.3   A box B sits in the back of a truck T as the truck slows down for a stop (the box remains motionless relative to the truck). What is the appropriate symbol for the horizontal force that the contact interaction between the box and the truck exerts on the truck?
- A. $\vec{F}_N^{B(T)}$
- B. $\vec{F}_N^{T(B)}$
- C. $\vec{F}_{SF}^{B(T)}$
- D. $\vec{F}_{SF}^{T(B)}$

(more)

E. $\vec{F}_{KF}^{B(T)}$

F.   Other (specify)

N7T.4   A child C pulls on a wagon W, using a string S; the wagon moves forward at a constant speed as a result. The third-law partner to the forward force exerted on the wagon is which of the following forces? (R = road.)

A. $\vec{F}_T^{S(W)}$

B. $\vec{F}_T^{W(S)}$

C. $\vec{F}_T^{W(C)}$

D. $\vec{F}_T^{C(W)}$

E. $\vec{F}_{KF}^{W(R)}$

F.   Other (specify)

N7T.5   A small car pushes on a disabled truck, accelerating it slowly forward. Each exerts a force on the other as a result of their contact interaction. Which vehicle exerts the *greater* force on the other?
A.   The car.
B.   The truck.
C.   Both forces have the same magnitude.
D.   The truck doesn't exert any force on the car.
E.   There is not enough information for a meaningful answer.

N7T.6   A physicist and a chemist are playing tug-of-war. For a certain length of time during the game, the participants are essentially at rest. During this time, each person pulls on the rope (which can be treated as an ideal string) with a force of 350 N. What is the tension on the rope?
A.   700 N
B.   350 N
C.   175 N
D.   Other (specify)

N7T.7   Object A ($m_A = 1.0$ kg) hangs at rest from an ideal string A connected to the ceiling. Object B ($m_B = 2.0$ kg) hangs at rest from an ideal string B connected to object A. The tension on string A is

A.   Twice the tension on string B
B.   3/2 times the tension on string B
C.   Equal to the tension on string B
D. $\frac{2}{3}$ the tension on string B
E.   Other (specify)

N7T.8   Two people are attempting to break a rope, which will break if the tension on the rope exceeds 360 N. If each person can exert a pull of 200 N,
A.   They can break the rope if they each take an end and pull.
B.   They can break the rope if they tie one end to the wall and both pull on the other.
C.   They can break the rope if they use either of the strategies above.
D.   They cannot break the rope.

N7T.9   A spring scale typically indicates the magnitude of the tension force exerted on its bottom hook. What will the scale read in each of the cases shown in the diagram? (*Hint:* Construct a free-body or free-particle diagram for the scale in each case.)
A.   49 N
B.   98 N
C.   186 N
D.   Other (specify)

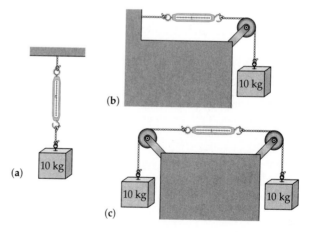

HOMEWORK PROBLEMS

Basic Skills

N7B.1   Which of the following force pairs are third-law partners? Explain your reasoning.
(a) The earth attracts a stone; the stone attracts the earth.
(b) A jet's engine thrusts it forward; drag pushes it back.
(c) You push on a box without moving it; the floor pushes back on the box.

N7B.2   Which of the following force pairs are third-law partners? Explain your reasoning.
(a) A mule pulls on a plow, moving it forward; the ground pulls backward on the plow.
(b) A team of dogs pulls a sled, moving it; the sled pulls backward on the dogs.
(c) Gravity tugs downward on a box sitting on the ground; the ground pushes up on the box.

N7B.3 In example N7.2, what would be the magnitude of the force that the blocks exerted on each other if blocks A and B had masses of 10 kg and 6.0 kg and we pushed on the blocks with a force of 4.0 N?

N7B.4 Find the acceleration of the system, the magnitude of the tension force on block B, and the difference in the magnitudes of the tension forces exerted by each end of the string in example N7.3 if the string has a mass of 100 g. Compare to the answers that we got before. Is this string even approximately ideal, in your opinion?

Free-Particle Diagrams and Third-Law Partners. For each of the situations described in problems N7B.5 through N7B.12, draw a separate, isolated free-particle diagram for each object involved (you may also draw free-body diagrams if this is helpful). Assign an appropriate symbol to each force vector, according to the conventions established in this chapter and in chapter N1. Indicate the approximate relative magnitudes of the forces by giving each arrow an appropriate length. Circle and link any third-law partners in the diagrams that you draw.

N7B.5 The moon orbits the earth. (Draw diagrams for both the moon and the earth.)

N7B.6 You jump up off the floor. (Draw a diagram for both you and the earth during the interval of time while you are still in contact with the floor and you are still accelerating upward.)

N7B.7 A little box sits on a bigger box. Both boxes are at rest. (Draw diagrams for both boxes.)

N7B.8 A little box sits on top of a bigger box sitting on an incline. Both boxes are at rest. (Draw diagrams for both boxes.)

N7B.9 A little box sits on top of a bigger box. The big box is sliding on a rough but level floor and as a result is slowing down. (Draw diagrams for both boxes.)

N7B.10 A tractor pulls a plow at a constant velocity in a field. (Draw diagrams for both the tractor and the plow.)

N7B.11 A person hangs from a helicopter by a rope as the helicopter begins to accelerate upward. (Draw diagrams for the person, the helicopter, and the rope.)

N7B.12 A small car pushes a large disabled truck so that both accelerate gently forward. (Draw diagrams for both the car and the truck. Ignore air resistance.)

Synthetic

The starred problems are particularly well suited for practice using the problem-solving framework.

N7S.1 Two teams of people are involved in a tug-of-war. Since the forces exerted by each team on the other are equal and opposite by the third law, how is it possible for either team to win? Explain carefully, using appropriate free-body diagrams, how one team *can* win.

(See problem N7S.1.)

N7S.2 A block with a substantial mass M is suspended from the ceiling by a light string A. An identical string B hangs from the bottom of the block. If you jerk suddenly on string B, it will break; but if you pull steadily on string B, string A will break. Using a force diagram of the block, carefully explain why.

*N7S.3 A 12,000-kg tugboat pushes on a 420,000-kg barge in still water. If the tug and barge accelerate at a rate of 0.20 m/s^2, what is the magnitude of the force that the water has to exert on the tugboat's propellers? What is the magnitude of the force that the tug exerts on the barge? Ignore friction.

*N7S.4 A 32-kg child puts a 15-kg box into a 12-kg wagon. The child then pulls horizontally on the wagon with a force of 65 N. If the box does not move relative to the wagon, what is the static friction force on the box?

*N7S.5 A 2500-kg helicopter lifts a 1500-kg crate suspended from its fuselage.
(a) Assume that the crate is suspended using steel cables of negligible mass. If the helicopter is accelerating upward at a rate of 2.0 m/s^2 (and the force that the downward-flowing air exerts on the crate is negligible), what is the magnitude of the force that the helicopter rotors must exert on the surrounding air, and what is the magnitude of the force the cables exert on the point where they are connected to the fuselage?
(b) Answer the same questions, assuming that the crate is suspended using thick chains whose total mass is 150 kg.
(c) What is the difference between the total forces exerted by the top and bottom ends of the chains in the situation described in part (b)?

(See problem N7S.5.)

*N7S.6   A 65-kg crate slides on a rough plane inclined up-
ward at an angle of 28°. The crate is hauled up the
plane by lightweight rope that goes parallel to the
incline and then over a pulley at the top of the incline. A worker standing below the pulley pulls
vertically downward on the rope. If the 55-kg
worker hangs his or her entire weight from the
rope, it is barely sufficient to move the crate up the
incline at a constant velocity of 1.2 m/s.
(a) What is the magnitude of the sliding friction
force on the box?
(b) What is the coefficient of kinetic friction between the box and the incline?

*N7S.7   A 85-kg crate sits in a 280-kg boat. What is the ac-
ceptable range of forces that a boat's propeller can
exert on the water if we want to ensure that the
crate remains motionless relative to the boat as the
boat accelerates from rest?

*N7S.8   A 82-kg worker clings to a lightweight rope going
over a lightweight, low-friction pulley. The other
end of the rope is connected to a 67-kg barrel of
bricks. If the worker is initially at rest 15 m above

the ground, how fast will he or she be moving
when he or she hits the ground?

*N7S.9   Consider the situation shown in figure N7.6, except
assume that the table is inclined at an angle of 22°,
with the pulley at the top of the incline. Does block
A slide up or down the incline? Assume that the
table is frictionless, $m_A = 0.75$ kg, $m_B = 0.25$ kg,
and that the masses of the pulley and string are
negligible.

*N7S.10 Imagine that a person is seated in a chair that is sus-
pended by a rope that goes over a pulley. The person holds the other end of the rope in his or her
hands, as shown in figure N7.7. Assume that the
combined mass of the person and chair is M.
(a) What is the magnitude of the downward force
the person must exert on the rope to raise the
chair at a constant speed? Express your answer
in terms of M and g. (*Hint:* The answer is not
Mg!)
(b) What is the magnitude of the required force
if the person is accelerating upward with
$\text{mag}(\vec{a}) = 0.10g$?

Figure N7.7
A drawing of the
situation discussed in
problem N7S.10.

Rich-Context

N7R.1   A mule is asked to pull a plow. The mule resists, explaining that "If I tug on the plow, Newton's third
law asserts that the plow will tug on me with an
equal and opposite force. Since these forces will
cancel each other out, it is obvious that we're not
going anywhere. Therefore, there is no point in trying." Carefully (but politely) explain to the mule
the error in its reasoning, and using appropriate
free-body or free-particle diagrams, explain why it
is possible for the mule to accelerate the plow.

N7R.2 A 0.62-kg glider slides frictionlessly on an 3-m air track inclined at an angle of 12°. The glider is connected to a lightweight string which passes over a low-friction pulley at the top of the air track. The other end of the string is connected to a spring scale (whose mass is 0.05 kg), which is connected to a 0.17-kg weight.

(a) If you hold the glider at rest, what will the scale read?

(b) If you then release the glider, will it accelerate up or down the incline?

(c) What will the scale read as the system accelerates?

N7R.3 Imagine that Chris, whose mass is 45 kg, has fallen into a crevasse in a glacier. Chris's partner Pat, whose mass is 65 kg, has thrown a rope to Chris, and intends to lift Chris out by holding on to the other end of the rope while sliding down the glacier's icy slope, which is tilted at an angle of 35° with respect to the horizontal. Assume that Pat has

used some camping equipment to cobble together a pretty good pulley to change the rope's direction at the lip of the crevasse. Also assume that Chris hangs freely when suspended below the lip of the crevasse. Is it possible for Pat to pull Chris out, considering likely coefficients of friction involved? Pat has a 30-kg backpack. Could this help?

Advanced

N7A.1 Imagine a train consisting of N frictionless cars following a locomotive that is accelerating the whole train forward with acceleration of magnitude a. Assume that the first car behind the locomotive has mass M. If the tension in the coupling at the rear of *each* car is 10% smaller than the tension in the coupling in the front of the car, what is the mass of each successive car as a fraction of M? What is the total mass of the cars in terms of M? [*Hint:* You might find it helpful to know that $1 + x + x^2 + \cdots + x^n = (1 - x^{n+1})/(1 - x)$.]

ANSWERS TO EXERCISES

N7X.1 Free-body diagrams for the elevator and the person look like this:

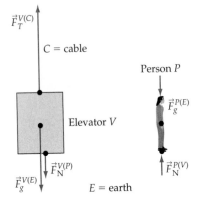

The only third-law partners appearing in this diagram are the normal forces $\vec{F}_N^{V(P)}$ and $\vec{F}_N^{P(V)}$ that the person and elevator floor exert on each other due to their mutual contact interaction. Note that $F_N^{P(V)}$ must be equal to the person's weight if the person is not accelerating vertically; similarly, $F_T^{V(C)}$ must be equal to the sum of $F_g^{V(E)}$ and $F_N^{V(P)}$. The latter force would not be there if there were no person in the elevator, so the tension on the cable has to increase by this amount (which happens to be equal to the person's weight) when the person is in the elevator (and there is no vertical acceleration). So, even though the person's weight acts on the *person*, not the elevator, the ef-

fect of this weight is communicated to the elevator (and ultimately to the cable) by the contact interaction between the person and the floor (in a manner constrained by Newton's second and third laws).

N7X.2 Plugging $F_N^{A(H)} = 9.0$ N into equation N7.6b, we get

$$a_x = \frac{9.0 \text{ N}}{6.0 \text{ kg}} \left(\frac{1 \text{ kg·m/s}^2}{1 \text{ N}} \right) = 1.5 \text{ m/s}^2 \qquad (N7.17)$$

Plugging this result into equation N7.6c, we find that $F_N^{A(B)} = F_N^{B(A)} = (2.0 \text{ kg})(1.5 \text{ m/s}^2) = 3.0$ N.

N7X.3 Solving equation N7.7c for $F_T^{B(S)}$ and using equation N7.7e, we get

$$m_B(a_z + g) = F_T^{B(S)} = F_T^{S(B)} \qquad (N7.18)$$

Plugging this into equation N7.7b, we get

$$m_S a_z = -m_S g + F_T^{S(A)} - m_B(a_z + g) \qquad (N7.19)$$

Solving this for $F_T^{S(A)}$ yields equation N7.8.

N7X.4 Plugging equation N7.8 into N7.7d and the result into N7.7a, we get

$$m_A a_z = -m_A g + F_T^{A(X)}$$
$$- (m_S a_z + m_S g + m_B a_z + m_B g) \qquad (N7.20)$$

Adding $m_S a_z + m_B a_z$ to both sides yields equation N7.9.

N7X.5 Plugging numbers into equation N7.11, we get

$$F_T^{B(S)} = (2.0 \text{ kg})(9.8 \text{ m/s}^2 + 0.98 \text{ m/s}^2) = 21.6 \text{ N} \tag{N7.21}$$

(Remember that $1 \text{ kg·m/s}^2 = 1 \text{ N}$.)

N7X.6 Adding $m_S g$ to both sides of equation N7.7b, we get

$$m_S(a_z + g) = +F_T^{S(A)} - F_T^{S(B)} \tag{N7.22}$$

But $F_T^{S(A)} = F_T^{A(S)}$ and $F_T^{S(B)} = F_T^{B(S)}$ according to Newton's third law. Plugging these into equation N7.22, we get

$$m_S(a_z + g) = +F_T^{A(S)} - F_T^{B(S)} \tag{N7.23}$$

which is equation N7.12.

N7X.7 Equation N7.10 now reads

$$a_z = \frac{54 \text{ N}}{5.001 \text{ kg}}\left(\frac{1 \text{ kg·m/s}^2}{1 \text{ N}}\right) - 9.8 \frac{\text{m}}{\text{s}^2} = 0.998 \text{ m/s}^2 \tag{N7.24}$$

Equation N7.11 now reads

$$F_T^{B(S)} = (2.0 \text{ kg})(9.8 \text{ m/s}^2 + 0.998 \text{ m/s}^2) = 21.6 \text{ N} \tag{N7.25}$$

remembering that $1 \text{ kg·m/s}^2 = 1 \text{ N}$. Note that this is essentially the same as before, and equation N7.12 reads

$$F_T^{A(S)} - F_T^{B(S)} = (0.001 \text{ kg})(10.8 \text{ m/s}^2) = 0.011 \text{ N} \tag{N7.26}$$

Note that $(0.011 \text{ N})/(21.6 \text{ N}) \approx 0.0005$, or 0.05%.

N8

Circularly Constrained Motion

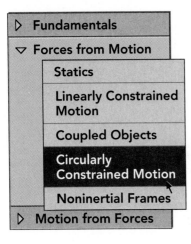
Chapter Overview

Introduction

We continue our examination of how we can determine forces from motion by examining the case in which an object is constrained to move in a circle, extending our work on uniform circular motion in chapter N2 to cover cases in which the object moves in a circle with nonconstant speed. This chapter lays important foundations for chapters N11 and N12.

Section N8.1: Uniform Circular Motion

In this section, we use a motion diagram to argue more carefully and rigorously than in chapter N2 that a particle that moves in a circle of radius R with constant speed v has an acceleration that points directly toward the center and has magnitude $\mathrm{mag}(\vec{a}) = v^2/R$. The section also notes that in such cases, we can write the particle's speed as

$$v = \frac{2\pi R}{T} \qquad \text{for circular motion at constant speed} \qquad \text{(N8.5)}$$

Purpose: This equation expresses the speed v of an object moving around a circle in terms of the circle's radius R and the time T it takes the object to go around the circle.

Limitations: This expression is true only if v is *constant*.

Section N8.2: Directionals

We will find it convenient in the future to express the direction of an object's acceleration using *directionals*. Chapter C2 introduced the idea of a directional as a symbol representing a pure direction. In this section, we will see that we can consider an arbitrary directional to be a vector with magnitude 1 (no units!). We can construct a directional $\hat{u}$ that points in the same direction as a given vector $\vec{u}$ as follows:

$$\hat{u} \equiv \frac{\vec{u}}{\mathrm{mag}(\vec{u})} \qquad \text{(N8.9)}$$

We define the directional $\hat{r}$ to be the direction of an object's position vector $\vec{r}$ from some origin. This directional is thus equivalent to the direction "directly away from the origin" at the particle's location. This directional is useful in a variety of contexts; in this chapter, we can use it to express the acceleration of an object moving with constant speed in a circle as follows:

$$\vec{a} = -\frac{v^2}{R}\hat{r} \qquad \text{(N8.12)}$$

Purpose: This equation describes the acceleration $\vec{a}$ of a particle (or an object's center of mass) that is moving in a circle of radius R with constant speed v.

Symbols: $\hat{r}$ is a directional meaning "directly away from the circle's center." (Since the acceleration in this case points *toward* the circle's center, it points in the $-\hat{r}$ direction, which is the point of the minus sign.)

Limitations: The particle (or object) must both move in a *circle* and have a *constant* speed.

Section N8.3: Nonuniform Circular Motion

In this section, we extend the argument presented in section N8.1 to cover the situation in which an object moves in a circle with a speed that is not necessarily constant. The result is

$$\vec{a} = \frac{dv}{dt}\hat{v} - \frac{v^2}{R}\hat{r} \qquad \text{(N8.19)}$$

Purpose: This equation describes the acceleration $\vec{a}$ of a particle (or an object's center of mass) that is moving in a circle of radius R with speed v.

Symbols: $\hat{r}$ is a directional meaning "directly away from the circle's center," and $\hat{v}$ is a directional pointing parallel to the particle's velocity. Note that dv/dt is the time derivative of the particle's *speed* v.

Limitations: The particle must move in a *circle*.

In circular motion the velocity directional $\hat{v}$ and the positional $\hat{r}$ are always perpendicular, so this equation indicates that the particle's or object's acceleration is constructed of the sum of two perpendicular component vectors in this case. Also, note that the first component vector points in the $+\hat{v}$ direction if the particle's speed is increasing ($dv/dt > 0$) and in the $-\hat{v}$ direction if its speed is decreasing ($dv/dt < 0$).

Section N8.4: Banking

This section discusses why an airplane must bank to make a turn, and it uses Newton's second law and equation N8.12 to calculate the banking angle. It also discusses why roads are often banked to make it safer for cars to navigate tight turns at high speeds.

Section N8.5: Examples

This section discusses how we can adapt the problem-solving framework developed in chapters N5 and N6 to solve circular motion problems. For example, in solving circular motion problems, it is useful to orient frame axes so that the circle lies in the plane defined by a suitable pair of coordinate axes, and orient one of the axes to point in either the $+\hat{r}$ or the $-\hat{r}$ direction. It is also useful to note that in most banking problems it looks as if you don't have enough information to solve the problem, but you can usually manipulate things so that unknown quantities cancel out. The chapter closes with examples that illustrate the use of the problem-solving framework.

N8.1   Uniform Circular Motion

We introduced the concept of *uniform circular motion* in chapter N2, where we saw that an object moving at a constant speed in a circle was accelerating toward the center of the circle and that the magnitude of this acceleration was plausibly equal to v^2/R, where v is the speed of the object and R is the radius of its circular trajectory. Our task in this section is to put these results on a firmer mathematical foundation, partly so that we can more easily go on to study *nonuniform* circular motion.

Consider a particle (or the center of mass of an object) moving in a circle of radius R at a constant speed v. Figure N8.1 shows a motion diagram for the particle as it travels past points 1, 2, and 3, and it displays how we can construct the change-in-velocity vector $\Delta\vec{v}$ from the average velocity vectors $\vec{v}_{12}$

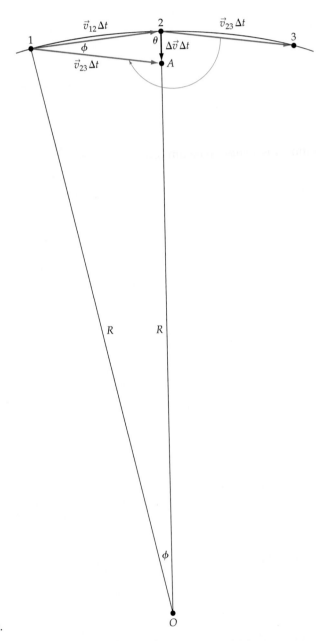

Figure N8.1

A motion diagram for an object moving at a constant speed in a circular path. Note that triangles A12 and 2O1 are similar triangles.

and $\vec{v}_{23}$. We can use this diagram to prove that the particle's acceleration as it passes point 2 points directly toward the circle's center and has a magnitude of $\text{mag}(\vec{a}) = v^2/R$ as follows.

Since the particle's speed is constant by hypothesis, the vectors $\vec{v}_{12}\,\Delta t$ and $\vec{v}_{23}\,\Delta t$ in the diagram have the same length. This means that triangles $A12$ and $2O1$ are similar: both are isosceles triangles with a common base angle θ. This means that the lengths of their sides must be proportional:

A proof of the expression for an object's acceleration when it moves in uniform circular motion

$$\frac{\text{mag}(\Delta\vec{v}\,\Delta t)}{\text{mag}(\vec{v}_{12}\,\Delta t)} = \frac{\text{mag}(\vec{v}_{12}\,\Delta t)}{R} \qquad \text{(N8.1)}$$

Now, $\text{mag}(q\,\vec{u}) = q\,\text{mag}(\vec{u})$ for any scalar q and vector $\vec{u}$. Therefore, if we divide both sides of equation N8.1 by Δt and define $v_{12} \equiv \text{mag}(\vec{v}_{12})$, we get

$$\frac{\text{mag}(\Delta\vec{v}/\Delta t)}{v_{12}} = \frac{v_{12}}{R} \qquad \Rightarrow \qquad \text{mag}\left(\frac{\Delta\vec{v}}{\Delta t}\right) = \frac{v_{12}^2}{R} \qquad \text{(N8.2)}$$

Now, the object's speed v in its circular path is the arclength between points 1 and 2, divided by Δt, where $v_{12} \equiv \text{mag}(\vec{v}_{12})$ is the length of the straight line between those points, divided by Δt. In the limit that $\Delta t \to 0$, points 1, 2, and 3 become infinitesimally separated, and the arclength and the straight-line distance between 1 and 2 become indistinguishable. Therefore,

$$\lim_{\Delta t \to 0}\left[\text{mag}\left(\frac{\Delta\vec{v}}{\Delta t}\right)\right] = \lim_{\Delta t \to 0}\frac{v_{12}^2}{R} = \frac{v^2}{R} \qquad \text{(N8.3)}$$

But the particle's acceleration $\vec{a}$ is *defined* to be $\lim_{\Delta t \to 0}(\Delta\vec{v}/\Delta t)$, so

$$\frac{v^2}{R} = \lim_{\Delta t \to 0}\left[\text{mag}\left(\frac{\Delta\vec{v}}{\Delta t}\right)\right] = \text{mag}\left(\lim_{\Delta t \to 0}\frac{\Delta\vec{v}}{\Delta t}\right) \equiv \text{mag}(\vec{a}) \qquad \text{(N8.4)}$$

We can also see from the diagram that the $\Delta\vec{v}\,\Delta t$ vector points directly toward the center of the circle, so the $\vec{a}$ vector (which is simply a scalar multiple of $\Delta\vec{v}/\Delta t$, even in the limit) must point toward the center. Q.E.D.

There is nothing special about point 2: the argument would be the same at *any* point along the particle's circular trajectory. Therefore, this result describes the particle's acceleration at *any* point along its path.

In uniform circular motion problems, we are often given the time T that it takes to go around the circle once instead of the object's speed. Since the object goes a distance of $2\pi R$ in this time, if its speed is constant, it is given by

$$v = \frac{2\pi R}{T} \qquad \text{for circular motion at constant speed} \qquad \text{(N8.5)}$$

A useful equation when we are given the time it takes to go around the circle

Purpose: This equation expresses the speed v of an object moving around a circle in terms of the circle's radius R and the time T it takes the object to go around the circle.

Limitations: This expression is true only if v is *constant*.

Exercise N8X.1

An object travels at a constant speed v once every 6.3 s around a circle of radius 3.0 m. What is v? The magnitude of its acceleration?

N8.2   Directionals

It is awkward to keep explaining in words that the direction of this vector is "toward the center of the circle." We can more compactly express this information by using a *directional*.

As discussed in chapter C2, a directional is a shorthand description of a pure direction. For example, the directional $\hat{x}$ means "in the $+x$ direction," so a vector written $\vec{v} = -(2.0 \text{ m/s})\hat{x}$ describes a velocity vector with a magnitude of 2.0 m/s pointing in the $-x$ direction.

We can consider a directional to be a vector

Now, in column-vector notation, we write a velocity vector whose magnitude is 2.0 m/s and which points the $-x$ direction as

$$\begin{bmatrix} v_x \\ v_y \\ v_z \end{bmatrix} = \begin{bmatrix} -2.0 \text{ m/s} \\ 0 \\ 0 \end{bmatrix} = -2.0 \text{ m/s} \begin{bmatrix} 1 \\ 0 \\ 0 \end{bmatrix} \tag{N8.6}$$

We see that the column vector [1, 0, 0] plays the same role in the expression above that the directional $\hat{x}$ plays in the expression $\vec{v} = -(2.0 \text{ m/s})\hat{x}$. This suggests that we can think of a directional as being a special case of a vector.

Indeed, let $\hat{u}$ be an *arbitrary* directional, and say that we can write some vector $\vec{w}$ in the form $\vec{w} = \pm w\hat{u}$, where $w = \text{mag}(\vec{w})$. If we consider $\hat{u}$ to be a vector and take the notation $\vec{w} = \pm w\hat{u}$ at face value, then taking the magnitude of both sides of this equation implies that

$$\text{mag}(\vec{w}) = \text{mag}(\pm w\hat{u}) = w\,\text{mag}(\hat{u}) \tag{N8.7}$$

where I used the fact that the magnitude of a scalar times a vector is the absolute value of the scalar times the magnitude of the vector (see section C2.6). But since $w = \text{mag}(\vec{w})$, we can divide both sides of this equation by w to get

... that has magnitude 1 (with no units)

$$1 = \text{mag}(\hat{u}) \tag{N8.8}$$

So, if we consider a directional to be a vector, then it *must* have magnitude 1 (with no units!). This is the essential characteristic of a directional vector, and for this reason, directionals are often called **unit vectors.** While the term *unit vector* describes the mathematical nature of $\hat{u}$, I prefer the term *directional* because it more clearly describes the *role* it plays in indicating a vector's direction.

How to construct a directional that indicates a given vector's direction

Note that we can construct a directional $\hat{u}$ that indicates the direction of an arbitrary vector $\vec{u}$ as follows:

$$\hat{u} \equiv \frac{\vec{u}}{\text{mag}(\vec{u})} \tag{N8.9}$$

Since any positive scalar multiple of a vector has the same direction as the vector, $\hat{u}$ as defined above has the same direction as $\vec{u}$. Moreover, if we take the magnitude of both sides of the equation above, we find that (as required)

$$\text{mag}(\hat{u}) = \text{mag}\left(\frac{\vec{u}}{\text{mag}(\vec{u})}\right) = \frac{1}{\text{mag}(\vec{u})}\text{mag}(\vec{u}) = 1 \tag{N8.10}$$

The $\hat{r}$ directional

Now that we have discussed directionals in some detail, let's get back to the issue of describing the direction of an object's acceleration when it is in uniform circular motion. A useful directional in a variety of contexts is

$$\hat{r} \equiv \frac{\vec{r}}{\text{mag}(\vec{r})} \tag{N8.11}$$

where $\vec{r}$ is the position vector of a certain point relative to some specified origin. The directional $\hat{r}$ evaluated at a given point therefore indicates the direction "directly away from the origin" at that point. If we define the circle's center to be the origin in situations involving circular motion, we can write the acceleration of an object in uniform circular motion as follows:

$$\vec{a} = -\frac{v^2}{R}\hat{r} \qquad (N8.12)$$

Purpose: This equation describes the acceleration $\vec{a}$ of a particle (or an object's center of mass) that is moving in a circle of radius R with constant speed v.

Symbols: $\hat{r}$ is a directional meaning "directly away from the circle's center." (Since the acceleration in this case points *toward* the circle's center, it points in the $-\hat{r}$ direction, which is the point of the minus sign.)

Limitations: The particle (or object) must both move in a *circle* and have a *constant* speed.

The acceleration vector $\vec{a}$ of an object moving in uniform circular motion

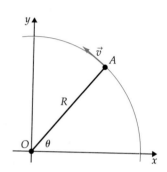

Figure N8.2

What are the components of the directional $\hat{r}$ for a particle passing through point A?

Exercise N8X.2

Imagine that a particle is passing through point A, shown in figure N8.2, as it moves in a circular trajectory of radius R. Write $\hat{r}$ as a column vector, expressing its components in terms of θ.

N8.3   Nonuniform Circular Motion

Now consider the case in which a particle (or the center of mass of an object) moves along a circular path with nonconstant speed. Figure N8.3 shows a motion diagram for a particle that is speeding up as it moves to the right. Note that we can consider the object's total displacement $\vec{v}_{23}\,\Delta t$ between times t_2 and t_3 to be the sum of the displacement $\vec{v}_{23a}\,\Delta t$ that it would have moved if its speed had remained the same, plus an increment $\vec{v}_{23b}\,\Delta t$ indicating how much farther it made it along the circle due to its increased speed. Therefore, the change in velocity vector $\Delta\vec{v}\,\Delta t$ that we would construct at point 2 is

How to find the acceleration vector $\vec{a}$ for nonuniform circular motion

$$\Delta\vec{v}\,\Delta t = \vec{v}_{23}\,\Delta t - \vec{v}_{12}\,\Delta t = \vec{v}_{23b}\,\Delta t + (\vec{v}_{23a}\,\Delta t - \vec{v}_{12}\,\Delta t) \qquad (N8.13)$$

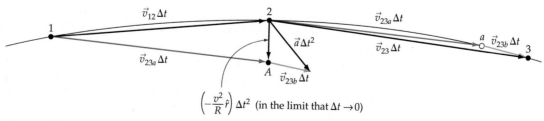

Figure N8.3

A motion diagram of a particle that follows a circular trajectory but speeds up with time. If the particle's speed were constant, it would make it to point a by time t_3, but because its speed has increased, it actually makes it to point 3 by that time.

If we divide both sides of this expression by Δt^2 and take the limit as $\Delta t \to 0$, we get

$$\vec{a} \equiv \lim_{\Delta t \to 0} \frac{\Delta \vec{v}}{\Delta t} = \lim_{\Delta t \to 0} \frac{\vec{v}_{23b}}{\Delta t} + \lim_{\Delta t \to 0} \frac{\vec{v}_{23a} \, \Delta t - \vec{v}_{12} \, \Delta t}{\Delta t^2} \qquad (N8.14)$$

But the rightmost quantity in the expression above is exactly the same as the vector $\Delta \vec{v} / \Delta t$ we evaluated in section N8.1, which we found to have magnitude v^2/R and point toward the circle's center in the limit that $\Delta t \to 0$. So equation N8.14 reduces to

$$\vec{a} = \lim_{\Delta t \to 0} \frac{\vec{v}_{23b}}{\Delta t} - \frac{v^2}{R} \hat{r} \qquad (N8.15)$$

To finish determining this acceleration vector, we need to evaluate the limit of $\vec{v}_{23b}/\Delta t$ as $\Delta t \to 0$. Notice that even for the generously large value of Δt shown in figure N8.3, the vectors $\vec{v}_{23a} \, \Delta t$, $\vec{v}_{23b} \, \Delta t$, and $\vec{v}_{23} \, \Delta t$ point pretty much along the same line. This approximation will become even better as Δt becomes small. This means that

$$\mathrm{mag}(\vec{v}_{23b}) \approx \mathrm{mag}(\vec{v}_{23}) - \mathrm{mag}(\vec{v}_{23a}) = \mathrm{mag}(\vec{v}_{23}) - \mathrm{mag}(\vec{v}_{12}) \quad (N8.16)$$

since $\mathrm{mag}(\vec{v}_{23a}) = \mathrm{mag}(\vec{v}_{12})$ by definition. Now each of these average velocities most closely represents the instantaneous velocity halfway through its respective interval, so $\Delta v \equiv \mathrm{mag}(\vec{v}_{23}) - \mathrm{mag}(\vec{v}_{12})$ very closely represents the change in the object's speed during a time interval of duration Δt centered on time t_2. All these approximations become increasingly exact as $\Delta t \to 0$. So

$$\lim_{\Delta t \to 0} \frac{\mathrm{mag}(\vec{v}_{23b})}{\Delta t} = \lim_{\Delta t \to 0} \frac{\Delta v}{\Delta t} \equiv \frac{dv}{dt} \qquad (N8.17)$$

where dv/dt is now the time derivative of the object's *speed*. Note also that as $\Delta t \to 0$, the direction of $\vec{v}_{23b}$ becomes increasingly horizontal, which is the same as the direction of the particle's velocity $\vec{v}$ at time t_2. Therefore, in the limit that $\Delta t \to 0$, we can write

$$\lim_{\Delta t \to 0} \frac{\vec{v}_{23b}}{\Delta t} = \frac{dv}{dt} \hat{v} \qquad (N8.18)$$

where $\hat{v} \equiv \vec{v}/v$ is a directional representing the direction of the particle's velocity at the time in question.

So the complete expression for the acceleration of a particle (or object's center of mass) moving in a circular trajectory with a nonuniform speed is

The acceleration of a particle (or an object's CM) that moves in a circle with varying speed

$$\vec{a} = \frac{dv}{dt} \hat{v} - \frac{v^2}{R} \hat{r} \qquad (N8.19)$$

Purpose: This equation describes the acceleration $\vec{a}$ of a particle (or an object's center of mass) that is moving in a circle of radius R with speed v.
Symbols: $\hat{r}$ is a directional meaning "directly away from the circle's center," and $\hat{v}$ is a directional pointing parallel to the particle's velocity. Note that dv/dt is the time derivative of the particle's *speed* v.
Limitations: The particle must move in a *circle*.

Note that this expression reduces to equation N8.12 when v is constant, because $dv/dt = 0$ in that case. Note also that if the particle is speeding up,

dv/dt is positive, so the acceleration vector "leans forward" in toward the direction of the particle's velocity; but if the particle is slowing down, dv/dt is negative and the acceleration vector "leans backward." Finally, note that since $\hat{v}$ is always perpendicular to $\hat{r}$ for circular motion, the magnitude of an object's acceleration in nonuniform circular motion is

$$\text{mag}(\vec{a}) = \sqrt{\left(\frac{dv}{dt}\right)^2 + \frac{v^4}{R^2}} \qquad \text{(N8.20)}$$

Exercise N8X.3

A car is traveling around a circular bend in the road, which has a radius of 450 m. At a certain instant, the car is traveling due west at a speed of 22 m/s and is slowing down at a rate of 1.5 m/s². What are the magnitude and direction of the car's acceleration at this instant?

N8.4   Banking

In the rest of this chapter, we will consider examples in which we use either equation N8.12 or equation N8.19 to infer things about the forces acting on objects in circular motion.

One interesting application relates to the phenomenon of *banking*. You perhaps know that when a pilot turns a plane, she or he has to bank the plane into the turn, that is, lower one wing and raise the other. Why?

Figure N8.4a shows a rear-view free-particle diagram of the plane. Two forces act vertically on the plane: its weight pulls it downward, and the lift from its wings pushes it upward. When the plane is in straight and level flight, these forces are directly opposite to each other and cancel each other.

We have seen, though, that a plane flying in a circular path has an acceleration toward the circle's center, which is *leftward* in figure N8.4b. Newton's second law tells us that the net force on the plane must also point in that direction. Where does this leftward net force come from? It can't come from the plane's engines, since the thrust force that they exert is always directly forward.

The easiest way to exert a leftward force on a plane is to tilt the wings at an angle θ so that the lift force that they exert (which acts perpendicular to

Why a plane banks when it turns

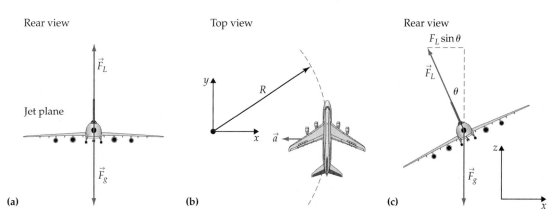

Figure N8.4
(a) A free-particle diagram (rear view) of a jet plane flying straight and level. (b) A top view of a plane in a circular turn to the left. (c) A free-particle diagram of the plane turning left.

the wings) has both an upward and a leftward component (see figure N8.4c). If the plane maintains its altitude as it turns, the upward part of the lift force must still balance the downward weight of the plane, leaving the leftward part as the net force on the plane. This net force is what causes the plane to accelerate away from its natural straight-line path into the circular path.

Example N8.1

Problem   A jet plane flies at a constant speed of 260 mi/h in a holding pattern that is a horizontal circle of radius 5.0 mi. What is the plane's banking angle?

Translation   A sketch of the plane's motion appears in figure N8.4b, and a free-particle diagram for the plane appears as in figure N8.4c.

Model   If the plane is moving in a *horizontal* circle, its acceleration in the z direction is zero: $a_z = 0$. On the other hand, because of its circular motion the plane must have a *nonzero* acceleration toward the circle's center: at the instant shown in figure N8.4b, $a_x = -v^2/R$ (negative because the acceleration is to the left, i.e., in the $-x$ direction as we have defined our coordinates) and $a_y = 0$.

Solution   Newton's second law then reads

$$\begin{bmatrix} -mv^2/R \\ 0 \\ 0 \end{bmatrix} = m\vec{a} = \vec{F}_g + \vec{F}_L = \begin{bmatrix} 0 \\ 0 \\ -mg \end{bmatrix} + \begin{bmatrix} -F_L \sin\theta \\ 0 \\ F_L \cos\theta \end{bmatrix} \qquad \text{(N8.21)}$$

If we multiply both sides of the top line of equation N8.21 by -1 and add mg to both sides of the bottom line of equation N8.21, we get

$$\frac{mv^2}{R} = F_L \sin\theta \qquad mg = F_L \cos\theta \qquad \text{(N8.22)}$$

This problem looks hopeless at first: while we know v and R, we do not know the plane's mass m, the magnitude of the lift force F_L, or the angle θ, meaning that we have more unknowns than equations! But it turns out that we can solve the problem anyway. If we divide the first equation by the second, the unknowns m and F_L divide out, and we are left with

$$+\frac{v^2}{Rg} = \frac{\sin\theta}{\cos\theta} = \tan\theta \qquad \Rightarrow \qquad \theta = \tan^{-1}\left(\frac{v^2}{Rg}\right) \qquad \text{(N8.23a)}$$

$$\theta = \tan^{-1}\left[\frac{(260 \text{ mi/h})^2}{(5.0 \text{ mi})[22 \text{ (mi/h)/s}]} \left(\frac{1 \text{ h}}{3600 \text{ s}}\right) \right] = 9.7° \qquad \text{(N8.23b)}$$

Exercise N8X.4

You may notice that the pilot when beginning a sharp turn will power up the engines somewhat if he or she wants to keep the plane from losing altitude. Why would the pilot do this?

A car doesn't need to bank

A car rounding a corner does *not* need to bank into the curve: when you turn the steering wheel to the left, the road exerts a static friction force on the

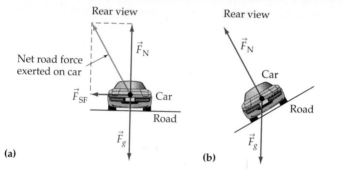

(a)   (b)

Figure N8.5
(a) A free-particle diagram of a car rounding a curve to the left on a level roadbed. Static friction has to provide the leftward force required to keep the car following the curve. (b) A free-particle diagram of a car on an ideally banked roadbed. Here the normal force alone provides the necessary sideward force.

tires to the left, which provides the leftward net force necessary to accelerate the car away from its natural straight-line motion (see figure N8.5a). The car does not need to lean to the left to do this. (In fact, if you consider the torques that have to be exerted on the car, you will find that the car actually leans somewhat to the *right* when turning to the left: see problem N8S.11.)

An engineer designing a road will often call for a tight curve to be banked at an angle. This is often easiest to see on freeway overpasses that curve in one direction or another. (It is even more obvious in racetracks and the tracks of roller coasters, which are often banked at very large angles.) If the banking angle θ is chosen just right for the radius of the curve and the typical speed of a car on that curve, then the tilted *normal* force acting on the car provides the required leftward acceleration (see figure N8.5b). This is advantageous because the car does not then have to *depend* on static friction to keep it traveling with the curve. Under bad weather conditions, the coefficient of static friction between the roadbed and the tires may become so low that the static friction force cannot keep the car following the curve. The normal force, on the other hand, is not affected by bad conditions. The car cannot move *into* a wet or icy roadbed any more easily than into a dry roadbed! Banking the curve thus cuts down on accidents in bad weather. (The advantages in the case of roller coasters are even more obvious!)

A person riding a bicycle, on the other hand, *does* have to lean into the curve as an airplane does. The reasons are superficially similar to the reason a plane has to bank, but a careful analysis requires techniques that we will develop in chapter N9 (see example N9.4).

Roads are sometimes banked for safety's sake

Bicyclists also have to lean into the curve

N8.5   Examples

How to choose an appropriate
reference frame

Examples N8.2 through N8.4 illustrate how we can adapt the problem-solving framework first developed in chapters N5 and N6 to solve various kinds of problems involving circular motion.

Circular motion problems, like problems involving linear acceleration, are generally easier to do if you orient the reference frame axes correctly. Usually the best orientation is such that one axis points along the line connecting the object to the center of its circular path. In the case of *uniform* circular motion, the object's acceleration will thus nicely lie along this axis direction, and the components of the net force in other axis directions will be zero. In the case of *nonuniform* circular motion, the two axis directions in the plane of the circle will then correspond in a simple way to the $\hat{r}$ and $\hat{v}$ directions of equation N8.19. Note that since these directions change as the object moves, the reference frame that you set up will only be useful at one instant. This is usually good enough to solve the problem, since other instants will be analogous.

Banking problems *seem* to have
too many unknowns

Most banking angle problems, like the airplane problem discussed in example N8.1, may look at first as though they have too many unknowns to solve. Press ahead anyway and see if you can get some unknowns to divide out.

In most other respects, circular motion problems are like any other constrained-motion problem: we take what we know about the object's motion (that it moves in a circle) and use that to determine unknown forces, banking angles, time to complete the circular path, and the like.

Example N8.2

Problem   A 1500-kg car travels at a constant speed of 22 m/s over the top of a hill whose cross section near the top is approximately a circle having an effective radius of 150 m. What is the magnitude of the normal force of the car just as it passes the top of the hill, and how does this compare to the car's weight?

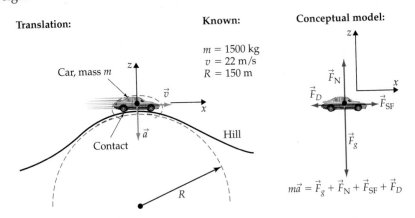

Prose Model   A free-particle diagram for the car is shown above and to the right ($\vec{F}_{\text{SF}}$ is from the tires' interaction with the road, and $\vec{F}_D$ is from the car's interaction with the air). The constraint on the car's motion is that we are told that it moves in a vertical circle (which we have defined to be the xz plane) at a constant speed. This means that the car's acceleration is downward (in the $-z$ direction in our reference frame) with a magnitude of $a = v^2/R$. This in turn means that $\vec{F}_{\text{SF}}$ from the interaction with the road must be directed *forward* (so that it cancels $\vec{F}_D$), and $\vec{F}_N$ must be smaller than $\vec{F}_g$ (as drawn) to yield a downward net force. The free-particle diagram is then consistent with the observed acceleration. We are assuming that the hill's cross section really *is* circular (it probably isn't quite), and that drag is the only significant opposing force.

Solution   Newton's second law in this case tells us that

$$\begin{bmatrix} 0 \\ 0 \\ -mv^2/R \end{bmatrix} = \begin{bmatrix} 0 \\ 0 \\ -mg \end{bmatrix} + \begin{bmatrix} 0 \\ 0 \\ +F_N \end{bmatrix} + \begin{bmatrix} +F_{SF} \\ 0 \\ 0 \end{bmatrix} + \begin{bmatrix} -F_D \\ 0 \\ 0 \end{bmatrix} \qquad (1)$$

The top line of this equation tells us that $F_{SF} = F_D$, which we knew intuitively. Solving the bottom line for F_N, we get

$$F_N = mg - \frac{mv^2}{R} = (1500 \text{ kg})(9.8 \text{ N/kg}) - (1500 \text{ kg})\frac{(22 \text{ m/s})^2}{150 \text{ m}}\left(\frac{1 \text{ N}}{1 \text{ kg·m/s}^2}\right)$$

$$= 14{,}700 \text{ N} - 4800 \text{ N} = 9900 \text{ N} \qquad (2)$$

Since we can read from this calculation that $mg = 14{,}700$ N, $F_N/F_g = 9900 \text{ N}/14{,}700 \text{ N} = 0.67$, so the normal force is two-thirds of the weight.

Evaluation   The units check out, and the magnitudes look reasonable here ($F_N < F_g$ is a good sign). The sign of F_N is also positive, which is appropriate for a vector magnitude (a negative result would indicate something seriously awry!).

Example N8.3

Problem   What is the banking angle needed to keep a 1500-kg car following a circular bend in the road of radius 330 m when it is going 25 m/s (≈ 58 mi/h) without requiring that *any* static friction force be exerted on the tires?

Translation   Rear- and top-view diagrams of the situation look like this:

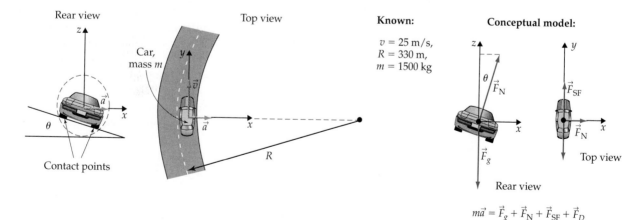

$$m\vec{a} = \vec{F}_g + \vec{F}_N + \vec{F}_{SF} + \vec{F}_D$$

Prose Model   Rear- and top-view free-particle diagrams for the car are shown to the right. The constraint here is that the car is moving in a horizontal circle at a constant speed. Its acceleration therefore points directly inward, which (as we have defined the frame axes) is in the $+x$ direction at the instant shown. This is consistent with the free-particle diagrams, which show the net force to point in that direction (due to the tilting of the normal force). We are assuming that the bend in the road is really circular, that it is level, and that the car's speed is constant. If this is so, then there is no component of acceleration in either the y direction or the z direction. I am also assuming that the force opposing the car's forward motion is mostly drag.

Solution   Newton's second law in this case implies that

$$\begin{bmatrix} +mv^2/R \\ 0 \\ 0 \end{bmatrix} = \begin{bmatrix} 0 \\ 0 \\ -mg \end{bmatrix} + \begin{bmatrix} +F_N\sin\theta \\ 0 \\ +F_N\cos\theta \end{bmatrix} + \begin{bmatrix} 0 \\ +F_{SF} \\ 0 \end{bmatrix} + \begin{bmatrix} 0 \\ -F_D \\ 0 \end{bmatrix} \quad (1)$$

The middle line of this equation tells us that $F_D = F_{SF}$, which is not particularly relevant to what we want to find. The bottom line, on the other hand, tells us that $mg = F_N\cos\theta$, whereas the first line tells us that $mv^2/R = F_N\sin\theta$. If we divide the second by the first, we get

$$\frac{mv^2}{mRg} = \frac{\cancel{F_N}\sin\theta}{\cancel{F_N}\cos\theta} \quad \Rightarrow \quad \tan\theta = \frac{v^2}{Rg} = \frac{(25\ \cancel{m/s})^2}{(330\ \cancel{m})(9.8\ \cancel{m/s^2})} = 0.19 \quad (2)$$

So $\theta = \tan^{-1}(0.19)$, implying that the road should be banked toward the center at an angle of $\theta = 11°$.

Evaluation   Note how the units work out so that $\tan\theta$ is correctly unitless, and θ is positive (consistent with the pictures) and its magnitude is reasonable ($> 45°$ would be unreasonable). Note also that the result does not depend on m but does depend on v. A car not going at exactly the right speed will need some x component of static friction to hold it on its path.

Example N8.4

Problem   A satellite in a circular orbit is a freely falling object that happens to be moving at just the right speed so that the gravitational force acting on it provides exactly the right acceleration to hold it in its circular path. How fast would an object have to travel to be in a circular orbit just above the earth's surface? How long would it take the object to go around the earth at this

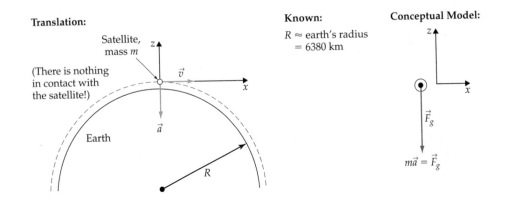

Translation:

Satellite, mass m

(There is nothing in contact with the satellite!)

Earth

$\vec{v}$

$\vec{a}$

R

Known:

$R \approx$ earth's radius
  $= 6380$ km

Conceptual Model:

$\vec{F}_g$

$m\vec{a} = \vec{F}_g$

speed? Ignore air friction (even though this is pretty absurd for something traveling near the earth's surface!).

Prose Model   A free-particle diagram of the object here is almost too trivial to draw: if we ignore air friction, the only force that can possibly act on the satellite is the force of gravity. A satellite in a circular orbit is (by definition) moving in a circular path. We will assume that its speed v is constant: if this is so, then the satellite's acceleration will have a magnitude of $a = v^2/R$ and will point toward the center of the earth. This is consistent with the net force drawn on the diagram, since it must also point toward the center of the earth. At the representative instant shown, then, both the net force and the satellite's acceleration have components only in the $-z$ direction (according to our frame).

Solution   At that instant, Newton's second law implies

$$\begin{bmatrix} 0 \\ 0 \\ -mv^2/R \end{bmatrix} = \begin{bmatrix} 0 \\ 0 \\ -mg \end{bmatrix} \tag{1}$$

Solving the last line of this equation for v^2 and then v, we get

$$v^2 = Rg \quad \Rightarrow \quad v = \sqrt{Rg} = \sqrt{(6{,}380{,}000 \text{ m})(9.8 \text{ m/s}^2)} = 7900 \text{ m/s} \tag{2}$$

This is pretty fast! (Note that the mass of the satellite is irrelevant.) Since the distance around the earth is $2\pi R$, the time T required for a satellite to cover this distance is

$$T = \frac{\text{distance}}{\text{speed}} = \frac{2\pi R}{v} = \frac{2\pi(6380 \text{ km})}{7.9 \text{ km/s}}\left(\frac{1 \text{ min}}{60 \text{ s}}\right) = 85 \text{ min} \ (= 1.4 \text{ h}) \tag{3}$$

Evaluation   Note how the units all work out in both cases, and v comes out positive (as a magnitude should). The speed is pretty fast, but this is correct (one of the reasons that space travel is difficult and expensive).

The problem with a satellite orbiting "just above" the earth's surface (in addition to the hazard that it might pose to airplanes and mountains) is that drag is *not* going to be negligible at 7.9 km/s. (Indeed, the space shuttle's wings glow red-hot from friction when it reenters the atmosphere at speeds comparable to this!)

TWO-MINUTE PROBLEMS

N8T.1   When a speeding roller-coaster car is at the bottom of a loop, the magnitude of the normal force exerted on the car's wheels due to its interaction with the track is
  A.   Greater than the weight of the car and its passengers.
  B.   Equal to the weight of the car and its passengers.
  C.   Less than the weight of the car and its passengers.

N8T.2   A child grips tightly the outer edge of a playground merry-go-round as other kids push on it to give it a dizzying rotational velocity. When the other kids let go, the horizontal component of the net force on the child points most nearly
  A.   Inward toward the center of the merry-go-round.
  B.   Outward away from the center of the merry-go-round.
  C.   In the direction of rotation.

(more)

A child using a playground merry-go-round
(see problem N8T.2).

D.   Nowhere: the horizontal component of the
      net force is zero.
E.   In some other direction (specify).

N8T.3 A car is traveling counterclockwise along a circular
bend in the road whose effective radius is 100 m. At
a certain instant of time, the car is traveling due
north, has a speed of 10 m/s, and is in the process of
increasing that speed at a rate of 1 m/s². The direc-
tion of the car's acceleration at that instant is most
nearly
A.   North
B.   Northeast
C.   East
D.   Northwest
E.   West
F.   Southwest
T.   Zero

N8T.4 A plane in a certain circular holding pattern banks
at an angle of 8° when flying at a speed of 150 mi/h.
If a second plane flies in the same circle at 300 mi/h,
what is its banking angle?
A.   A bit less than 16°.
B.   Exactly 16°.
C.   A bit more than 16°.
D.   A bit less than 32°.
E.   Exactly 32°.
F.   A bit more than 32°.
T.   Answer depends on the planes' masses.

N8T.5 Car 1 with mass m rounds a curve of radius R traveling at a constant speed v. Car 2 with mass $2m$
rounds a curve of radius $2R$ traveling at a constant
speed $2v$. How does the magnitude F_2 of the sideward static friction force acting on car 2 compare
with the magnitude F_1 of the sideward static friction force acting on car 1?
A. $F_1 = 4F_2$
B. $F_1 = 2F_2$
C. $F_1 = F_2$
D. $F_2 = 2F_1$

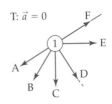

Figure N8.6
(For problems N8T.6 through N8T.8.) A mass
(bob) swings from the end of a string. At
point 1, the bob is at the extreme point of
the swing and thus is instantaneously at rest.
At point 3, the bob is directly below its
suspension point and has its maximum
speed.

E. $F_2 = 4F_1$
F.   Other (specify)

N8T.6 (See figure N8.6.) Which one of the arrows to the
right most closely indicates the direction of the
bob's acceleration when it is at point 1?

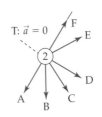

N8T.7 (See figure N8.6.) Which one of the arrows to the
right most closely indicates the direction of the
bob's acceleration when it is at point 2?

N8T.8 (See figure N8.6.) Which one of the arrows to the
right most closely indicates the direction of the
bob's acceleration when it is at point 3?

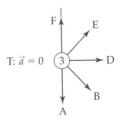

(more)

HOMEWORK PROBLEMS

Basic Skills

N8B.1 A car traveling at a constant speed of 50 mi/h travels around a curve. An accelerometer in the car measures its sideward acceleration to be 0.1g. What is the effective radius of the curve? [*Hint:* $g = 22$ (mi/h)/s.]

N8B.2 A plane is traveling in a circular path 32 km in diameter. It is banking at an angle of 12°, which indicates that its sideward acceleration is 2.1 m/s². What is its speed?

N8B.3 An airplane banks at an angle of 11° while flying in a level circle at 320 mi/h. What is the radius of its circular path?

N8B.4 A car is traveling counterclockwise along a circular bend in the road with an effective radius of 200 m. At a certain instant, the car's speed is 20 m/s, but it is slowing down at a rate of 1.0 m/s². What is the magnitude of the car's acceleration?

N8B.5 The position $\vec{r}$ of a certain object relative to the origin has components [3.0 m, 5.0 m, −2.0 m]. What are the components of the unit vector $\hat{r}$ that points in the same direction?

N8B.6 The position $\vec{r}_{B(A)}$ of object B relative to object A has the components [7 m, −12 m, 0 m]. What are the components of the unit vector $\hat{r}_{B(A)}$ that points in the same direction?

Synthetic

The starred problems are especially well suited for practicing the use of the problem-solving framework.

N8S.1 In the limit that $R \to \infty$, a circular trajectory becomes a straight line. Argue carefully that in this limit, equation N8.19 becomes what you'd expect for linear motion with nonconstant speed.

*N8S.2 A stunt driver drives a car over the top of a hill having a cross section that can be approximated by a circle of radius 250 m. What is the greatest speed the car can reach before it leaves the road at the top of the hill?

*N8S.3 A car travels at a constant speed of 23 m/s through a small valley whose cross section is like a circle of radius 310 m. What is the magnitude of the normal force on the car, expressed as a multiple of the car's weight?

*N8S.4 A child places a lunch box on the rim of a playground merry-go-round that has a radius of 2.0 m ($\approx$ 6 ft). If the merry-go-round goes once around

every 6.0 s, what is the speed of the box? What must be the coefficient of static friction between the box and the merry-go-round if the box is to stay on?

*N8S.5 The acceleration of gravity near the surface of the moon has about one-sixth the magnitude that it does on earth. The radius of the moon is 1740 km. How long would a satellite orbiting just above the moon's surface take to go once around the moon? (The moon has no atmosphere!)

*N8S.6 Imagine that you are designing a circular curve in a highway that must have a radius of 330 ft and will carry traffic moving at 60 mi/h.
(a) At what angle should the roadbed be banked for maximum safety?
(b) If the roadway is *not* banked, what would the necessary coefficient of static friction between the tires and the asphalt road have to be to keep a car on the road? Is such a coefficient reasonable? (*Hint:* 1 m = 3.3 ft, 1 m/s = 2.24 mi/h.)

*N8S.7 A 150-lb student rides a ferris wheel that rotates at a constant rate. At the highest point, the seat exerts a normal force of magnitude 110 lb on the student. What would the magnitude of this normal force be at the lowest point? (*Hint:* An object with mass 1 kg weighs 2.2 lb.)

*N8S.8 A ball of mass m is tied to one end of a string of length L, the other end of which is fixed to the ceiling. The ball is then set in motion at a constant speed in a horizontal circle of radius $R < L$ around the axis that the string would make if the ball were to hang at rest (the string thus makes a constant angle θ with the vertical direction). Determine how long it takes the ball to go around the circle once in terms of m, g, L, and θ.

*N8S.9 As you are riding in a 1650-kg car, you approach a hairpin curve in the road whose radius is 50 m. The roadbed is banked inward at an angle of 10°.
(a) Suppose the road is very icy, so that the coefficient of static friction is essentially zero. What is the maximum speed at which you can go around the curve?
(b) Now suppose that the road is dry and that the static friction coefficient between the tires and the asphalt road is 0.6. What is the maximum speed at which you can safely go around the curve?

*N8S.10 At a certain instant of time, a 1200-kg car traveling along a curve 250 m in radius is moving at a speed of 10 m/s (22 mi/h) but is slowing down at a rate of 2 m/s². Ignoring air friction, what is the total static friction force on the car as a fraction of its weight at that instant?

*N8S.11 Consider a car turning a corner to the left. Using the methods of chapter N5, show that in order to balance the torques on the car that seek to rotate the car around an axis going through its center of mass along the direction of its motion, the roadbed has to exert a greater normal force on the right wheels of the car than on the left wheels. Since this normal force will be transmitted to the car's body through the springs of the car's suspension, this means that the suspension springs on the right will have to compress more than those on the left, and thus the car will lean to the right, as asserted in section N8.4.

A car leans *away* from a curve (see problem N8S.11).

*N8S.12 An unpowered roller-coaster car starts at rest at the top of a hill of height H, rolls down the hill, and then goes around a vertical loop of radius R.

A roller coaster with loop-de-loop
(see problem N8S.12).

Determine the minimum value for H required if the car is to stay on the track at the top of the loop. (*Hints:* At the top of the loop, the car is upside down. If it is in contact with the track, though, the contact interaction will exert a normal force on the car perpendicular to the track and *away* from the track, since the normal force is a compression force. You may find it helpful to use conservation of energy here.)

N8S.13 This problem explores a purely mathematical method for deriving equation N8.12. A particle moving in a circle of fixed radius R around an origin point O has a position vector that can be written $\vec{r}(t) = R\hat{r}$, where $\hat{r}$ is the direction of $\vec{r}$, which changes with time.

(a) In a coordinate system oriented so the particle's path lies in the xy plane, show that at any given instant

$$\hat{r} = \begin{bmatrix} \cos\theta \\ \sin\theta \\ 0 \end{bmatrix} \qquad (\text{N8.24})$$

where the angle θ is the angle that the object's position vector makes with the x axis at that instant. Note that θ varies with time as the object moves around the circle.

(b) By taking the time derivative of both sides of $\vec{r}(t) = R\hat{r}$ in column-vector form and using the chain rule (see appendix NA), show that

$$\vec{v}(t) = R\frac{d\theta}{dt}\begin{bmatrix} -\sin\theta \\ \cos\theta \\ 0 \end{bmatrix} \qquad (\text{N8.25})$$

(c) Argue from the expression above that

$$v(t) = R\left|\frac{d\theta}{dt}\right| \qquad (\text{N8.26})$$

and that we can consider the column vector in brackets in equation N8.25 to be the directional $\hat{v}$.

(d) Note that equation N8.26 implies that if the particle's speed is constant, then $|d\theta/dt|$ must be a constant. *Assuming* this, take the time derivative of both sides of equation N8.25 to show that

$$\vec{a}(t) = -\frac{v^2}{R}\hat{r} \qquad (\text{N8.27})$$

Rich-Context

N8R.1 "Rotor" is a ride found in many amusement parks that consists of a hollow cylindrical room (roughly 8 ft in radius) that rotates around a central vertical axis. Riders enter the room and stand against the canvas-covered wall. The room begins to rotate, and when a certain speed is reached, the floor of the room drops away, revealing a deep pit. The riders do not fall, though: they are supported by a static friction force exerted by the person's contact interaction with the wall. Estimate the rate at which the room should rotate (in revolutions per minute) to safely pin the riders to the wall. (You may have to make some estimates.)

People enjoying Rotor (see problem N8R.1).

N8R.2   You are the technical consultant for a car-chase sequence in an action movie. In a certain part of the scene, the director wants a car to round a certain curve while braking from 66 mi/h to rest before it travels 290 ft along the curve. You measure the radius of the curve to be 590 ft. Is this scene possible? Defend your response and suggest an alternative scene if it is not possible.

Advanced

N8A.1   Equations N8.24 through N8.26 in problem N85.13 apply even when the particle's speed in its circular trajectory is *not* constant. Read that problem carefully, then use the product rule to calculate the time derivative of equation N8.25, and argue that your result is consistent with

$$\vec{a}(t) = \frac{dv}{dt}\hat{v} - \frac{v^2}{R}\hat{r} \qquad (\text{N8.28})$$

when $d\theta/dt$ is *not* necessarily constant. (*Hint:* You will have to separately discuss the cases in which $d^2\theta/dt^2 > 0$ and $d^2\theta/dt^2 < 0$. Show that in both cases, the term involving $d^2\theta/dt^2$ reduces to the first term on the right side of equation N8.28.)

ANSWERS TO EXERCISES

N8X.1   By equation N8.5, we have

$$v = \frac{2\pi R}{T} = \frac{2\pi (3.0 \text{ m})}{6.3 \text{ s}} = 3.0 \text{ m/s} \qquad (\text{N8.29})$$

So the magnitude of its acceleration is

$$a = \frac{v^2}{R} = \frac{(3.0 \text{ m/s})^2}{3.0 \text{ m}} = 3.0 \text{ m/s}^2 \qquad (\text{N8.30})$$

N8X.2   We can write the particle's position at point *A* as a column vector as follows:

$$\vec{r} = \begin{bmatrix} R\cos\theta \\ R\sin\theta \\ 0 \end{bmatrix} \qquad (\text{N8.31})$$

Therefore, the directional $\hat{r}$ is

$$\hat{r} = \frac{\vec{r}}{r} = \frac{1}{R}\begin{bmatrix} R\cos\theta \\ R\sin\theta \\ 0 \end{bmatrix} = \begin{bmatrix} \cos\theta \\ \sin\theta \\ 0 \end{bmatrix} \qquad (\text{N8.32})$$

N8X.3   The radial part of this car's acceleration has a magnitude of $v^2/R = (22 \text{ m/s})^2/(450 \text{ m}) = 1.1 \text{ m/s}^2$. We

are told that the car is slowing down at the rate of 1.5 m/s², so the component of the acceleration in the direction of motion is $dv/dt = -1.5 \text{ m/s}^2$. The magnitude of the total acceleration (according to the pythagorean theorem) is

$$a = \sqrt{(1.1 \text{ m/s}^2)^2 + (-1.5 \text{ m/s}^2)^2} = 1.85 \text{ m/s}^2 \qquad (\text{N8.33})$$

Note that $\vec{a}$ points somewhat backward here. The angle ϕ that $\vec{a}$ makes with the backward direction is

$$\phi = \tan^{-1}\left|\frac{v^2/R}{dv/dt}\right| = \tan^{-1}\left|\frac{1.1 \text{ m/s}^2}{1.5 \text{ m/s}^2}\right| = 36° \qquad (\text{N8.34})$$

N8X.4   Notice that in figure N8.4a, the whole lift force goes to supporting the plane: $mg = F_L$. In figure N8.4c, though, we see that the plane is supported by only the vertical component of the lift force: $mg = F_L \cos\theta$. This means that the lift force must increase in magnitude as the plane banks to keep its vertical component equal in magnitude to the plane's weight. The easiest way to make the wings exert greater lift is to increase the speed of the plane through the air.

Noninertial Frames

Chapter Overview

Introduction

Throughout this subdivision, we have been using Newton's second law and an object's observed motion to infer the forces acting on that object. However, this approach can backfire if we are observing the object's motion in a noninertial reference frame, because it can lead us to infer forces that do not physically exist. This chapter discusses this problem in depth.

Section N9.1: Fictitious Forces

In certain situations, such as in an airplane accelerating for takeoff, we seem to experience forces that are linked to the motion of the plane and do *not* arise from physical interactions with other objects. The purpose of this chapter is to explain why such forces are in fact *inventions of our imagination*, and that we can adequately explain what we observe *without* inventing such forces, if we use an appropriate reference frame.

Section N9.2: The Galilean Transformation

The first step in the process is to describe mathematically how observations in one reference frame S are linked to observations in another frame S'. If we orient these frames' axes so that they point in the same direction in space, then a straightforward argument yields the following transformation equations:

$$\vec{v}'(t) = \vec{v}(t) - \vec{\beta}(t) \qquad (N9.2)$$

Purpose: This equation allows us to compute an object's velocity $\vec{v}'$ as measured in the S' frame if we know its velocity $\vec{v}$ in the S frame and the velocity $\vec{\beta}$ of the S' frame relative to the S frame.
Limitations: All speeds must be much smaller than that of light.
Notes: We call this the **galilean velocity transformation equation.**

$$\vec{a}'(t) = \vec{a}(t) - \vec{A}(t) \qquad (N9.7)$$

Purpose: This equation allows us to compute an object's acceleration $\vec{a}'$ in a frame S' given the same object's acceleration $\vec{a}$ in frame S and the acceleration $\vec{A}$ of frame S' relative to frame S ($\vec{A} \equiv d\vec{\beta}/dt$).
Limitations: The speeds of the objects and frames involved must be much less than the speed of light.

Section N9.3: Inertial Reference Frames

Newton's first law states that an isolated object has zero acceleration. An **inertial reference frame** is a frame in which Newton's first law is observed to hold. In

noninertial reference frames, violation of the first law means that *none* of Newton's laws apply.

We can test whether a frame is inertial by placing an object at rest relative to that frame, isolating it from external effects, and seeing whether it remains at rest. This is tricky in a gravitational field (since an object cannot be isolated from the effects of gravity), but we can imagine an object (such as a puck floating on an air table) that is at least isolated from external effects in the horizontal plane.

Equation N9.7 implies that if such **first-law detectors** establish that frame S is inertial, then another frame S' will also be inertial if and only if its acceleration $\vec{A}$ relative to S is zero. A frame attached to the ground is nearly inertial, so any frame that accelerates significantly relative to it will be noninertial.

Section N9.4: Linearly Accelerating Frames

In a frame attached to a plane accelerating for takeoff, a force appears to press objects backward. However, one comes to this conclusion only by inappropriately applying Newton's second law in the noninertial plane frame. If we analyze the situation in the ground frame, we see that the observed effects are easily explained if we understand that the objects are simply trying to remain at rest as the plane accelerates forward.

Section N9.5: Circularly Accelerating Frames

Similarly, the "centrifugal" force you seem to feel in a turning car arises from your attempt to use Newton's second law to explain why objects appear to be thrown to the right when your car turns left. In the ground frame, though, we see that the objects are really attempting to travel in a straight line while the car frame veers left. Again in the *ground* frame, we can explain what we observe without invoking forces unconnected to physical interactions. Such forces are *not real*.

Section N9.6: Using Fictitious Forces

Having said that, we know that in some situations, analysis using a noninertial frame is much easier than using an inertial frame. In such cases, equation N9.7 implies that

$$m\vec{a}' = -m\vec{A} + \vec{F}_1 + \vec{F}_2 + \cdots \qquad \text{(N9.12)}$$

Purpose: We can use this version of Newton's second law in an noninertial frame *if* we know that frame's acceleration $\vec{A}$ relative to an inertial frame.
Symbols: m is the mass of an object, $\vec{a}'$ is its acceleration in the noninertial frame, and $\vec{F}_1, \vec{F}_2, \ldots$ are real forces acting on the object.
Limitations: The relative speed of the two frames and the speed of the object relative to each must be much smaller than the speed of light.

We call the term $-m\vec{A}$ a **frame-correction force** (or *inertial force*): this force acts as if it were an additional gravitational force applied to the object. Note that this "force" does not physically exist: the *term* exists in this equation only to correct for our using a noninertial frame. Use this equation only if (1) an analysis in an inertial frame is very difficult and (2) you clearly indicate you are using a NONINERTIAL frame in your analysis.

Section N9.7: Freely Falling Frames and Gravity

Equation N9.12 implies that since all objects fall with the same acceleration $\vec{g}$ in an external gravitational field, the frame-correction and gravitational forces in a freely falling frame cancel out. This is the newtonian explanation for why we can ignore external gravitational fields in a frame floating in space (as claimed in chapter C4).

In Einstein's theory of general relativity, however, freely falling frames are *really* inertial, and frames at rest on the surface of the earth are not! The section and selected homework problems discuss this issue in greater detail.

N9.1   Fictitious Forces

Examples of *fictitious forces* in daily life

According to the newtonian model, *all* physical forces express the interaction of two objects, and all are either long-range forces or contact forces. The latter are easy to recognize since you can easily see when two objects are touching each other, and the only long-range forces that operate over macroscopic distances are gravitational and electromagnetic forces.

Yet in certain situations, forces appear to act that do *not* fit these categories. For example, imagine yourself in a jet accelerating for takeoff. When the pilot opens up the throttle, a magical force seems to appear that pushes you backward in your seat. The more extreme the plane's acceleration, the stronger that this apparent force seems to be. When the plane lands and the pilot applies the brakes and reverses the thrust on the engines, the reverse happens: a mysterious force appears that tugs you forward. If you are in an automobile and the driver suddenly applies brakes sharply, you feel the same kind of force pulling you forward.

As a car turns a corner, a magical force seems to press you against the side of the car away from the turn. Again, this force seems to be associated with the car's acceleration, because we know that a car going around a curve is accelerating toward the center of the curve. As the car's acceleration increases (i.e., the tighter the curve or the higher the car's speed), the magnitude of this force appears to increase as well.

These alleged forces do not fit in any of the categories described in chapter N1. They are obviously not contact forces (nothing that you are touching presses you back into the plane seat during takeoff). Such forces feel like gravitational forces, but they cannot *be* gravitational (an extremely large and massive object does not magically appear behind the plane when the pilot opens the throttle). Nor are they electrostatic or magnetic (you do not become electrically charged or magnetized when your car turns a sharp corner). These alleged forces do not in fact seem to be a consequence of the presence of *any* external object.

Our task: to explain the effects without the forces

Even though these forces "feel" quite real, *I claim that these alleged forces are in fact inventions of your imagination.* Ladies and gentlemen of the jury, I will show you beyond a shadow of a doubt that it is possible within the context of the newtonian model to explain *all* the described *effects* of these forces *without* assuming that these forces really exist. Since the argument for the existence of these alleged forces is based *entirely* on the circumstantial evidence of their effects, an alternative explanation of those effects makes the case for the existence of these **fictitious forces** disappear. Let me begin my case.

N9.2   The Galilean Transformation

Fictitious forces are associated with *accelerating* reference frames

The first step in understanding this situation is to recognize that these forces only seem to arise when you are riding in something (such as a plane or car) that is *accelerating* relative to the earth's surface. While being pushed back into your chair is something you might expect during an airplane takeoff, you would be very surprised (even terrified) if a mysterious force were to spontaneously push you back into your chair while you were sitting at home reading a book!

When you are riding in a plane, car, or even a playground merry-go-round, you automatically and unconsciously use your surroundings (the cabin of the plane, the frame of the car, or the structure of the merry-go-round) as your frame of reference, and you judge your motion and the motion of objects around you in terms of that frame. When your jet or car is

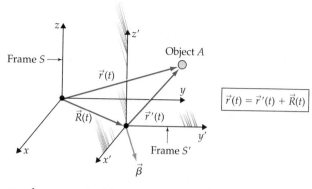

Figure N9.1
Two reference frames S and S'. Frame S' moves relative to S at a velocity $\vec{\beta}$, as shown. The positions of object A as measured in the two reference frames are related by the vector equation in the box.

$$\vec{r}(t) = \vec{r}'(t) + \vec{R}(t)$$

cruising at a constant speed, no magical forces appear: only when the jet or car changes its velocity do strange things seem to happen. *The presence of these forces has something to do with observing motion from within an accelerating reference frame.*

The next step is to find a means of mathematically connecting observations made in one reference frame with observations made in another. Figure N9.1 shows two abstract reference frames S and S' (the latter is read "S prime"). Frame S' is moving with respect to S at some (possibly time-dependent) relative velocity $\vec{\beta}$ (the Greek letter beta here refers to the "boost" in velocity required to take you from frame S to frame S'). Let's also take advantage of our ability to choose the orientation of reference frames to set them up so that their axes point in the same directions (this makes the analysis easier). Assume that we measure the position of object A as a function of time in both reference frames: let $\vec{r}(t)$ be its position vector as measured in frame S at time t, and let $\vec{r}'(t)$ be its position as measured in S' at the same instant, and let the position of S' relative to S at that instant be $\vec{R}(t)$. As shown in figure N9.1, the definition of vector addition means that the mathematical relationship between $\vec{r}(t)$, $\vec{r}'(t)$, and $\vec{R}(t)$ at any time (no matter how S' is moving) is given by

$$\vec{r}(t) = \vec{r}'(t) + \vec{R}(t) \qquad \text{(N9.1)}$$

Transformation equation for positions

In words, this equation tells us that the object's position in frame S is the vector sum of its position in frame S' and the position of S' relative to S.

If we solve this equation for $\vec{r}'(t)$, take the time derivative of both sides, and note that $d\vec{R}/dt$ is the same as the velocity $\vec{\beta}$ of S' relative to S, we get

$$\vec{v}'(t) = \vec{v}(t) - \vec{\beta}(t) \qquad \text{(N9.2)}$$

(Galilean) transformation equation for velocities

Purpose: This equation allows us to compute an object's velocity $\vec{v}'$ as measured in the S' frame if we know its velocity $\vec{v}$ in the S frame and the velocity $\vec{\beta}$ of the S' frame relative to the S frame.
Limitations: All speeds must be much smaller than that of light.
Notes: We call this the **galilean velocity transformation equation.**

Does this make sense? Let's consider some examples.

Example N9.1

Problem Imagine that you are on a train traveling in the $+x$ direction at a speed of 25 m/s with respect to the ground. If you throw a baseball in the $+x$ direction at a speed of $+12$ m/s relative to the train, what is the ball's *x*-velocity relative to the ground? *Intuitively,* you might say that the ball's velocity with respect to the ground should be the *sum* of the train's velocity with respect to the ground and the ball's velocity with respect to the train, that is, 37 m/s. Is this correct?

Translation and Model   To apply equation N9.2 or N9.3 correctly, we have to do two things: (1) determine what the reference frames are and which is *S* and which is *S'*, and (2) determine which of the stated velocities correspond to which of the symbolic quantities $\vec{v}$, $\vec{v}'$, and $\vec{\beta}$.

The first of these steps is actually a fairly arbitrary decision: as long as we keep everything straight, we would get the same ultimate answer no matter which frame (ground or train) we took to be frame *S*. But note that $\vec{\beta}$ is defined to specify the velocity of frame *S'* *relative to S*. In the problem description, the train's velocity is specified relative to the ground, so it is *convenient* to let *S'* be the train frame and *S* be the ground frame, because then we know immediately from the problem description that $\vec{\beta} = (+25 \text{ m/s})\hat{x}$.

We are told that the ball moves at a speed of 12 m/s in the *x* direction with respect to the train, so this velocity must be $\vec{v}'$, since the train is the "primed" frame. We are trying to find the velocity of the ball with respect to the ground, so $\vec{v}$ is the unknown quantity.

Solution   Solving equation N9.2 for $\vec{v}$, we get

$$\vec{v} = \vec{v}' + \vec{\beta} = \begin{bmatrix} +12 \text{ m/s} \\ 0 \\ 0 \end{bmatrix} + \begin{bmatrix} +25 \text{ m/s} \\ 0 \\ 0 \end{bmatrix} = \begin{bmatrix} +37 \text{ m/s} \\ 0 \\ 0 \end{bmatrix} \qquad (N9.3)$$

So the ball's velocity with respect to the ground is $(+37 \text{ m/s})\hat{x}$, as expected.

Example N9.2

Problem   Imagine that you are in a train traveling in the +*x* direction at a speed of 35 m/s relative to the ground, and you observe a car traveling in the same direction at a speed of 29 m/s relative to the ground. What is the car's speed relative to you?

Translation and Model   Again, we are *given* that the train's velocity with respect to the ground is 35 m/s in the *x* direction. If we take the train to be frame *S'* and the ground to be frame *S*, then this means that $\vec{\beta} = (+35 \text{ m/s})\hat{x}$. We are given that the velocity of the car in the ground frame is $\vec{v} = (+29 \text{ m/s})\hat{x}$. The velocity $\vec{v}'$ of the car in the train frame is the unknown quantity.

Solution   Equation N9.2 then directly implies that

$$\vec{v}' = \vec{v} - \vec{\beta} = \begin{bmatrix} 29 \text{ m/s} \\ 0 \\ 0 \end{bmatrix} - \begin{bmatrix} 35 \text{ m/s} \\ 0 \\ 0 \end{bmatrix} = \begin{bmatrix} -6 \text{ m/s} \\ 0 \\ 0 \end{bmatrix} \qquad (N9.4)$$

This means that the car will appear to drift backward at a speed of 6 m/s relative to the train.

Evaluation   If you visualize the situation, perhaps you will agree that this has to be right.

Example N9.3

Problem   An airplane is flying with a velocity of 75 m/s due north relative to the air. If the wind is blowing 12 m/s west relative to the ground, what is the plane's velocity (magnitude and direction) relative to the ground?

Translation and Model   Let's take the ground to be frame S and the air to be frame S'. According to the description of the situation, the air is moving relative to the ground at 12 m/s west. If both our frames are oriented in the usual way relative to the earth's surface, this means that $\vec{\beta} = [-12 \text{ m/s}, 0, 0]$. The plane's velocity with respect to the air is $\vec{v}' = [0, +75 \text{ m/s}, 0]$. We want to find the plane's velocity with respect to the ground, which is $\vec{v}$.

Solution   Solving equation N9.2 for $\vec{v}$, we then get

$$\vec{v} = \vec{v}' + \vec{\beta} = \begin{bmatrix} 0 \\ +75 \text{ m/s} \\ 0 \end{bmatrix} + \begin{bmatrix} -12 \text{ m/s} \\ 0 \\ 0 \end{bmatrix} = \begin{bmatrix} -12 \text{ m/s} \\ +75 \text{ m/s} \\ 0 \end{bmatrix} \quad \text{(N9.5)}$$

The magnitude of this velocity is

$$v = \sqrt{v_x^2 + v_y^2 + v_z^2} = \sqrt{(-12 \text{ m/s})^2 + (75 \text{ m/s})^2 + 0} = 76 \text{ m/s} \quad \text{(N9.6a)}$$

The angle that this velocity makes with the y (north) axis is

$$\theta = \tan^{-1}\left(\frac{12 \text{ m/s}}{75 \text{ m/s}}\right) = 9° \quad \text{(N9.6b)}$$

Since v_x is negative, the plane is flying at 76 m/s, 9° *west* of north.

Now, if we take the time derivative of both sides of equation N9.2, we get

$$\vec{a}'(t) = \vec{a}(t) - \vec{A}(t) \quad \text{(N9.7)}$$

Transformation equation for accelerations

Purpose: This equation allows us to compute an object's acceleration $\vec{a}'$ in a frame S' given the same object's acceleration $\vec{a}$ in frame S and the acceleration $\vec{A}$ of frame S' relative to frame S ($\vec{A} \equiv d\vec{\beta}/dt$).
   Limitations: The speeds of the objects and frames involved must be much less than the speed of light.

Equation N9.7 implies that if frame S' moves at a *constant velocity* with respect to S (so that $\vec{A} = 0$), the object's acceleration is the same in both frames:

$$\vec{a}'(t) = \vec{a}(t) \quad \text{if } S' \text{ moves at a constant velocity relative to } S \quad \text{(N9.8)}$$

We will build our case regarding fictitious forces on an analysis of equations N9.7 and N9.8.

Exercise N9X.1

A train is moving at a constant velocity of 23 m/s in the $+x$ direction. A child throws a ball out the back end of the train with a velocity of 3 m/s in the $-x$ direction relative to the train. What is the velocity of the ball with respect to the ground?

Exercise N9X.2

A ship is moving at a constant velocity of 8 m/s due north. A sailboat nearby has a velocity of 6 m/s due west. What is the velocity of the sailboat relative to the ship?

A sailboat and ship (exercise N9X.2).

Exercise N9X.3

Imagine that you are in an elevator whose vertical acceleration is 3.0 m/s^2 upward. You drop a ball that falls with a downward acceleration of $g = 9.8$ m/s^2 in the frame of the earth. What is the ball's acceleration in the frame of the elevator?

N9.3   Inertial Reference Frames

Imagine that an object is completely isolated from external interactions. This means that $\vec{F}_{\text{net}} = 0$, and by Newton's second law, the object should have *zero* acceleration. Newton's first law says this even more directly: *an isolated object moves at a constant velocity.* Now, either an object is isolated from external interactions, or it is not: you do not have to use a reference frame to determine whether an object is massive or whether it is electrically charged or whether it touches something else. Observers in all reference frames should therefore agree as to whether a given object is isolated.

Imagine that a universally accepted isolated object *is* observed to move at a constant velocity (zero acceleration) in some frame S. Equation N9.7 implies that all *other* frames can be divided into two categories. In those frames that move at a constant velocity with respect to S (so that their acceleration $\vec{A}$ relative to S is zero), the isolated object will *still* be measured to have zero acceleration and thus be observed to obey Newton's first and second laws. Such frames are called **inertial reference frames** (because Newton's first law

An isolated object doesn't obey Newton's laws in a noninertial reference frame

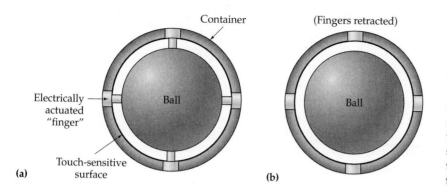

Container   (Fingers retracted)

Electrically
actuated
"finger"

Ball   Ball

Touch-sensitive
surface

(a)   (b)

Figure N9.2
(a) A cross-sectional view of a floating-ball first-law detector. Electrically actuated "fingers" hold the ball initially at rest in the spherical container. (b) After the fingers are retracted, the ball should continue to float at rest in the container if the container frame is inertial.

is sometimes called the *law of inertia*). In those frames that are accelerating relative to frame S, the isolated object will be measured to have a nonzero acceleration $\vec{a}' = -\vec{A}$, meaning that Newton's first and second laws *fail* to work in such reference frames. Such frames are called **noninertial reference frames.**

Let me emphasize again that *Newton's first and second laws do not apply in noninertial reference frames:* such frames are "bad" for analyzing motion in the newtonian model. If we want to apply Newton's laws to analyze an object's motion, we must do the analysis in an *inertial* reference frame.

We can use a **first-law detector** to distinguish an inertial frame from a noninertial frame *without* determining the frame's motion with respect to something else. Figure N9.2 shows such a detector. Electrically actuated "fingers" hold an electrically uncharged and nonmagnetic ball in the center of a spherical container from which the air has been removed (figure N9.2a). When the fingers are retracted, the ball is (at least momentarily) at rest and completely isolated from contact and electromagnetic interactions (figure N9.2b). An isolated object at rest will *remain* at rest if the frame is inertial. If the frame to which the container is attached is noninertial, though, the ball will accelerate relative to that frame and thus the container (see equation N9.10). This can easily be detected, because if the ball drifts away from rest in any direction, it will eventually hit the container wall, where touch-sensitive sensors can register the violation of the first law (and trigger the "fingers" to reset the ball).

Actually, the detector as just described will only distinguish inertial from noninertial frames in deep space (far away from any gravitating objects) because there is no other way to isolate the internal ball from gravity. We might adapt our detector for operation in a gravitational field by changing the spherical container to a cylindrical one and allowing the ball to drop from rest. Since the ball's acceleration should be perfectly vertical by definition, any sideward deviation that would cause the ball to strike a container wall will indicate that the frame of the container is noninertial. Also, since all objects fall with the same acceleration, the ball *should* take a well-defined amount of time to fall the length of the cylinder. If it arrives at the bottom early or late, the frame is not inertial.

A more practical way to make a first-law detector for use in a gravitational field is to replace the ball with a puck that floats on a cushion of air above a flat, level air table. This puck is isolated from external interactions that might affect its *horizontal* motion, so if we place it at rest at the center of the table, it should *remain* at rest. If we observe the puck to accelerate away from the center, the frame in which the table sits is not inertial. This detector will not register violations of the first law due to vertical accelerations, but it is easier to imagine and construct than a dropped-ball detector.

Constructing an idealized first-law detector

A floating-puck first-law detector

Formal definitions of *inertial* and *noninertial* frames

Imagine now attaching first-law detectors in a number of places throughout a reference frame. A frame is defined to be *inertial* if *no* detector in the frame registers any violation of Newton's first law; the frame is *noninertial* otherwise.

If we apply this definition to a frame attached to the surface of the earth, we find that such a frame is *not* perfectly inertial (because the earth is rotating). The earth rotates fairly slowly, though, so the deviations from perfection are *usually* negligibly small. Unless otherwise stated, we will assume that a reference frame attached to the earth's surface is sufficiently inertial for our purposes.

N9.4   Linearly Accelerating Frames

What you experience in a linearly accelerating frame

Consider again the example of an airplane accelerating for takeoff. Imagine that you hold a floating-puck first-law detector in your lap. Then the pilot turns on the engines, and the plane begins to accelerate down the runway. You feel a force pushing you back into your seat. You will also see the puck accelerate toward the rear of the plane. In the plane frame, then, it certainly looks as if some magical force is acting on you and the puck (see figure N9.3a), pulling everything backward. Yet the puck is horizontally *isolated*, and therefore there is no way for an external force to act on you and the puck. How can we resolve this paradox?

The problem is that we are inappropriately trying to make Newton's second law work in an accelerating frame. Let's look at the motion of the puck from the perspective of the ground frame. When the pilot turns up the engines, the plane begins to accelerate. The puck is horizontally isolated, so if it was initially at rest relative to the ground, it will remain at rest. As the plane accelerates forward relative to the ground, the air table, on the other hand, is carried with the plane forward with respect to the ground and thus with respect to the puck, making the puck appear to accelerate backward with respect to the table. In the inertial frame of the ground, no "magical forces" are needed to explain the puck's behavior (see figure N9.3b).

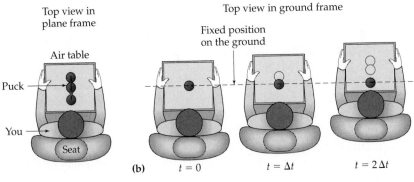

(a)   **(b)** $t = 0$ $t = \Delta t$ $t = 2\Delta t$

Figure N9.3
(a) In a plane accelerating for takeoff, a puck in a first-law detector accelerates to the rear of the plane relative to the box. (b) When viewed in the ground frame, the puck remains at rest. It only accelerates relative to the air table because the *table* is accelerating forward with you and the plane. (Note that I have displaced successive images of the situation to the right for clarity's sake. The plane is accelerating straight forward and is not moving to the right.)

We can also see this from equation N9.7. Let the ground frame be frame S and the plane frame be frame S'. Since the puck has zero net physical force on it, its acceleration will be zero in the inertial ground frame S ($\vec{a} = 0$). If the plane has a nonzero acceleration with respect to the ground frame ($\vec{A} \neq 0$), then the puck's acceleration measured in the plane frame is

$$\vec{a}' = \vec{a} - \vec{A} = 0 - \vec{A} = -\vec{A} \qquad \text{(N9.9)}$$

An explanation using equation N9.7

(This applies to *any* object not accelerating in the S frame.) This tells us that the puck's acceleration as measured in the plane is equal in magnitude and opposite in direction to the plane's acceleration relative to the ground. This is what we would expect if the puck had no horizontal acceleration relative to the ground.

The point is that we do not need a "magical rearward force" to explain the behavior of the puck: we can explain its behavior just fine in the ground frame *without* appealing to such a force. This alleged force therefore vanishes if we analyze the situation while using a proper inertial reference frame: *it is not real.*

"Wait," you say. "If this force is not real, what causes the puck to drift backward?" Think! The air table is being accelerated *forward* along with the plane, as observed in the ground frame. If the puck were to remain at rest relative to the table, it would *also* have to accelerate forward relative to the ground frame. But there is no force acting on the puck that can accelerate it forward. Therefore, it is unable to keep up with the table as the latter accelerates forward.

Resolving intuitive objections to this explanation

"But," I hear you cry, "we can *feel* this backward force. Therefore, it *must* be real!" Not so, I say. What you *feel* is the airplane seat pressing you forward. In an *inertial* reference frame, if your seat presses you forward but you don't accelerate forward, it must be because some other force is pressing you backward. But in the noninertial plane frame, this logic does not apply, because Newton's laws do not apply. In the inertial frame of the ground, the seat presses you forward not because something else is pressing you backward, but because the seat *has* to press you forward to accelerate you along with the plane (figure N9.4).

Part of the reason that you seem to really *feel* this force is that parts of your body have to flex and stretch to accelerate other parts of your body

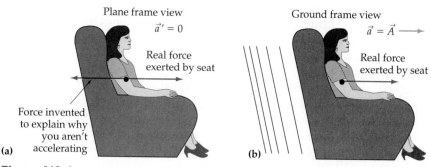

Figure N9.4

(a) In an accelerating plane, you seem to be at rest and thus have zero acceleration. You feel the chair pressing forward on you, so you assume that there must be a force pressing back on you to make the net force on you zero. (b) But the net force on you really is not zero: the chair has to push forward on you to accelerate you along with the plane. When you analyze things in the ground frame, there is no need for a backward force.

along with the plane. For example, if you hold your hand in front of you, it feels as if something were pushing it toward you. But what is really going on is that you have to flex the muscles in your arm to push your hand forward so that it accelerates along with the plane. You feel this muscular action and intuitively interpret it as a response to something pushing your hand toward you.

In short, when you feel the seat pressing you forward, you intuitively *invent* a force pressing you backward, because otherwise your lack of acceleration in the plane frame and the way that your body feels would be inexplicable. In doing this, though, you are illegally applying Newton's second law in an accelerated reference frame (an offense punishable throughout the universe by nonsensical answers!). In the inertial reference frame of the ground, such a force (1) cannot be made consistent with the idea that *forces express interactions* and (2) is entirely *unnecessary* to explain what we observe if we use an inertial frame. Therefore, from the point of view of the newtonian model, this force is *not real*.

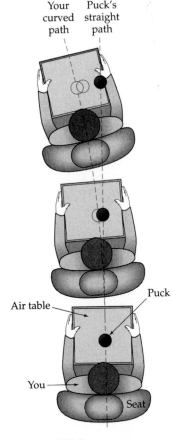

Your curved path   Puck's straight path

Air table

Puck

You

Seat

Figure N9.5
The puck in a floating-puck first-law detector seems to accelerate to the right as you ride in a car turning left because the car and you accelerate to the left under the puck while it moves straight forward.

Exercise N9X.4

If you are in a car that suddenly slows down, you seem to feel a magical force acting to draw you forward. What is *really* going on? In a serious accident, people without seat belts can go through the windshield. If they are not being thrown forward by this magical force, how is this possible?

N9.5   Circularly Accelerating Frames

Now imagine being in a car that is turning a corner or riding on a playground merry-go-round. The outward force that you seem to feel under such circumstances is such a vivid and common part of daily experience that it even has a colloquial English name: **centrifugal force**. This force is so deeply embedded in our prenewtonian intuition that even some science writers and precollege textbook authors discuss this force as if it were real.

We can analyze what you experience in the car turning a corner by using the approach we used in section N9.4. When viewed in the inertial ground frame, as the car turns left (say), a floating puck continues to move forward in the direction of the car's original velocity, as shown in figure N9.5. In the car's frame, the air table is accelerating toward the left, so the puck seems to accelerate right relative to the table. In the car's frame, it *looks* as if the puck is pushed outward (to the right) by some centrifugal force. What is *really* happening, though, is that the table (along with the car) is accelerating left "underneath" the puck.

Your body, like the puck, would move forward in a straight line if the seat belt or the side of the car did not exert a leftward force on you to accelerate you along with the car. In the car's frame, you do not *seem* to be accelerating, so you might interpret the leftward force exerted by the belt or car wall as being due to a rightward force pulling you outward. But you are not *allowed* to use Newton's second law in the accelerating frame of the car, so this inference is *false*.

Really, we have two choices in both of these situations. We could in principle *insist* that Newton's laws apply in *all* reference frames. This would require us to invent forces in some reference frames that are not due to contact or any other interaction with an external object. Moreover, these forces seem to exist at some times and not at others, and exist in one reference frame (e.g., the plane) but not exist in another (e.g., the ground). It seems to me that forces that either exist or don't exist depending on one's arbitrary choice of reference frame have an extremely poor claim on being "real."

It is better, simpler, and entirely sufficient to simply assert that Newton's laws *don't apply* in noninertial reference frames and insist that we use only *inertial* reference frames when analyzing motion by using Newton's laws. We have no physical evidence for these alleged forces except for the effects that they seem to produce. We can entirely explain these effects *without* the use of these alleged forces by using an inertial reference frame. These alleged forces are, therefore, not real in any useful sense of the word, and we call them *fictitious* forces.

Ladies and gentlemen of the jury, I have shown beyond a shadow of a doubt that motion can be easily interpreted and analyzed in an inertial reference frame without invoking the magical forces others claim to exist. Are these forces real? Only one verdict is logical, sensible, and right. Do your duty. I rest my case.

Closing argument

Exercise N9X.5

When you are on a rapidly rotating playground merry-go-round, you have to hold on tightly, using the full strength of your arms to keep your body from flying off. If centrifugal forces are not real, why do you have to exert so much strength to keep yourself at rest relative to the merry-go-round?

N9.6   Using Fictitious Forces

Now that you *completely understand* that the mysterious forces that appear to arise in a noninertial reference frame are totally fictitious and without any basis in physical reality, I will reluctantly admit that in a few cases, a given situation is much easier to analyze in a noninertial reference frame and therefore it is useful to *pretend* that these fictitious forces exist (as long as one completely understands what one is doing).

Imagine that we have an inertial frame S and a noninertial frame S' that is accelerating relative to S with acceleration $\vec{A}$. Imagine that real physical forces $\vec{F}_1, \vec{F}_2, \ldots$ act on a certain object of mass m. Newton's second law applies in the inertial frame S, so in that frame, we have

$$m\vec{a} = \vec{F}_\text{net} = \vec{F}_1 + \vec{F}_2 + \cdots \qquad (N9.10)$$

Equation N9.7 states that the object's acceleration in the noninertial S' frame is $\vec{a}' = \vec{a} - \vec{A}$, implying that $\vec{a} = \vec{a}' + \vec{A}$. Plugging this into equation N9.10 yields

$$m\vec{a}' + m\vec{A} = \vec{F}_1 + \vec{F}_2 + \cdots \qquad (N9.11)$$

The $m\vec{A}$ term in this equation should *rightly* be considered to be a correction to the acceleration side of Newton's second law to enable us to use the acceleration $\vec{a}'$ measured in the noninertial frame. But if we subtract $m\vec{A}$ from both sides, we can put this term on the force side of the equation and *pretend* that it is just another force acting on the object:

A version of Newton's second law that we can use in a noninertial frame

$$m\vec{a}' = -m\vec{A} + \vec{F}_1 + \vec{F}_2 + \cdots \qquad \text{(N9.12)}$$

Purpose: We can use this version of Newton's second law in a noninertial frame *if* we know that frame's acceleration $\vec{A}$ relative to an inertial frame.
Symbols: m is the mass of an object, $\vec{a}'$ is its acceleration in the noninertial frame, and $\vec{F}_1, \vec{F}_2, \ldots$ are real forces acting on the object.
Limitations: The relative speed of the two frames and the speed of the object relative to each must be much smaller than the speed of light.

Let's call this "force" $-m\vec{A}$ a **frame-correction force** to emphasize that we are supplying this "force" *only* to compensate for measuring the object's acceleration in a noninertial reference frame. Such a "force" is also often called an *inertial force*. This "force" has the same mathematical form as the gravitational force $\vec{F}_g = m\vec{g}$: both are directly proportional to the object's mass. Indeed, in the noninertial frame, an object behaves *exactly as if it were experiencing an additional gravitational force* whose gravitational field vector is $\vec{g}_{FC} = -\vec{A}$.

We use the frame-correction "force" in situations where an analysis in an inertial frame is awkward. Example N9.4 illustrates such a situation.

Example N9.4

Problem   A bicyclist goes around a curve of radius R at a speed v. Why does the bicyclist have to lean into the curve? What is the cyclist's lean angle?

Model   We *could* analyze this situation in the inertial frame of the ground, but the analysis is complicated by the fact that the center of mass of the bike and rider is accelerating and both the bike and rider are also rotating in that frame. In the NONINERTIAL frame of the bike, however, this situation reduces to a straightforward statics problem. Figure N9.6b shows a free-body diagram of the situation as viewed in this frame. Note that since the bike's acceleration $\vec{A}$ relative to the ground points in the $-x'$ direction and has magnitude $A = v^2/R$, the frame-correction force $\vec{F}_{FC}$ in the bike's frame points in the $+x'$ direction and has magnitude mv^2/R, where m is the combined mass of the bike and rider.

Figure N9.6
(a) A bicyclist leaning into a curve.
(b) A free-body diagram for the bicyclist, as we would construct it in the NONINERTIAL frame of the bike. The right-hand diagram shows the two forces that exert nonzero torques around point O.

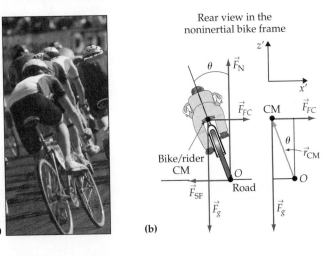

In this NONINERTIAL frame, the bike is at rest and not rotating, so both the net force and the net torque on the system must be zero. Let's choose the point of contact between the tire and road to be the origin O for computing torques. Both $\vec{F}_N$ and $\vec{F}_{SF}$ exert zero torque around O (since both are applied *at* point O). Both $\vec{F}_{FC}$ and $\vec{F}_g$ behave as if they are applied to the center of mass of the bike and rider, which we will say has position $\vec{r}_{CM}$ relative to O (see the diagram on the right side of figure N9.6). Using the right-hand rule, we see that $\vec{F}_g$ exerts a torque directly toward us in figure N9.6 while $\vec{F}_{FC}$ exerts a torque in the opposite direction. These torques will therefore add to zero (and thus the net torque on the bike will be zero) if their magnitudes are equal.

Solution   It is easiest to compute these torque magnitudes by using the fact that $\text{mag}(\vec{r} \times \vec{F}) = r_\perp F$, where $r_\perp$ is the magnitude of the vector component of $\vec{r}$ perpendicular to $\vec{F}$. We can see on the diagram that this component has a magnitude of $r_{CM} \sin\theta$ in the case of $\vec{F}_g$ and $r_{CM} \cos\theta$ in the case of $\vec{F}_{FC}$, so to balance the torques, we must have

$$(r_{CM} \sin\theta) F_g = (r_{CM} \cos\theta) F_{FC} \qquad \text{(N9.13)}$$

Canceling r_{CM} from both sides and using $F_g = mg$ and $F_{FC} = mv^2/R$, we get $mg \sin\theta = (mv^2/R) \cos\theta$, and dividing both sides by $mg \cos\theta$ yields

$$\tan\theta = \frac{\sin\theta}{\cos\theta} = \frac{v^2}{Rg} \quad \Rightarrow \quad \theta = \tan^{-1}\left(\frac{v^2}{Rg}\right) \qquad \text{(N9.14)}$$

Evaluation   Note that the angle here is the same as we computed for a banking airplane in example N8.1. We see that the rider must lean into the curve so that the net torque on the bike is zero in the bike's NONINERTIAL frame. (Another way that we could express this is to say that in the bike frame, the rider feels an *effective* total gravitational force of $\vec{F}_g + \vec{F}_{FC}$ that points at an angle θ outward from directly down, and that the rider must position his or her center of mass directly "above" the tire's contact point with the ground in this effective gravitational field to keep from falling over.)

Exercise N9X.6

Redo example N9.4, using the system's center of mass as the origin O for computing torques, and show that you *still* have to lean at the angle given by equation N9.14 to balance torques. [*Hint:* Use equation N9.12 to argue that $\text{mag}(\vec{F}_N) = mg$ and $\text{mag}(\vec{F}_{SF}) = \text{mag}(\vec{F}_{FC})$.]

Use equation N9.12 only in situations in which an analysis in an inertial frame would be horrendously complex. Even then, indicate very clearly in your solution that you are *deliberately* using a noninertial frame (as I did in example N9.4 by capitalizing NONINERTIAL), and denote the frame-correction "force" using the symbol $\vec{F}_{FC}$ to make it clear that you understand its nature.

When to use frame-correction forces

N9.7   Freely Falling Frames and Gravity

In chapter C4, we discussed the fact that when we use a freely falling reference frame, we can *ignore* the external gravitational fields in which the frame falls. We are now in a position to understand *why*. Consider an inertial frame S

A newtonian explanation of why we can treat a falling frame as if it were inertial

in a gravitational field $\vec{g}$, and consider a frame S' that is freely falling in that field. (*Freely falling* in this context means that no significant forces *other* than gravity act on the frame.) Since all objects fall with the same acceleration $\vec{g}$ in a gravitational field, the acceleration of frame S' relative to S will be $\vec{A} = \vec{g}$.

Now imagine that we analyze the motion of an object of mass m that is subject to the force of gravity $\vec{F}_g = m\vec{g}$ as well as additional forces $\vec{F}_1, \vec{F}_2, \ldots$. According to equation N9.12, the object's motion as observed in the falling frame S' will obey the equation

$$m\vec{a}' = -m\vec{A} + \vec{F}_g + \vec{F}_1 + \vec{F}_2 + \cdots = -m\vec{g} + m\vec{g} + \vec{F}_1 + \vec{F}_2 + \cdots$$

$$\Rightarrow \quad m\vec{a}' = \vec{F}_1 + \vec{F}_2 + \cdots \tag{N9.15}$$

where in the second step I used the fact that $\vec{A} = \vec{g}$. Since the frame-correction force cancels the gravitational force, the final equation has the same form as Newton's second law *ignoring* that gravitational force: an object in a freely falling frame thus behaves as if Newton's second law applied *ignoring the gravitational field in which the frame falls*. This is why we could treat freely falling or "floating" frames as if they were inertial. From the perspective of newtonian mechanics, however, such frames are *not* inertial frames, and the effective erasure of the consequences of real external gravitational interactions is a symptom of that.

Einstein's alternative explanation: the *equivalence principle*

However, in 1907, Albert Einstein presented an alternative take on this phenomenon. He pointed out that if one is placed in a freely falling enclosed room, equation N9.15 implies that no *mechanical* experiment that one could perform entirely within the room could tell whether that room is freely falling in a uniform gravitational field or is a genuine inertial frame floating in deep space far from any gravitating objects. Even the first-law detectors

The space shuttle, when it is orbiting the earth, is a freely falling reference frame. Einstein's principle of equivalence states that such a frame is physically equivalent to a frame floating in deep space, far from any gravitating bodies.

we discussed in section N9.3 would certify that such a frame was inertial. To express his point in analogy to a famous cliche about ducks, we might say, "If it looks like an inertial frame and physically behaves as an inertial frame, it *is* an inertial frame." Einstein boldly hypothesized that a freely falling frame in a uniform gravitational field is *completely equivalent* to an idealized inertial frame floating in deep space: there is *no* experiment that one could perform that would distinguish the two. By extension, since a frame at rest on the surface of the earth is accelerating *upward* relative to any freely falling frame, then (to the extent that we can consider the earth's gravitational field to be uniform) a frame at rest on the earth's surface is physically indistinguishable from an *accelerating* frame in deep space.

Einstein made a variety of testable predictions based on this hypothesis (some of which are discussed in the homework problems). These predictions have been experimentally tested and have all been found to be true.

This hypothesis, called the **equivalence principle,** is the foundation of Einstein's theory of general relativity. In that theory, what we normally think of as gravity is a *fictitious* force (!) that appears in a frame at rest on the surface of the earth only because such a frame is *accelerating* upward relative to genuinely inertial (i.e., freely falling) frames in the earth's vicinity. (The theory's main task, therefore, is to explain why *inertial* frames near the earth are *accelerating* rather than at rest relative to the earth's surface.)

TWO-MINUTE PROBLEMS

N9T.1 In a Western movie, a person shoots an arrow backward from a fleeing horse. If the velocity of the horse relative to the ground is 13 m/s west and the arrow's velocity relative to the horse is 38 m/s east, what is the arrow's velocity with respect to the ground?
A. 41 m/s east
B. 41 m/s west
C. 25 m/s east
D. 25 m/s west

N9T.2 A blimp has a velocity of 8.2 m/s due west relative to the air. There is a wind blowing at 3.5 m/s due north. The speed of the balloon relative to the ground is
A. 11.7 m/s
B. 8.9 m/s
C. 7.4 m/s
D. 4.7 m/s
E. Some other speed (specify)

N9T.3 An elevator moves downward with an acceleration of 6.2 m/s². A ball dropped from rest by a passenger will have what *downward* acceleration relative to the elevator?
A. 3.6 m/s²
B. 6.2 m/s²
C. 9.8 m/s²
D. 16.0 m/s²
E. Some other acceleration (specify)

N9T.4 When you go over the crest of a hill in a roller coaster, a force appears to lift you up out of your seat. This is a fictitious force, true (T) or false (F)?

N9T.5 If your car is hit from behind, you are suddenly pressed back into the seat. The normal force that the seat exerts on you is a fictitious force, T or F?

N9T.6 You are pressed downward toward the floor as an elevator begins to move upward. The force pressing you down is a fictitious force, T or F?

N9T.7 A beetle (black dot on the top view shown at the right) sits on a rapidly rotating turntable. The table rotates faster and faster, and eventually the beetle loses its grip. What is its subsequent trajectory relative to the ground (assuming it cannot fly)?

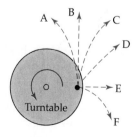

N9T.8 When is an elevator a reasonably good inertial reference frame?

(more)

   A.   Never, under any circumstances.

   B.   Always, since it is attached to the surface of the earth.

   C.   It is except when it is changing speed.

   D.   Other conditions (describe).

N9T.9   Is a freely falling elevator an accurate inertial reference frame?

   A.   No, such a frame is accelerating toward the earth.

   B.   It is not a *real* inertial frame, but we can treat it as one if we ignore the gravitational forces on objects inside.

   C.   Yes.

   D.   It depends (explain).

HOMEWORK PROBLEMS

Basic Skills

N9B.1   A person shoots a tranquilizer dart at a fleeing bear. If the dart's speed is 33 m/s relative to the ground and the bear is fleeing at 9 m/s relative to the ground, what is the dart's speed relative to the bear?

N9B.2   In a certain baseball game, a well-hit line drive moves almost horizontally westward at 125 mi/h. The center fielder runs toward the ball at a speed of 15 mi/h. What is the ball's speed relative to the fielder?

N9B.3   An airplane flies due north at a speed of 145 km/h relative to the ground. If there is a wind blowing east at 15 km/h, what is the plane's speed relative to the air?

N9B.4   Two cars approach an intersection, one traveling north at 18 m/s and the other traveling west at 14 m/s. What is the cars' relative speed?

N9B.5   A boat originally traveling at a speed of 5.2 m/s due east hits a sandbar and comes to rest in 1.3 s. What is the approximate average acceleration (magnitude and direction) of a piece of slippery ice sitting on a table in the ship's galley

   (a) relative to the ground and

   (b) relative to the ship? Explain your answers.

N9B.6   Two cars in a race start from rest. Both cars accelerate forward, car A at 12 (mi/h)/s and car B at 9 (mi/h)/s. What is the acceleration (magnitude and direction) of car B relative to car A?

N9B.7   Imagine that you toss a baseball to a friend while you are both in the back of a bus moving due west at a constant speed 15 m/s. While the baseball is in the air, its acceleration with respect to the ground is 9.8 m/s^2 downward (since only the force of gravity acts on it once it leaves your hand). What is the ball's acceleration relative to the bus? Explain.

N9B.8   A car originally moving at 25 mi/h hits a brick wall and comes to rest in 0.12 s. As the car comes to rest, about how fast will an unbelted passenger accelerate

   (a) relative to the ground and

   (b) relative to the windshield? Explain your response. (Ignore friction.)

Synthetic

N9S.1   A boat moves directly toward a dock 2.0 km due west across a river flowing 1.0 m/s due south. How fast does the boat have to travel relative to the water to make it across in 20 min?

N9S.2   An airplane flies due north at 250 mi/h relative to the ground in air that is moving east at 25 mi/h. At what angle is the plane pointed relative to north? What is its speed relative to the air?

N9S.3   An airplane accelerates for takeoff, reaching a speed of 150 mi/h in 30 s. A 65-kg person feels pressed back into his or her seat during this time.

   (a) What is the person's acceleration relative to the ground?

   (b) What is the magnitude of the forward force that the seat must exert on the person during this time?

   (c) What is the person's acceleration in the plane frame?

   (d) If you were to naively apply Newton's second law in this frame, what is the magnitude of the fictitious force that you infer must be pushing the person backward?

N9S.4   A 35-kg child is holding on for dear life on a playground merry-go-round that has a radius of 2.5 m and is turning at a rate of one turn every 5.0 s.

   (a) What is the child's acceleration relative to the ground?

   (b) What is the magnitude of the force acting on the child through the child's arms?

   (c) What is the child's acceleration relative to the place on the merry-go-round where the child is standing?

   (d) If you were to naively apply Newton's second law to this situation, what is the magnitude of the apparent centrifugal force acting on the child?

A child hanging onto a playground merry-go-round (problem N9S.4).

A centrifuge for training astronauts (see problem N9S.7).

N9S.5 An elevator is accelerating upward at 1.8 m/s². Someone is standing on a bathroom scale in the elevator, and it reads 180 lb. Use equation N9.12 to calculate the person's true weight.

N9S.6 A test tube is placed in a centrifuge. When the centrifuge rotates at full speed, the test tube's bottom is about 38 cm from the axis of rotation. The material near the bottom of the test tube experiences an apparent effective gravitational force 18 times larger than normal. How many times does the centrifuge rotate per second?

Scientists often use centrifuges to create a stronger-than-normal effective gravitational field for separating materials in solution (see problem N9S.6).

N9S.7 Astronauts train for the high accelerations they experience during launch by using a centrifuge that consists of a little room suspended at the end of a long boom that is rotated in a horizontal plane. If the total apparent effective gravitational field strength in the room is supposed to have a magnitude of $5g$, and the distance between the room's center and the center of rotation is 10 m, about how long should it take the boom to rotate once?

N9S.8 Imagine that the ball in a floating-ball detector can travel only 1.0 cm before hitting the side of its container. Imagine also that we have figured out a way to cancel exactly the true effects of gravity. If we place our detector on the earth's equator, how long will it take for the ball (after it is released from rest) to reach the side, and in which direction will it drift? Compare this to the time that it would take in a frame that accelerates from rest to walking speed (about 1 m/s) in 1 min.

N9S.9 An object's inertial mass m_I expresses its resistance to being accelerated by a given force: $F_{net} = m_I a$ according to Newton's second law. An object's gravitational mass m_G expresses the strength of the gravitational force the object feels in a given gravitational field $\vec{g}$: $\vec{F}_g = m_G \vec{g}$. These are completely distinct ideas; and even though Newton recognized that one could explain much by assuming that $m_G = m_I$, newtonian mechanics offers no explanation about why these quantities should be equal or even related. However, one of the consequences of the equivalence principle is that an object's inertial mass must be *exactly* equal to its gravitational mass, no matter what the object is made of. Carefully explain why this must be so. (*Note:* Recent experiments show that these quantities are experimentally equal to roughly 13 significant digits.)

N9S.10 One of the consequences of the principle of equivalence is that light should bend in a gravitational field. To see this, imagine a laboratory in deep space that is being accelerated by a rocket engine so that its acceleration relative to inertial frames is upward (relative to the laboratory's floor) and has a magnitude of A. Imagine that a beam of light enters a hole in one side of the laboratory traveling in an initially horizontal direction relative to the floor. If the laboratory frame were inertial, the light would travel in a straight line, and thus would hit the laboratory's opposite wall at the same height as the hole. However, the laboratory is accelerating upward, so during the time that it takes the light to move across the laboratory, the laboratory's floor will move upward relative to the beam, which will make it appear to curve toward the floor in the laboratory frame. This will ultimately cause the beam to hit the far wall a distance d below the expected position.

(a) If the laboratory's width is L, what is d in terms of A and the speed of light c?

(b) The principle of equivalence implies that a laboratory at rest on the surface of the earth is physically equivalent to a laboratory in deep space whose acceleration relative to inertial frames is $A = g$. Making a reasonable estimate for the size of the laboratory, make an estimate of the distance through which a light beam will be deflected by the earth's gravitational field. Would this be easy to detect? (*Note:* In 1919, astronomers observing an eclipse measured the deflection of starlight passing near the obscured sun and verified the prediction of general relativity that a gravitational field would bend light.)

Rich-Context

N9R.1   You are kidnapped and put blindfolded in an elevator at the ground floor of a building. As the

elevator starts, you notice that your weight seems to increase by 10% for 3 s, remain normal for 24 s, then decrease by 10% for 3 s. You are then taken out and put into a locked room. What is the approximate floor number your room is on?

N9R.2   A freely falling frame does not behave *precisely* as an inertial frame because $\vec{g}$ is not precisely constant near a gravitating object.[†] For example, $\vec{g}$ near the earth always points directly toward the earth's center, which is in different directions at different points. Consider a boxcar that is 10 m long and that is freely falling toward the earth (assume the boxcar is oriented horizontally). Two marbles float initially at rest at opposite ends of the boxcar. Estimate the marbles' acceleration relative to each other. (*Hint:* $\sin\theta \approx \tan\theta \approx \theta$ if θ is measured in radians and $\theta \ll 1$.)

Advanced

N9A.1   A jet flies a distance L and back, first against and then with a wind with speed u. One might think that since the wind will slow the jet one way and speed it equally the other way, the effects would cancel and the trip time would be $2L/v$, where v is the jet's speed relative to the air. Show that the round-trip time is *actually* $(2L/v)(1 - u^2/v^2)^{-1}$.

ANSWERS TO EXERCISES

N9X.1   Here the train is the S' frame, and the ground is the S frame. We are given that $\beta_x = +23$ m/s and $v'_x = -3$ m/s, so solving the x component of equation N9.2 for v_x implies that $v_x = \beta_x + v'_x = +20$ m/s.

N9X.2   Implicitly, the speeds of the boats given in the problem are both relative to the ground. We choose the ground to be the S frame and the ship to be the S' frame, and the sailboat to be the object whose motion is examined in both frames. Thus we are given that $\vec{v} = 6$ m/s west, while $\vec{\beta} = 8$ m/s north, and we want to find $\vec{v}'$. Using equation N9.2, we find that $\vec{v}' = [-6$ m/s, -8 m/s, $0]$. This vector has a magnitude of 10 m/s and a direction 37° west of south.

N9X.3   Let the elevator and the ground be the S' and S frames, respectively. We are given $A_z = +3.0$ m/s² and that the ball's acceleration in S is $a_z = -9.8$ m/s²,

we want to find its acceleration a'_z in S'. The z component of equation N9.7 tells us that $a'_z = a_z - A_z = -12.8$ m/s².

N9X.4   The "magical force" that seems to draw a person forward is simply the tendency of the person's body to continue moving at a constant velocity while the car accelerates backward. If the person is not held to the car by a seat belt, the person will continue to move through the windshield as the car comes to rest. Thus the person is not "thrown through the windshield" so much as the windshield is thrown backward by the collision past the person's body as the latter moves naturally forward.

N9X.5   You have to hold on tightly in order to exert the inward force on your body required to keep it accelerating in the circular path of the rim of the merry-go-round!

[†]*Note:* A *uniform* gravitational field would be *completely erased* in a freely falling frame. According to general relativity, a uniform gravitational field is therefore as fictitious as a centrifugal force. However, the *real* gravitational fields created by real physical objects are always nonuniform, and therefore they cannot be completely erased even by going to a freely falling frame. In general relativity, what we normally consider "gravity" is a fictitious force that arises from using a noninertial reference frame; for general relativity, what is *real* about gravity is the "tidal effects" of the gravitational field that remain and are observable even in a freely falling frame. Einstein was able to develop a logically coherent theory of general relativity partly because the equivalence principle eventually allowed him to focus on what was *real* about gravitation instead of being distracted by the more obvious but fictitious part.

N9X.6 *Model:* In this case, the gravitational and frame-correction forces exert zero torque around the system's center of mass, so we need to calculate the torque contributed by the normal force and the static friction force. These forces contribute torques in opposite directions, so their torques will cancel if they have the same magnitude.

The net force on the system in the bike frame must be zero, so we must have

$$0 = \begin{bmatrix} 0 \\ 0 \\ +F_N \end{bmatrix} + \begin{bmatrix} -F_{SF} \\ 0 \\ 0 \end{bmatrix} + \begin{bmatrix} 0 \\ 0 \\ -mg \end{bmatrix} + \begin{bmatrix} +mv^2/R \\ 0 \\ 0 \end{bmatrix}$$

(N9.16)

implying that $F_N = mg$ and $F_{SF} = mv^2/R$. In this case, the position vector $\vec{r}$ of the point where both these forces are applied relative to the origin is the inverse of $\vec{r}_{CM}$ in figure N9.6, so the magnitudes of its components perpendicular to $\vec{F}_N$ and $\vec{F}_{SF}$ are $r \sin\theta$ and $r \cos\theta$, respectively.

Solution: Therefore, setting the magnitudes of the torques exerted by these forces equal, we find that

$$F_N r \sin\theta = F_{SF} r \cos\theta$$

$$\tan\theta = \frac{F_{SF}}{F_N} = \frac{\cancel{m}v^2/R}{\cancel{m}g} = \frac{v^2}{Rg}$$

(N9.17)

Evaluation: This is what we got before.

N10 Projectile Motion

Chapter Overview

Introduction

This chapter launches a four-chapter subdivision exploring practical situations where we can use our knowledge about forces acting on an object to determine the object's motion. The three situations we will examine in depth are projectile motion (this chapter), oscillatory motion (chapter N11), and orbital motion (chapters N12 and N13).

Section N10.1: Weight and Projectile Motion

As noted in unit C, we define an object's weight to be the force that gravity exerts on that object. Weight is not the same as mass, but at a given position in a gravitational field, an object's weight is proportional to its mass

$$\vec{F}_g = m\vec{g} \tag{N10.1}$$

where the constant of proportionality $\vec{g}$ at that location is called the **gravitational field vector** at that location: $\vec{g}$ expresses the strength and direction of the gravitational field at that point. At points near the earth's surface, $\mathrm{mag}(\vec{g}) = 9.8 \text{ m/s}^2 = 22 \text{ (mi/h)/s}$. An object is **freely falling** if its weight is the *only* force acting on it. Newton's second law implies that a freely falling object's acceleration is $\vec{a} = \vec{g}$: for this reason, $\vec{g}$ is often called the **acceleration of gravity.**

We call an object a **projectile** and its motion **projectile motion** if the following conditions apply:

1. The object remains close enough to the earth's surface so that $\mathrm{mag}(\vec{g}) \approx$ constant.
2. The object's trajectory is short enough that the direction of $\vec{g}$ is constant.

Note that $\mathrm{mag}(\vec{g})$ varies by less than $\pm 1\%$ within 30 km of the earth's surface, and the direction of $\vec{g}$ varies by less than $1°$ during a 100-km trajectory.

If the drag forces acting on an object are negligible compared to its weight, we say that its motion is **simple projectile motion.**

Section N10.2: Simple Projectile Motion

In the case of simple projectile motion, integrating both sides of $d\vec{v}/dt \equiv \vec{a}(t) = \vec{g}$ and $d\vec{r}/dt \equiv \vec{v}(t)$ with respect to time yields

The basic equations describing simple projectile motion

$$\begin{bmatrix} v_x(t) \\ v_y(t) \\ v_z(t) \end{bmatrix} = \begin{bmatrix} v_{0x} \\ v_{0y} \\ -gt + v_{0z} \end{bmatrix} \tag{N10.6}$$

$$\begin{bmatrix} x(t) \\ y(t) \\ z(t) \end{bmatrix} = \begin{bmatrix} v_{0x}t + x_0 \\ v_{0y}t + y_0 \\ -\frac{1}{2}gt^2 + v_{0z}t + z_0 \end{bmatrix} \tag{N10.7}$$

Purpose: These equations express how the components $[v_x, v_y, v_z]$ and $[x, y, z]$ of a simple projectile's velocity $\vec{v}$ and position $\vec{r}$ vary with time t.

Symbols: $g \equiv \text{mag}(\vec{g})$ is the gravitational field strength, and $[v_{0x}, v_{0y}, v_{0z}]$ and $[x_0, y_0, z_0]$ are the components of the object's initial velocity and position at time $t = 0$, respectively.

Limitations: Gravity must be the only significant force acting on the object, and $\vec{g}$ must be essentially constant in magnitude and direction during the object's trajectory.

Section N10.3: Some Basic Implications

Here are some of the important implications of equations N10.6 and N10.7:

1. The horizontal components of motion are unaffected by gravity.
2. An object's vertical motion is independent of its horizontal motion.
3. An object's trajectory is confined to a plane (which we usually choose to be the xz plane).

One can find the time t when a simple projectile reaches the peak of its trajectory by setting its vertical velocity component $v_z(t)$ to zero and solving for t. Similarly, one can determine when a simple projectile returns to the ground by solving $z(t) = 0$ for t. Since the latter is a quadratic equation, you will get two solutions for t, but one is usually unphysical (it yields a value for t that is before or after the time interval when the object is a simple projectile).

Section N10.4: A Projectile Motion Framework

In solving simple projectile motion problems, it is important to do the following things. The *translation* section of a problem solution should include clear definitions of the reference frame, the origin, and initial conditions, and it should define symbols for important instants of time as well as other quantities. In the conceptual motion part, it is important to establish *when* the simple projectile model applies in this case as well as to state any approximations and assumptions you are making. In the solution part, you should use the column-vector versions of equations N10.6 and N10.7, recalling that each component equation is independent.

Section N10.5: Drag and Terminal Speed

The drag force on a reasonably large object moving through air at a reasonably large speed v is usually accurately modeled by

$$F_D = \tfrac{1}{2}C\rho A v^2 \qquad \text{(N10.10)}$$

where C is the drag coefficient, ρ is the density of air, and A is the object's cross-sectional area. An object falling vertically from rest will eventually stop accelerating when it reaches a speed where the drag is equal to the object's weight: this **terminal speed** is

$$v_T = \sqrt{\frac{2mg}{C\rho A}} \qquad \text{(N10.11)}$$

Purpose: This equation describes the maximum speed v_T that an object dropped from rest will reach when falling through air.

Symbols: A is the object's cross-sectional area, m is its mass, C is the object's drag coefficient, ρ is the density of air, and $g \equiv \text{mag}(\vec{g})$ is the gravitational field strength.

Limitations: This equation does not apply to specks of dust and other tiny objects.

An object falling from rest at time $t = 0$ will behave as a simple projectile for times $t \ll t_T \equiv v_T/g$, but will essentially reach its terminal speed after a time $t \approx 2t_T$.

N10.1   Weight and Projectile Motion

One of the fundamental forces of nature that we experience daily is the *force of gravity*, the force that incessantly and insistently tugs us toward the center of the earth. Since this force is such a common part of our experience, it is worth looking at in some detail.

As we've said before, in physics we define an object's *weight* $\vec{F}_g$ to be *the force that gravity exerts on that object*. Note that this force expresses the interaction between the object and whatever nearby body (usually the earth) is massive enough to create an appreciable gravitational field. In section C3.4, we saw that near the earth's surface, the weight force on an object is given by

Defining the gravitational field vector $\vec{g}$

$$\vec{F}_g = m\vec{g} \qquad \text{(N10.1)}$$

where $\vec{g}$ is a *vector* that points toward the center of the earth and whose magnitude is $g = 9.80\ \text{N/kg} = 9.80\ \text{m/s}^2$ (since $1\ \text{N} = 1\ \text{kg·m/s}^2$). We can take this equation as the *definition* of $\vec{g}$. Since $\vec{g}$ expresses the effect of the gravitational interaction on an object of given mass at a certain point in space, and since $\vec{g}$ has the same value for all objects at that point, it expresses something basic about the character of the gravitational field that the earth creates at that point. We therefore call $\vec{g}$ the **earth's gravitational field vector** at that point. (We can define $\vec{g}$ near other celestial objects in a similar way.)

If the *only* force on an object is its weight $\vec{F}_g$, we say that it is **freely falling.** As we saw in section N4.4, Newton's second law tells us that since the object's weight is strictly proportional to its mass, its acceleration is

The acceleration of a freely falling object is $\vec{g}$

$$\vec{a} = \frac{\vec{F}_{\text{net}}}{m} = \frac{\vec{F}_g}{m} = \frac{m\vec{g}}{m} = \vec{g} \qquad \text{(N10.2)}$$

implying that *every* object at a given point in space (independent of its mass, composition, or other characteristics) will fall with the *same* acceleration equal to the value of $\vec{g}$ at that point. Since the value of $\vec{g}$ at a point characterizes the common acceleration of all falling objects at that point, people sometimes call it the **acceleration of gravity** (even though it obviously isn't *gravity* that is accelerating!). I will use the term *gravitational field vector* not only because it more accurately describes the physical meaning of $\vec{g}$, but also because it underlines its analogy with the *electric field vector* $\vec{E}$ that we will discuss in unit E.

The simple projectile motion model

As long as (1) an object remains "sufficiently close" to the surface of the earth and (2) its trajectory is "sufficiently short" that the curvature of the earth is not significant, the magnitude and direction of $\vec{g}$ (and thus of the object's weight $\vec{F}_g$) will be approximately constant. If in addition (3) the object is "not significantly affected" by other forces except possibly air drag, then we call it a **projectile** and its motion **projectile motion.** When (4) we *also* neglect drag, we say that the object's motion is **simple projectile motion.** Newton's second law implies that such an object will have constant acceleration $\vec{a} = \vec{g}$.

When is drag important?

What do these restrictions mean in practice? Let's look at each in turn, starting with the last. A *freely falling* object is strictly defined to be an object that is not influenced by anything *except* its own weight. In practice, it is hard to isolate falling objects from the effects of air friction (drag). In some cases, these effects can be substantial: it is clear that a feather does not fall in the same way that a steel ball does. The statement that an object is "freely falling" thus could only strictly apply to objects in a vacuum. However, the description is *approximately* true in any case in which the drag force is small compared to its weight.

Whether this approximation is good or not depends on the object's *mass, shape, size,* and *speed.* For example, we would expect two objects of the same size and shape and moving at the same speed to experience the same drag. However, if one object is more massive than the other, the same drag force will change its velocity *less* in a given time interval (according to Newton's second law) and thus will affect its motion less. Thus a steel ball will behave more like a freely falling object than a ping-pong ball of the same size. An object's shape is also important: for example, a sheet of paper behaves more like a freely falling object if it is compressed into a tight ball than if it is left flat, because more air interacts with the flat sheet than with the ball. Finally, the speed of an object is important: higher speeds produce higher drag forces. So the approximation works best if the object is small, massive, and moving relatively slowly.

What does it mean for an object to be "sufficiently close" to the earth's surface? We have seen that the effect of the earth's gravity diminishes with distance from the earth's center. However, you have to travel a significant distance vertically before variations in $\vec{g}$ become noticeable. Empirically, the variation in the value of g is smaller than $\pm 1\%$ within a range of ± 30 km relative to sea level. Even at altitudes of several hundred kilometers (roughly the altitude where the space shuttle flies), the value of g is only 2% to 3% smaller than it is at sea level. So depending on what we would consider a significant variation in the value of g to be, "sufficiently close" could range from within a few kilometers to within hundreds of kilometers of the earth's surface.

When is an object "sufficiently close" to the earth?

The other restriction is that the object's trajectory be "sufficiently short" compared to the earth's curvature. This is because $\vec{g}$ points toward the earth's *center,* so if the object moves far enough horizontally, the direction of $\vec{g}$ could change significantly (relative to the distant stars, say). The earth is pretty large, though! Since the earth's circumference is about 40,000 km and the change in the direction of $\vec{g}$ for an object going around the earth will be $360°$, we see that we have to travel more than 100 km for the direction of $\vec{g}$ to change by more than $1°$. So we can consider the direction of $\vec{g}$ to be constant as long as the object's trajectory is not longer than 100 km or so.

When is its trajectory "sufficiently short"?

Exercise N10X.1

Imagine that the drag force on a 1.5-in. sphere moving through air at a speed of 5 m/s is 0.009 N. What fraction is this of the weight of a 2.5-g ping-pong ball of this size? How about a 220-g steel ball?

Exercise N10X.2

A rocket is launched near the equator and travels about 1200 km due east. By about how much does the direction of $\vec{g}$ at the rocket's position change (relative to the distant stars) during the rocket's flight?

N10.2   Simple Projectile Motion

So, when the drag on an object is negligible and $\vec{g}$ is relatively constant in magnitude and direction, we can model the object as a *simple projectile* and take its acceleration to be $\vec{a} = \vec{g} = $ constant. Since this *is* a constant, we can easily integrate this by using the methods of section N4.5 to find the falling object's velocity and position as a function of time. Integrating the acceleration

Integrating $\vec{a}(t)$ to find $\vec{v}(t)$ and $\vec{r}(t)$

to find the velocity (see equation N4.18), we find that

$$\vec{v}(t) - \vec{v}(0) = \int_0^t \vec{a}(t)\,dt = \int_0^t \vec{g}\,dt = \vec{g}\int_0^t dt = \vec{g}(t-0) = \vec{g}t$$

$$\Rightarrow \quad \vec{v}(t) = \vec{g}t + \vec{v}(0) \tag{N10.3}$$

Similarly, integrating the velocity to find the position (see equation N4.21), we get

$$\vec{r}(t) - \vec{r}(0) = \int_0^t \vec{v}(t)\,dt = \int_0^t [\vec{g}t + \vec{v}(0)]\,dt = \vec{g}\int_0^t t\,dt + \vec{v}(0)\int_0^t dt$$

$$= \vec{g}\left(\tfrac{1}{2}t^2 - \tfrac{1}{2}0^2\right) + \vec{v}(0)(t-0) = \tfrac{1}{2}\vec{g}t^2 + \vec{v}(0)t$$

$$\Rightarrow \quad \vec{r}(t) = \tfrac{1}{2}\vec{g}t^2 + \vec{v}(0)t + \vec{r}(0) \tag{N10.4}$$

Note how I have treated both $\vec{g}$ and $\vec{v}(0)$ as simple constants in the integration process.

In practical component language, if we define the z axis to be vertically upward and the x and y axes to be horizontal, then $\vec{g} = [0, 0, -g]$. (The z component of $\vec{g}$ is negative because $\vec{g}$ points vertically downward. Note that $g \equiv \text{mag}(\vec{g})$ is always a positive number, no matter how $\vec{g}$ points.) If we also define

$$\vec{v}(0) \equiv \begin{bmatrix} v_{0x} \\ v_{0y} \\ v_{0z} \end{bmatrix} \quad \text{and} \quad \vec{r}(0) \equiv \begin{bmatrix} x_0 \\ y_0 \\ z_0 \end{bmatrix} \tag{N10.5}$$

then we can express equations N10.3 and N10.4 in column-vector form as follows:

The basic equations describing simple projectile motion

$$\begin{bmatrix} v_x(t) \\ v_y(t) \\ v_z(t) \end{bmatrix} = \begin{bmatrix} 0t \\ 0t \\ -gt \end{bmatrix} + \begin{bmatrix} v_{0x} \\ v_{0y} \\ v_{0z} \end{bmatrix} \quad \Rightarrow \quad \begin{bmatrix} v_x(t) \\ v_y(t) \\ v_z(t) \end{bmatrix} = \begin{bmatrix} v_{0x} \\ v_{0y} \\ -gt + v_{0z} \end{bmatrix}$$

$$\tag{N10.6}$$

$$\begin{bmatrix} x(t) \\ y(t) \\ z(t) \end{bmatrix} = \begin{bmatrix} \tfrac{1}{2}0t^2 \\ \tfrac{1}{2}0t^2 \\ -\tfrac{1}{2}gt^2 \end{bmatrix} + \begin{bmatrix} v_{0x}t \\ v_{0y}t \\ v_{0z}t \end{bmatrix} + \begin{bmatrix} x_0 \\ y_0 \\ z_0 \end{bmatrix}$$

$$\Rightarrow \quad \begin{bmatrix} x(t) \\ y(t) \\ z(t) \end{bmatrix} = \begin{bmatrix} v_{0x}t + x_0 \\ v_{0y}t + y_0 \\ -\tfrac{1}{2}gt^2 + v_{0z}t + z_0 \end{bmatrix} \tag{N10.7}$$

Purpose: These equations express how the components $[v_x, v_y, v_z]$ and $[x, y, z]$ of a simple projectile's velocity $\vec{v}$ and its position $\vec{r}$ vary with time t.

Symbols: $g \equiv \text{mag}(\vec{g})$ is the gravitational field strength, and $[v_{0x}, v_{0y}, v_{0z}]$ and $[x_0, y_0, z_0]$ are the components of the object's initial velocity and position at time $t = 0$, respectively.

Limitations: Gravity must be the only significant force acting on the object, and $\vec{g}$ must be essentially constant in magnitude and direction during the object's trajectory.

We will use equations N10.6 and N10.7 very often in what follows.

Exercise N10X.3

According to equation N10.7, how long would it take for an object to fall a vertical distance of 4.9 m from rest? How would this result change if the object initially had a horizontal x-velocity of 5.0 m/s?

N10.3   Some Basic Implications

Equations N10.6 and N10.7 imply three basic but important things about simple projectile motion that are worth underlining. First, notice that *gravity affects only the z component of a simple projectile's motion*. In particular, equation N10.6 tells us that the horizontal part of the projectile's original velocity is completely unaffected by the presence of gravity: as it falls, a simple projectile will maintain whatever original horizontal velocity it had. For example, a package dropped from a plane will continue to move horizontally at the same rate as the plane (see figure N10.1a).

Gravity affects only the vertical component of an object's motion

Second, *the component equations in either equation N10.6 or equation N10.7 are completely independent of each other*. In particular, note that the projectile's vertical motion is completely unaffected by its horizontal velocity. This means, for example, that an object dropped from rest will have the same vertical motion as an object with a large initial horizontal velocity, as shown in figure N10.1b.

The component equations are independent

Third, since a projectile's horizontal motion is unaffected by gravity, the projection of a projectile's motion on the horizontal (xy) plane will always be a straight line. It is thus generally possible to orient our reference frame so that its x axis lies in the direction of motion [i.e., so that $v_y(t) = v_{0y} = 0$]. If we do this, the projectile's motion lies entirely in the xz plane, which simplifies things somewhat. I will commonly do this in the examples that follow.

We can always define the horizontal motion to be along the x axis

Most actual projectiles are launched from near the ground and follow a trajectory that looks like an arc, going first upward and then curving back downward to the ground. How do we determine when the object has reached its maximum altitude and when it reaches the ground?

When does a (simple) projectile reach its peak?

If the initial velocity of a projectile is at all upward, then $v_{0z} > 0$. The bottom line of equation N10.6 tells us that subsequently $v_z(t) = -gt + v_{0z}$, so the projectile's z-velocity starts positive, but decreases linearly through zero and eventually becomes negative, as shown in figure N10.2. I claim that the

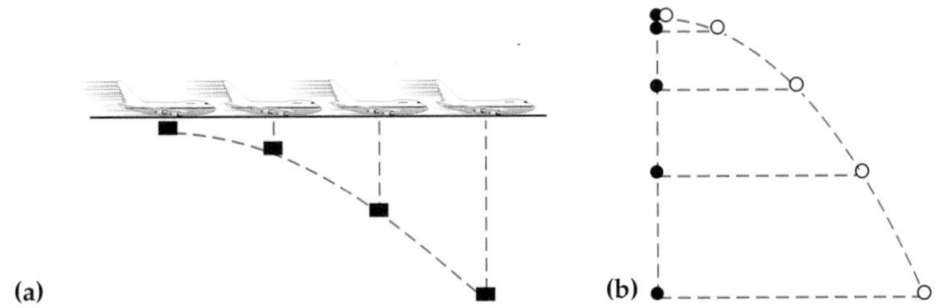

Figure N10.1
(a) A projectile's horizontal motion is unaffected by gravity: as an object falls, it will continue to move forward at whatever horizontal velocity it had originally. (b) A projectile's vertical motion is independent of its horizontal motion: objects fall vertically at the same rate no matter how they are moving horizontally.

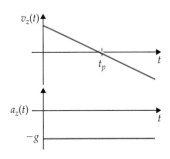

Figure N10.2
Graphs of a projectile's z-velocity and z-acceleration near the peak of its trajectory.

projectile is at its peak when its z-velocity passes through zero. Why? At an instant when $v_z(t)$ is positive, the projectile is still going upward, and so it will reach a still higher position later. At an instant where $v_z(t)$ is negative, it is moving downward and thus was at a higher position earlier. Only at the exact instant where $v_z(t) = 0$ is it neither climbing to nor coming down from a higher position: this must be the peak. Thus to find t_p ($\equiv$ time of the peak), we solve

$$0 = v_z(t_p) = -gt_p + v_{0z} \quad \Rightarrow \quad t_p = \frac{v_{0z}}{g} \qquad \text{(N10.8)}$$

[Note, by the way, that even though the projectile's z-velocity is passing through zero at this point, its z-acceleration remains nonzero and equal to $-g$ always. This can be seen clearly in figure N10.2: even as the value of $v_z(t)$ passes through zero, its slope is always constant and negative.]

If we have set up our reference frame so that the ground is at $z = 0$, then finding the time t_h ($\equiv$ time when the projectile hits the ground) is simply a matter of solving equation $z(t_h) = 0$ for t_h by using the quadratic equation:

$$0 = z(t_h) = -\tfrac{1}{2}gt_h^2 + v_{0z}t_h + z_0 \quad \Rightarrow \quad t_h = \frac{v_{0z} \pm \sqrt{v_{0z}^2 + 2gz_0}}{g} \qquad \text{(N10.9)}$$

When does a (simple) projectile hit the ground?

Exercise N10X.4

Check that the last result in equation N10.9 is correct.

Note that this equation yields *two* solutions. This is because if the projectile had been freely falling for all time and had sufficient energy to get up beyond $z = 0$ at all, then its parabolic trajectory would cross $z = 0$ *twice*, once going up and once going down. Usually we will be more interested in the latter solution. Note in particular that since the projectile is *not* freely falling before it is launched, neither equation N10.8 nor equation N10.9 applies for times before the launch time; and so if either equation yields such a solution, this solution is not physically meaningful.

Exercise N10X.5

Are there values of v_{0z} and z_0 for which equation N10.9 yields *no* solution? If so, explain in physical terms *why* there is no solution.

Other implications

Other implications of equations N10.6 and N10.7 include the *parabolic* shape of a projectile's trajectory in simple projectile motion (see problem N10S.10) and the fact that the horizontal distance a projectile launched from level ground will go before returning to the ground is greatest if it is launched at a 45° angle (see problem N10S.11).

N10.4 A Projectile Motion Framework

Projectile motion problems are somewhat simpler in structure than the constrained-motion problems we have done in previous chapters. We already know the basic model we will use (the simple projectile model) and the master equations we will use (equations N10.6 and N10.7). Even so, there are things we can and should do to increase our chances of solving the problem correctly.

The drawing that you create for the translation step of a projectile motion solution should display

The translation step

1. The initial position of the object of interest and its subsequent trajectory.
2. A labeled arrow indicating the object's initial velocity $\vec{v}_0$.
3. Reference frame axes with a clearly specified origin.
4. Labels defining symbols for other relevant quantities (particularly initial and final positions, important instants of time, and perhaps velocities).
5. A list of symbols with known values.

Establishing an appropriate reference frame is very important for simple projectile motion problems because equations N10.6 and N10.7 are expressed in terms of components that have meaning only in the context of a well-defined reference frame with a clearly specified origin. Also, time t is a very important variable in equations N10.6 and N10.7, so it is important to define symbols for important instants of time. This is often difficult to do on the diagram, so short prose descriptions of the variables are often a good way to define time symbols.

The point of the *conceptual model* in any problem is to assess the applicability and limitations of the model that we are applying. With simple projectile motion problems, the basic model is clear, so the main things that the conceptual model part of the solution should describe are the approximations needed to make the situation fit the simple projectile model (and how good these approximations are likely to be) and the time interval during which the model applies.

The conceptual model step

The latter helps us recognize whether a mathematically possible answer is in fact physically absurd. There is no efficient way to deal with these issues by using a diagram, but a very brief prose model is usually sufficient. The conceptual model should end with a statement of the master equation(s) (equations N10.6 and/or N10.7) with any unknown quantities circled.

When you solve the problem, remember that each row in equations N10.6 and N10.7 is independent and must be satisfied simultaneously (as in a conservation of momentum problem). It is common in projectile problems to solve one of these independent equations for a time and then plug the result into a different equation to compute another quantity of interest.

The solution step

Examples N10.1 and N10.2 illustrate the process.

Example N10.1

Problem   Your friend, an unemployed actor, has a chance at a role in a Western movie if he can learn to slide a mug precisely down the length of a saloon bar. After practicing a lot, he thinks he can do it. But when asked to perform in front of the producer, he is a bit nervous, and the mug goes off the end of the bar (which is 1.4 m high) at a speed of 2.5 m/s. Does the mug compound your friend's problem by hitting the producer's foot, which is 2.0 m beyond the end of the bar?

Translation   A sketch of the situation appears on the next page. Note that $t = 0$ is the instant that the mug leaves the edge of the bar; t_h is the instant when the mug hits the floor.

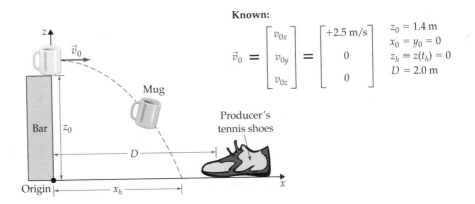

Known:

$$\vec{v}_0 = \begin{bmatrix} v_{0x} \\ v_{0y} \\ v_{0z} \end{bmatrix} = \begin{bmatrix} +2.5 \text{ m/s} \\ 0 \\ 0 \end{bmatrix}$$

$z_0 = 1.4$ m
$x_0 = y_0 = 0$
$z_h \equiv z(t_h) = 0$
$D = 2.0$ m

Model   A mug is both small and relatively massive, so the simple projectile model should pretty accurately describe the situation for the short time the mug is in the air after it leaves the bar at $t = 0$ until it hits the floor (or the producer's leg) at $t = t_h$. I am also treating the mug as a point particle, essentially ignoring its size. If the mug's final position x_h turns out to be very close to D, I may need to do the calculation again, making some guesses as to the mug's actual size. I'm also assuming that the mug's initial velocity is exactly horizontal. Equation N10.7 tells us that

$$\begin{bmatrix} x_h \\ y_h^{\;0} \\ z_h \end{bmatrix} = \begin{bmatrix} x(t_h) \\ y(t_h) \\ z(t_h) \end{bmatrix} = \begin{bmatrix} v_{0x}(t_h) + x_0^{\;0} \\ v_{0y}t_h^{\;0} + y_0^{\;0} \\ -\frac{1}{2}gt_h^2 + v_{0z}t_h^{\;0} + z_0 \end{bmatrix} \qquad (1)$$

The first and last rows of this equation provide two useful equations in the two unknowns t_h and x_h.

Solution   We know that $z(t_h) = 0$ (since the mug hits the floor at $t = t_h$ by definition), so the third row of equation 1 implies that

$$0 = z_h = -\tfrac{1}{2}gt_h^2 + z_0 \quad \Rightarrow \quad \tfrac{1}{2}gt_h^2 = z_0 \quad \Rightarrow \quad t_h = \pm\sqrt{\frac{2z_0}{g}} \qquad (2)$$

Since the mug is not a projectile before $t = 0$, the negative solution is not relevant. Plugging the positive result into the first line of equation 1, we get

$$x_h = v_{0x}t_h = v_{0x}\sqrt{\frac{2z_0}{g}} = (2.5 \text{ m/s})\sqrt{\frac{2(1.4 \text{ m})}{9.8 \text{ m/s}^2}} = 1.3 \text{ m} \qquad (3)$$

Evaluation   This has the right units and the right sign (the mug should hit to the right of $x = 0$) and seems plausible. So the mug misses hitting the producer, whose shoes are $D = 2.0$ m from the end of the bar. (Let's just hope that the mug also does not splash the producer's shoes.)

Example N10.2

Problem   During a soccer game, you kick the ball so that it leaves the ground at an angle of 33° traveling at a speed of 15 m/s. The ball reaches its maximum height just as it passes the goalie, who is standing well in front of the net. Is the ball high enough at this point to be out of the goalie's reach?

Translation   A drawing of the situation appears below. Let $t = 0$ be the instant the ball leaves the foot and t_p be the time when the ball reaches its peak. Note that the peak is defined to be where $v_z = 0$.

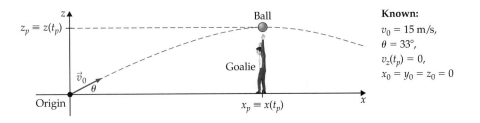

Known:
$v_0 = 15$ m/s,
$\theta = 33°$,
$v_z(t_p) = 0$,
$x_0 = y_0 = z_0 = 0$

Model   To use the simple projectile model, we have to assume that air resistance is negligible: this may not be a very good assumption for this rapidly moving, fairly lightweight ball. If the model is good at all, it applies after the ball leaves the ground at $t = 0$ until it hits the ground or is caught: assume this occurs after $t = t_p$. I also am going to assume that the goalie can catch the ball if z_p is less than about 3.0 m (which is ≈ 9.5 ft): this is about how high I think I could reach (with a bit of a jump). Equations N10.6 and N10.7 in this case read

$$\begin{bmatrix} v_x(t_p) \\ v_y(t_p) \\ v_z(t_p) \end{bmatrix} = \begin{bmatrix} v_0 \cos \theta \\ 0 \\ -gt_p + v_0 \sin \theta \end{bmatrix} \quad (1) \qquad \begin{bmatrix} x(t_p) \\ y(t_p) \\ z(t_p) \end{bmatrix} = \begin{bmatrix} (v_0 \cos \theta)(t_p) + x_0^{\;0} \\ y_0^{\;0} \\ -\frac{1}{2}gt_p^2 + (v_0 \sin \theta)t_p + z_0^{\;0} \end{bmatrix} \quad (2)$$

The top and bottom rows of these equations provide four useful equations in our four circled unknowns, so we can solve.

Solution   Since we know that $v_z(t_p)$ is zero, we can solve the bottom line of equation 1 for t_p and then plug it into the bottom line of equation 2 to determine the value of $z_p \equiv z(t_p)$ that we are particularly interested in. Doing this, we get

$$0 = v_z(t_p) = -gt_p + v_0 \sin \theta \quad \Rightarrow \quad t_p = \frac{v_0 \sin \theta}{g} \qquad (3)$$

$$\Rightarrow \quad z_p = z(t_p) = -\frac{1}{2}gt_p^2 + v_0(\sin \theta)(t_p) = -\frac{1}{2}g\left(\frac{v_0 \sin \theta}{g}\right)^2 + \frac{(v_0 \sin \theta)^2}{g}$$

$$= \left(-\frac{1}{2} + 1\right)\frac{v_0^2 \sin^2 \theta}{g} = \frac{v_0^2 \sin^2 \theta}{2g}$$

$$= \frac{(15 \text{ m/s})^2 \sin^2(33°)}{2(9.8 \text{ m/s}^2)} = 3.4 \text{ m} \qquad (4)$$

Evaluation   This result has the right units, the right sign (the peak is above the ground!), and a reasonable magnitude. The kick is almost certainly out of the goalie's reach, though, so it's a gooaal!

N10.5 Drag and Terminal Speed

Up to now, we have been considering *simple* projectile motion, in which the only significant force acting on an object is its weight. Realistically, though, projectiles moving through air experience some drag. How would including drag affect our predictions about a projectile's motion?

In section N6.5, we saw that the drag force on a large object with cross-sectional area A moving at a reasonably large speed v through air has a magnitude of

The drag force on a typical object moving through air

$$F_D = \tfrac{1}{2}C\rho Av^2 \qquad (N10.10)$$

where ρ is the density of air (which is about $1.2\ \mathrm{kg/m^3}$ at room temperature and standard pressure) and C is a unitless constant (called the *drag coefficient*) that depends on an object's shape.

Predicting a falling object's qualitative behavior

First consider the simplest possible situation: an object falling vertically from rest. Without much mathematical analysis, we can pretty easily determine qualitatively what will happen. At first (because v is very small) the magnitude of the drag force is negligible, and the object falls essentially freely downward (see figure N10.3a). But as its downward speed increases, the drag force also increases, decreasing the net force on the object and thus its downward acceleration (figure N10.3b). The downward speed of the object will continue to increase, however, until the drag force becomes essentially equal to the object's weight (figure N10.3c). As the object approaches the speed where this happens, the net force on it approaches zero. When the net force is essentially zero, the object no longer accelerates, but continues to fall with constant downward velocity. We call the magnitude of this final constant downward velocity the object's **terminal speed** v_T. You can easily show that its value must be

The terminal speed of an object falling with drag

$$v_T = \sqrt{\dfrac{2mg}{C\rho A}} \qquad (N10.11)$$

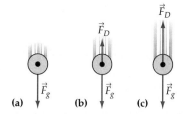

Figure N10.3
(a) Just after an object begins to fall, the magnitude of the drag force on it is very small. (b) As it begins to pick up downward speed, the drag force grows in magnitude. (c) When the magnitude of the drag force equals the magnitude of the object's weight, the object does not continue to accelerate, but instead maintains a *constant* downward velocity.

The acceleration of an object falling with drag

Exercise N10X.6

Verify equation N10.11. (*Hint:* when is $F_D = F_g$?)

Exercise N10X.7

A person's terminal speed in air is typically about 60 m/s. If so, what is the value for CA for a falling person? (Assume that $m \approx 60$ kg.)

More mathematically, if we define the z axis to be positive upward, then the z component of Newton's second law implies that

$$ma_z = F_{\mathrm{net},z} = -mg + \frac{1}{2}C\rho Av_z^2 = -mg + mg\left(\frac{C\rho A}{2mg}\right)(v_z^2) = -mg\left(1 - \frac{v_z^2}{v_T^2}\right)$$

$$\Rightarrow \quad a_z = -g\left(1 - \frac{v_z^2}{v_T^2}\right) \qquad (N10.12)$$

(Note that the drag force term is positive because it acts *upward* on an object falling downward.) You can see directly from this equation that the vertical acceleration becomes smaller and smaller until it becomes essentially zero as $|v_z|$ approaches the terminal speed v_T.

The next step would be to integrate both sides of this equation with respect to time to find v_z as a function of t. But here we run into a roadblock. To *find* v_z as a function of t, we have to integrate the right side of equation N10.12, and to do that, we already have to *know* how v_z depends on t!

This illustrates a typical problem that arises when we attempt to use forces to determine motion. *In principle,* once we know the forces acting on an object, we can use Newton's second law to find its acceleration $\vec{a}(t)$ and then integrate $\vec{a}(t)$ with respect to time to find the object's velocity $\vec{v}(t)$ and position $\vec{r}(t)$ as functions of time. *In practice,* however, the expression for an object's acceleration almost always refers to the object's unknown velocity and/or position, making a straightforward integration of the acceleration impossible. For realistic problems, either we have to resort to some kind of trick to perform the integration, or we have to solve the problem numerically (either constructing a trajectory diagram by hand or using Newton or a similar program to do the job). Problem N10A.2 discusses the trick needed to solve this particular problem mathematically.

But we can in fact make some useful estimates without much more mathematics. One very basic question we might like to answer is this: about how long will it take a falling object to reach its terminal velocity? If there is no drag force acting on the object, the z component of equation N10.6 implies that an object falling from rest satisfies $v_z = -gt$, so the time it would take a drag-*free* object to reach speed v_T would be

$$t_T = \frac{v_T}{g} \tag{N10.13}$$

Now, drag will cause the real object to fall more slowly than a free object, so we would *not* expect it to actually reach the terminal speed by this time. However, the time required is also not likely to be many times this value: surely after, say, 10 times the time required for a drag-free object to reach v_T, the real object must finally be falling at pretty close to this speed as well.

The exact mathematical solution to this problem is shown in figure N10.4. One can see that by time t_T, the object has only reached a speed of about $\frac{3}{4}v_T$, but the object essentially reaches its terminal speed after a time $t \approx 2t_T$. For very early times ($t < 0.2t_D$ or so), the object's downward speed increases at essentially the free-fall rate of $v_z = -gt$, so this provides a good estimate of the time period during which the simple projectile model might apply.

Exercise N10X.8

For a human being, $v_T \approx 60$ m/s. About how long would a skydiver have to fall until $v_z \approx v_T$?

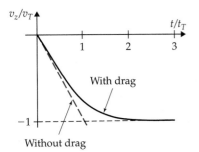

Figure N10.4

A graph of the mathematical solution for the z-velocity of an object falling from rest with drag (this solution is discussed in problem N10A.2). The straight line shows that at very early times ($t \ll t_T$) the object's downward speed increases at approximately the free-fall rate. However, the object essentially reaches its terminal speed by $t \approx 2t_T$, where $t_T = v_T/g$.

How long will it take this person to reach terminal speed? (See exercise N10X.8.)

If the object is *not* falling from rest, the problem becomes even *more* difficult. Here is where using the Newton program becomes a virtual necessity. The important issue to understand here is that even in very straightforward circumstances, it may not be easy to solve Newton's second law for the trajectory of an object (a problem we will struggle with for the rest of the unit).

TWO-MINUTE PROBLEMS

N10T.1 You are driving 5 ft or so behind a pickup truck (don't actually try such a stupid tailgating stunt, of course!). A jostled crate tips off the back of the truck with only a very small backward velocity. The crate will not hit your car until after it hits the road, regardless of your speed, true (T) or false (F)? (Ignore air resistance.)

N10T.2 A person standing in the cabin of a jet plane drops a coin. This coin hits the floor of the cabin at a point directly below where it was dropped (as seen in the cabin) no matter how fast the plane is moving, T or F?

N10T.3 A tennis ball is dropped from rest at the exact same instant and height that a bullet is fired horizontally. Which hits the ground first (ignoring air resistance)?
   A.   The bullet hits first.
   B.   The ball hits first.
   C.   Both hit at the same time.

N10T.4 As a projectile moves along its parabolic trajectory, which of the following remain constant (ignoring air resistance, and defining z axis to point upward)?
   A.   Its speed.
   B.   Its velocity.
   C.   Its x-velocity and y-velocity.
   D.   Its z-velocity.
   E.   Its acceleration.
   F.   Its x-velocity, y-velocity, and acceleration.
   T.   Some other combination of the given quantities.

N10T.5 Imagine that we throw a baseball with an initial speed of 12 m/s in a direction 60° upward from the horizontal. What is the baseball's speed at the peak of its trajectory? (*Hint:* You do not need to do a *lot* of calculating here.)
   A.   12 m/s
   B.   10.4 m/s
   C.   6 m/s
   D.   3 m/s
   E.   0 m/s
   F.   Other (specify)

N10T.6 Imagine that you serve a tennis ball with an initial speed of 10 m/s in a direction 10° below the horizontal. What is its speed at the peak of its trajectory?
   A.   10 m/s
   B.   9.8 m/s
   C.   1.7 m/s
   D.   0 m/s
   E.   There is no "peak" to this tennis ball's trajectory
   F.   Other (specify)

N10T.7 Imagine that you throw a tennis ball vertically into the air. At the exact top of its trajectory it is at rest. What is the magnitude of its acceleration at this point?
   A. 9.8 m/s^2
   B. -9.8 m/s^2
   C. $0 < a < 9.8 \text{ m/s}^2$
   D.   0
   E.   Other (specify)

N10T.8 Two balls have the same size and surface texture, but one is twice as heavy as the other. How many times larger is the terminal speed of the more massive ball falling through air than that of the lighter ball?
   A.   The balls fall with the *same* speed in air.
   B.   The massive ball's terminal speed is $[2]^{1/2}$ times larger than the other's.
   C.   The massive ball's terminal speed is 2 times larger than the other's.
   D.   The massive ball's terminal speed is 4 times larger than the other's.
   E.   The massive ball's terminal speed is some other multiple of the other's (specify).

N10T.9 Two balls have the same weight and surface texture, but one has twice the diameter of the other. How many times larger is the terminal speed of the smaller ball falling through air than that of the bigger ball?
   A.   The balls fall with the *same* speed in air.
   B.   The smaller ball's terminal speed is $[2]^{1/2}$ times larger than other's.

(more)

C. The smaller ball's terminal speed is 2 times larger than the other's.
D. The smaller ball's terminal speed is 4 times larger than the other's.

E. The smaller ball's terminal speed is some other multiple of the other's (specify).

HOMEWORK PROBLEMS

Basic Skills

N10B.1 A ballistic missile traveling 3200 km between its launch point and final destination freely falls during much of its trajectory, most of which is above the atmosphere. Is this missile a projectile or not? Explain.

N10B.2 A package dropped from an airplane flying at an altitude of 1.0 km will take how long to reach the ground? (Ignore air resistance.)

N10B.3 How long will it take a stone thrown with a vertical velocity of 22 m/s to reach the peak of its trajectory?

N10B.4 If a fireworks rocket has an initial upward speed of 58 m/s when launched, for how long will it coast upward before reaching its peak? (Ignore air resistance and assume that the rocket engines burn out very shortly after launch.)

N10B.5 If an object is launched from the ground with an initial z-velocity of 25 m/s, how much time will pass before it returns to the ground? (Ignore air resistance.)

N10B.6 If an object is launched from a point 10 m above the ground with an initial z-velocity of 25 m/s, how much time will pass before it returns to the ground? (Ignore air resistance.)

N10B.7 Estimate the terminal speed for a ping-pong ball whose diameter is ≈ 1.5 in. and whose mass is ≈ 2.5 g. (For a sphere C is roughly 0.5.)

Synthetic

The starred problems are particularly well suited for practicing the use of the problem-solving framework.

N10S.1 Estimate the speed at which the drag on a 150-g steel ball becomes equal to about 1% of its weight. (The density of steel is about 7900 kg/m^3, and C for a sphere is about equal to 0.5.)

N10S.2 (a) Estimate the drag coefficient for a spread-eagled person falling through air, using data given section N10.5.
(b) Roughly how far has a person fallen by the time the person reaches terminal speed, within ±10% or so? (*Hint:* Look at figure N10.4. Remember that an integral is equal to the area under the curve of the integrand.)

*N10S.3 Your Frisbee is lodged in a tree 12 m above your head.
(a) How fast would you have to throw a baseball to dislodge the Frisbee?
(b) Check your work, using a calculation based on conservation of energy.
(c) Which approach is easier to use in this kind of problem?

*N10S.4 A police officer is chasing a burglar across a rooftop. Both are running at a speed of 6.0 m/s. Before the burglar approaches the edge of the roof, the burglar needs to make a decision about whether to jump the gap to the next building, whose roof is 6.2 m away but 3.5 m lower. Will the burglar be able to land on the next building on his or her feet (assuming that the burglar's initial velocity is 6.0 m/s horizontally when the jump begins)?

*N10S.5 You are standing on a cliff 32 m tall overlooking a flat beach. The edge of the water is 25 m from the base of the cliff. How fast would you have to throw a stone horizontally to reach the water?

*N10S.6 In serving a tennis ball, the server hits the ball horizontally from a height of about 2.2 m. The ball has to travel over the net (0.9 m high) a distance of 12 m away. What is the minimum initial speed that the tennis ball can have if it is to make it over the net?

Serving a tennis ball (see problem N10S.6).

*N10S.7 You are the pilot of a Coast Guard rescue plane. Your job is to drop a package of emergency supplies to a person floating in the ocean. If your plane is flying directly toward the person at an elevation of 520 m at a speed of 85 m/s (about 190 mi/h), about how far ahead of the person should you release the package, if it is to land near the person? Assume that the package's motion is not significantly affected by air friction (probably not a very good assumption here).

*N10S.8 A batter hits a fly ball with an initial velocity of 37 m/s at an angle of 32° from the horizontal. How much time does the outfielder have to get to the appropriate position to catch the ball?

*N10S.9 During volcanic eruptions, chunks of solid rock can be blasted out of a volcano: these projectiles are called *volcanic blocks*. Imagine that during an eruption of Mount Saint Helens (a volcano in Washington state) a block lands near an observing station that is located 2.5 km east of and 900 m below the summit. Assume that air drag is negligible and that the block was ejected from the summit at a 35° angle from the horizontal. Find
(a) its initial speed and
(b) its time of flight.
(c) Imagine that this particular block is a cube of rock about 1 m on a side: such a block would have a mass of about 3000 kg. Is drag really negligible for this block at the launch speed you calculated for part (a)?

N10S.10 The formula for a parabola in the zx plane is

$$z(x) = Ax^2 + Bx + C \qquad (N10.14)$$

where A, B, and C are constants. Prove that the trajectory of a simple projectile has this form.

N10S.11 The range R of a projectile is the horizontal distance that it travels between its launch and the time it hits the ground.
(a) Show that if the projectile is launched from the ground, its range will be

$$R = \frac{2v_0^2 \sin\theta \cos\theta}{g} \qquad (N10.15)$$

(b) Prove that a projectile launched from the ground with a given initial speed v_0 will travel the farthest if it is launched at an angle of 45° with respect to the horizontal. (*Hint:* There exists a trigonometric identity that says that $\sin 2\theta = 2\sin\theta \cos\theta$.)

N10S.12 Consider dropping a baseball from rest at an altitude of 300 m. Assume that the baseball has a drag coefficient of 0.3, a mass of 0.145 kg, and a radius of 3.7 cm (roughly the real values for a baseball).

(a) What is the terminal speed v_T of this baseball?
(b) Calculate the characteristic time $t_T = v_T/g$.
(c) Set up the Newton program to model this situation. Let's agree that the ball has essentially reached its terminal velocity when its acceleration falls below $g/20$. About how long must the baseball fall to reach terminal velocity according to this criterion? Express your answer as a multiple of the characteristic time t_T. (*Hints:* Equation N10.12 might be helpful as you set up the program. A time step of 0.1 s works pretty well.)

N10S.13[†](a) Use the Newton program to calculate the trajectory of a batted baseball that starts at $y = 1$ m (where y is the vertical axis) and has an initial velocity of 50 m/s in a direction that is 37° above the horizontal (note that $\cos 37° = \frac{4}{5}$, $\sin 37° = \frac{3}{5}$). Use a time step of 0.2 s. Assume zero drag. Look at a close-up view of the construction to verify that the (blue) acceleration arrows are indeed vertical. (1) How far does the baseball travel horizontally? (2) What is the baseball's height above the ground at the peak of its trajectory? (3) How long is it in flight? Print out a copy of the trajectory, and record your answers on the printout.
(b) Now add drag, using equation N10.10 to model the drag. Note that the drag force always acts opposite to the object's velocity. Note also that the Newton program requires you to enter terms in the expression for the ball's *acceleration*, so you need to solve Newton's second law for acceleration to determine the values of these terms. The radius of a real baseball is 3.7 cm, its mass is 0.145 kg, and its drag coefficient is about 0.3. Use the same initial conditions as in part (a). Look at a close-up view and verify that the acceleration arrows are no longer vertical, but tip backward, as might be expected when there is a backward force on the ball. (1) How far does the baseball travel now? (2) What is its peak height? (3) How long does it take to reach the ground? (4) Is drag an important effect here? Write your responses to these questions on a printout of the trajectory, and show how you calculated the coefficient for the drag acceleration term that you typed into the slot in the Setup window.
(c) The baseball could also be spinning. It turns out that a baseball spinning in the correct way (so that the bottom part of the ball moves in the same direction as the ball's center of mass) experiences a lift force in addition to the drag force. This lift force has a magnitude of $F_L = kfv^2$, where f is the frequency (in rotations

[†]Thanks to Phil Krasicky of Cornell University for developing this application of the Newton program. See A. F. Rex, "The Effect of Spin on the Flight of Batted Baseballs," *American Journal of Physics,* **53:** 1073–1075 (1985) for greater discussion of the general issue.

per second) of the ball's rotation, v is the speed of the ball's center of mass, and k is a constant whose experimental value is roughly $1.1 \times 10^{-5} \, \text{N} \cdot \text{s}^3/\text{m}^2$. The direction of the force is perpendicular to the velocity on the upward side (this is exactly vertically upward *only* if the ball's velocity is exactly horizontal, so *don't* set the direction of the acceleration term contributed by this force to be in the y-direction). Add this effect into the simulation, assuming the baseball spins at 20 turns per second. (1) How far does the ball travel now? (2) What is its peak height? (3) How long does it take to reach the ground? (4) Could such a spin turn a long fly ball into a home run, do you think? Record your answers on a printout of the trajectory, and show how you calculated the acceleration coefficient for the spin lift term that you typed into the Setup window.

(d) Finally, check out what happens if you vary the size of the time step (this is something one should always do with computer models). (1) What happens if you choose Δt equal to 0.1 s? How about 0.05 s? Does the distance the ball travels, its peak height, or the time of flight change much? (2) What if you choose something larger, such as 0.5 s or 1 s? (3) What do you think is the largest time step you could choose to get answers that are within 2% or so of the true result? (Don't do new printouts here: just record your answers on one of the previous printouts.)

N10S.14 A soccer ball has a mass of about 0.45 kg and a radius of about 10 cm.

(a) Do some runs, using the Newton program to determine the optimal angle ($\pm 1°$) for the goalie to kick the ball to achieve the maximum range in still air with a launch speed of 30 m/s. Use a time step of 0.05 s, and do *not* ignore drag.

The optimal angle when drag is not significant is 45° (see problem N10S.11), so start with that angle.

(b) Explain physically why the optimal angle should be *less* than 45° when drag is significant.

(c) On the other hand, does choosing the optimal angle (as opposed to 45°) in this case give you much improvement in the range? Submit printouts justifying your conclusions.

Rich-Context

N10R.1 Abel Knaebble, professional stunt man and amateur physicist, wants to jump the Grand Canyon in his new, super-streamlined motorcycle (which is completely immune to the effects of air friction). He simply plans to ride his motorcycle off one rim, so his initial velocity will be entirely horizontal. Assume that the width of the canyon is 8.0 km at the proposed jump site, that the mass of his fully loaded cycle (including Abel himself) is 380 kg, and that the rim of the canyon from which he will jump is 320 m higher than the far side.

(a) How fast will he need to take off to make it to the other side?

(b) Discuss some of the physical impracticalities of this proposed stunt. (The more the better, but discuss at least two of the impracticalities you consider to be the most important.)

N10R.2 An essentially spherical boulder the size of several houses sits precariously on a slope above a village, as shown in figure N10.5. If this boulder ever gets loose, will it hit the village? (*Hint:* First estimate the speed with which the boulder will roll off the cliff. The total center-of-mass plus rotational kinetic energy of a rolling spherical object is $\frac{7}{10}mv^2$: see chapter C9.)

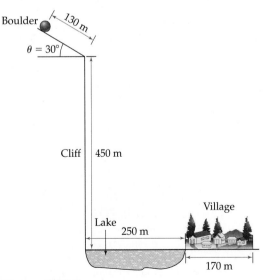

Figure N10.5
A village at risk? (See problem N10R.2.)

What is the optimal kick angle? (See problem N10S.14.)

Advanced

N10A.1 Prove that the ranges of two projectiles launched from the ground with the same initial speed v_0 at an angle of $45° \pm \phi$ are the same. [*Hints:* See problem N10S.11. Note also that $\sin(\pi/2 + \phi) = \cos\phi$.]

N10A.2 One can integrate equation N10.12 as follows.
(a) Using the fact that $a_z = dv_z/dt$, show that you can rearrange equation N10.12 to read

$$\frac{dv_z}{1 - v_z^2/v_T^2} = -g\,dt \qquad \text{(N10.16)}$$

(b) Take the integral of both sides of this equation, evaluating the integral on the left from $v_z(0)$ to $v_z(t)$ and the integral on the right from 0 to t.

(Feel free to take advantage of the labors of some brave and selfless mathematician by looking up the integral of the left side in a table of integrals. You should find that the result involves an inverse hyperbolic tangent of v_z/v_T.)

(c) Rearrange the result to show that

$$\frac{v_z(t)}{v_T} = \tanh\left(\frac{-gt}{v_T}\right) \qquad \text{(N10.17)}$$

This is the function graphed in figure N10.4.

(d) Integrate equation N10.17 to find an expression for the z-position of the object as a function of time. Again, feel free to look up the integral you need in a table of integrals. (You should get something involving the natural logarithm of a hyperbolic cosine.)

ANSWERS TO EXERCISES

N10X.1 You should find that F_D/mg is about 37% for the ping-pong ball and about 0.42% for the steel ball.

N10X.2 Since $\vec{g}$ points toward the center of the earth, the change in the angle of $\vec{g}$ will be simply equal to the angle relative to the center of the earth that the trajectory spans (see figure N10.6). If the trajectory has a horizontal range of $s = 1200$ km and the radius of the earth is $R = 6380$ km, the angle the trajectory spans is about $\theta = s/R = 0.19\,\text{rad} = 11°$.

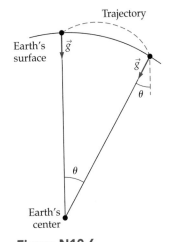

Figure N10.6
The change in the angle of $\vec{g}$ during a trajectory that is a significant fraction of the earth's radius.

N10X.3 It would take an object 1.0 s to fall 4.9 m from rest; this answer is independent of the horizontal velocity.

N10X.4 The quadratic formula implies that

$$\text{If}\quad ax^2 + bx + c = 0 \quad \text{then} \quad x = \frac{-b \pm \sqrt{b^2 - 4ac}}{2a} \qquad \text{(N10.18)}$$

In the case at hand, $a = -g/2$, $b = v_{0z}$, and $c = z_0$. Plugging these into the above yields equation N10.9.

N10X.5 If z_0 is negative and $2g|z_0| > v_{0z}^2$, then the quantity inside the square root is negative and the equation has no solution. Having z_0 being negative means that the projectile begins below $z = 0$. Conservation of energy means that the projectile in this case must have an initial kinetic energy of $\frac{1}{2}mv_{0z}^2 > mg|z_0|$ $\Rightarrow v_{0z}^2 = 2g|z_0|$ just to get up to $z = 0$. Therefore, if $v_{0z}^2 < 2g|z_0|$, the projectile *never* makes it up to $z = 0$, and thus there *should be* no real solutions to the equation.

N10X.6 The object reaches its terminal velocity when the drag acting on it is equal in magnitude to its weight, so

$$mg = \tfrac{1}{2}C\rho A v_T^2 \quad \Rightarrow \quad v_T^2 = \frac{2mg}{C\rho A} \qquad \text{(N10.19)}$$

Taking the square root of this yields equation N10.11.

N10X.7 By the equation above, $CA = 2mg/\rho v_T^2$. Estimating $m = 60$ kg, we get $CA = 0.27\,\text{m}^2$. If $C \approx 0.5$, then the cross-sectional area of the person is about $0.54\,\text{m}^2$, which is plausible (a person is about 1.5 m tall and about 0.4 m wide $= 0.60\,\text{m}^2$).

N10X.8 Since $g \approx 10$ m/s², it would take 6 s for an object that is freely falling to reach a speed of 60 m/s. If an object falling with drag took about twice as long to actually reach that terminal speed, then a skydiver would have to fall for about 12 s to reach the terminal speed. As a check, note that when $t = 12$ s,

$$\frac{t}{t_D} = \frac{gt}{v_T} = \frac{(10\,\text{m/s}^2)(12\,\text{s})}{60\,\text{m/s}} = 2 \qquad \text{(N10.20)}$$

which is the place on figure N10.4 where the object approximately reaches its terminal speed.

N11

Oscillatory Motion

Chapter Overview

Introduction

This is the second in the series of chapters exploring how we can use Newton's second law and knowledge of the forces acting on an object to predict its motion. In this chapter, we explore the important case of *oscillatory motion*, where an object moves in one dimension in response to a force that seeks to push it back toward an equilibrium point.

Section N11.1: A Mass on a Spring

Consider an object with mass m connected to a fixed point with an ideal spring, and imagine that the mass moves in one dimension along the x axis on a level, frictionless surface. If $x = 0$ is the object's position when the spring is relaxed, then we can show from the spring potential energy formula that

$$F_{\text{Sp},x} = -k_s x \qquad (N11.2)$$

    Purpose: This expression describes the x-force $F_{\text{Sp},x}$ that an ideal spring exerts on an object whose x-position is x.
    Symbols: k_s is the spring's **spring constant.**
    Limitations: The spring must be ideal, and the object must be in equilibrium at position $x = 0$. This equation also assumes that the object is moving in one dimension.
    Note: This equation is called **Hooke's law.**

Section N11.2: Solving the Equation of Motion

We call any object moving in one dimension whose *net* x-force has the form of equation N11.2 a **simple harmonic oscillator (SHO).** Newton's second law then implies that the object's position obeys the **(simple) harmonic oscillator equation:**

$$\frac{d^2x}{dt^2} = -\omega^2 x \qquad (N11.4)$$

    Purpose: This equation describes the consequences of Newton's second law for an object of mass m experiencing a *net* force given by Hooke's law.
    Symbols: x is the mass's x-position relative to its equilibrium position, t is time, and ω is a constant $= (k_s/m)^{1/2}$, where k_s is the constant appearing in Hooke's law.
    Limitations: It assumes that the object moves in *one dimension* and that the net force on the object is given by Hooke's law.

One can show by substitution that the following function satisfies the simple **differential equation** given by equation N11.4:

$$x(t) = A\cos(\omega t + \theta) \qquad\qquad \text{(N11.6)}$$

Purpose: This equation describes the most general solution to the harmonic oscillator equation.

Symbols: $x(t)$ is the oscillating object's x-position at time t, A is a constant with units of distance called the oscillation's **amplitude**, θ is a constant with units of angle called the oscillation **initial phase**, and ω is a constant with units of angle/time called the oscillation's **phase rate.**

Limitations: This equation assumes that the harmonic oscillator equation adequately describes the motion of the object.

Notes: Again, $\omega \equiv (k_s/m)^{1/2}$, where m is the object's mass and k_s is the spring constant appearing in Hooke's law. The values of A and θ are determined by the object's position and velocity at time $t = 0$.

The oscillation's **period** (the time required to complete 1 cycle) is $T = 2\pi/\omega$, and the oscillation's frequency (the number of oscillations per unit time) is $f = \omega/2\pi$. The SI unit for the latter quantity is the **hertz,** abbreviated as Hz, where 1 Hz = 1 cycle/s.

Section N11.3: The Oscillator as a Model

Any oscillating object behaves as a simple harmonic oscillator if:

1. It has an **equilibrium position** at which the net force on the object is zero.
2. It experiences a **restoring force** if it is displaced from equilibrium.
3. The magnitude of the restoring force is directly proportional to the displacement.

For sufficiently small displacements, almost any restoring force satisfies these conditions, so almost any oscillating system will behave as a SHO in that limit.

Section N11.4: A Mass Hanging from a Spring

Even though the net force on an object hanging from a spring includes gravity, if we orient our x axis vertically and shift the origin so that $x = 0$ is where the net force (including gravity) is zero, then the force law for displacements from this new origin becomes $F_{\text{net},x} = -k_s x$. Therefore, such an object is a simple harmonic oscillator.

Section N11.5: An Analogy to Circular Motion

Whether our oscillating object moves vertically or horizontally, its x-position is the same as the x coordinate of an object fixed to the edge of a steadily rotating circular disk whose radius R is equal to the oscillation amplitude A. We conventionally think of the disk as rotating counterclockwise, and we take counterclockwise angles to be positive. An oscillation's initial phase θ then corresponds to the angle of the disk at $t = 0$ relative to its orientation where x is maximum.

Mathematically, one can solve for A and θ from initial conditions by noting that $v_x(t) = dx/dt = -A\omega\sin(\omega t + \theta)$, setting $x(0)$ and $v_x(0)$ equal to the initial conditions, and solving these two equations for A and θ. The disk analogy often helps one to keep signs straight.

Section N11.6: The Simple Pendulum

A **simple pendulum** consists of an object (the pendulum **bob**) swinging from an inextensible massless string of length L. Newton's second law implies that the angle ϕ that the pendulum makes with the vertical obeys the differential equation that in the limit of small angles reduces to $d^2\phi/dt^2 = -(g/L)\phi$. This is the harmonic oscillator equation with ϕ replacing x and g/L replacing ω^2. Therefore, the solution to this equation is $\phi(t) = A\cos(\omega t + \theta)$ with $\omega = (g/L)^{1/2}$, and the pendulum's period is

$$T = \frac{2\pi}{\omega} = 2\pi\sqrt{\frac{L}{g}} \qquad \text{for small angles} \qquad \text{(N11.31)}$$

N11.1   A Mass on a Spring

Imagine an ideal, massless spring with one end connected to a fixed point (such as a wall) and the other connected to a movable object with mass m, as shown in figure N11.1a. Assume the object is free to slide in one dimension on a horizontal frictionless surface. In this case, the object's weight will be exactly canceled by the normal force due to its interaction with the surface, and the net force acting on the object will be the force supplied by the spring.

We conventionally define the x axis to coincide with the line along which the object moves. Let r be the magnitude of the object's position vector relative to the fixed point at a given time, and let r_0 be the same when the spring is relaxed. If we define our reference frame origin so that $x = r - r_0$, then $x = 0$ corresponds to the object's position when the spring is relaxed, and $|x|$ expresses the distance that the spring is either stretched or compressed.

According to equation C8.25, the force exerted by an ideal spring is

$$\vec{F}_{Sp} = -k_s(r - r_0)\hat{r} \tag{N11.1}$$

where k_s is the **spring constant** that characterizes the spring's stiffness and $\hat{r}$ is the direction in which r increases. In our coordinate system, $x = r - r_0$, and $\hat{r}$ is the $+x$ direction, so the x component of the force exerted on the object by the spring is given by the simple linear formula

$$F_{Sp,x} = -k_s x \tag{N11.2}$$

This equation is called **Hooke's law** after the British scientist (a contemporary of Newton) who first stated it. Note that the force acts in the negative x direction when x is positive and in the positive x direction when x is negative: in both cases this tends to push the object back toward $x = 0$. This is completely consistent with what we know qualitatively about the behavior of springs.

Conventional definition of reference frame (margin note)

The force law for a spring (Hooke's law) (margin note)

Exercise N11X.1

The units of the spring constant k_s were given in chapter C7 as joules per meter squared (J/m^2). These units were appropriate when we were using k_s to calculate potential energies. What would be the appropriate units for k_s in the context of equation N11.1? Show that your units are equivalent to joules per meter squared.

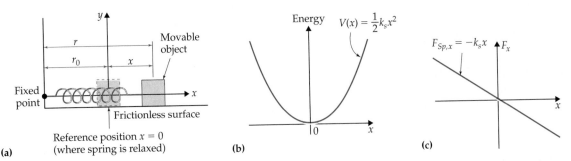

Figure N11.1

(a) The archetype of the simple harmonic oscillator: a movable object allowed to move in one dimension that is connected to a fixed point by an ideal, massless spring whose relaxed length is r_0. (b) A graph of the system's potential energy as a function of x. (c) A graph of the x-force on the movable object as a function of x.

Exercise N11X.2

If a spring whose relaxed length is 5.0 cm exerts a force of magnitude 3.0 N when stretched to 8.0 cm, what is its spring constant k_s?

N11.2   Solving the Equation of Motion

We call any object moving in one dimension whose net x-force has the form of equation N11.2 a **simple harmonic oscillator (SHO).** In *any* such situation, Newton's second law reads

$$F_{\text{Sp},x} = ma_x \quad \Rightarrow \quad a_x = \frac{F_{\text{Sp},x}}{m} = \frac{-k_s x}{m} = -\omega^2 x \qquad \text{(N11.3)}$$

where $\omega^2 \equiv k_s/m$. (The reason I have defined the *square* of ω to equal k_s/m will become clear shortly.) Since $a_x = dv_x/dt = d^2x/dt^2$, equation N11.3 becomes

$$\frac{d^2 x}{dt^2} = -\omega^2 x \qquad \text{(N11.4)}$$

    Purpose: This equation describes the consequences of Newton's second law for an object of mass m experiencing a *net* force given by Hooke's law.
    Symbols: x is the mass's x-position relative to its equilibrium position, t is time, and ω is a constant $= \sqrt{k_s/m}$, where k_s is the spring constant appearing in Hooke's law.
    Limitations: This equation assumes that the object moves in *one dimension* and that the net force on the object is given by Hooke's law.

The harmonic oscillator equation

This important equation is called the **(simple) harmonic oscillator equation** or **SHO equation.**

    How can we find the object's motion in this case? When we studied projectile motion in chapter N4, we simply integrated both sides of Newton's second law twice to find the projectile's position as a function of time (see section N4.4). We cannot do the same thing in this case, because in order to integrate both sides of equation N11.4 with respect to time, we have to know how x depends on time. Unfortunately, this is what we are trying to *find*. As I mentioned in section N10.5 on drag, this commonly occurs when we are trying to determine an object's motions from forces, and we ultimately have to find some trick to get around this problem.

    The harmonic oscillator equation is an example of a simple **differential equation.** A differential equation sets a function, $x(t)$ in this case, in an equation having terms that also involve derivatives of the function. Differential equations cannot generally be solved in any straightforward manner: often the only method is to guess what the solution is, plug the guess into the differential equation, and check to see whether the guessed solution satisfies the equation. If the guess works, you have solved the equation; if not, you try another guess. (A course in differential equations basically makes you a more intelligent guesser!)

Our trick here: guess the solution and see if it works

So the trick we will use here is to guess a possible solution and see if it works. What kind of an intelligent guess can we make in this case? We know that an object on the end of a spring will oscillate, so we expect $x(t)$ to be some function that repeats in time, something like $\sin(bt)$ or $\cos(bt)$, where b is some constant. The differential equation tells us that the second time derivative of our $x(t)$ function should be equal to a negative constant times that same function. Both $\sin(bt)$ and $\cos(bt)$ have that characteristic as well! Note that (see equation NA.24 in appendix NA on derivatives)

$$\frac{d}{dt}\sin(bt) = b\cos(bt) \qquad \frac{d^2}{dt^2}\sin(bt) = \frac{d}{dt}b\cos(bt) = -b^2\sin(bt)$$

$$\text{(N11.5a)}$$

$$\frac{d}{dt}\cos(bt) = -b\sin(bt) \qquad \frac{d^2}{dt^2}\cos(bt) = -\frac{d}{dt}b\sin(bt) = -b^2\cos(bt)$$

$$\text{(N11.5b)}$$

So if we set $x(t)$ equal to either one of these functions, we could satisfy the harmonic oscillator equation $d^2x/dt^2 = -\omega^2 x$ as long as we identify the constant b as being equal to $\omega = \sqrt{k_s/m}$.

The equation $x(t) = \sin(\omega t)$ actually can't be right, as $x(t)$ has units of meters while the sine function always produces a unitless number by definition (it is the ratio of two sides of a right triangle). So a more viable solution has to be something like $x(t) = A\sin(\omega t)$, where A is a constant with units of meters.

However, as we have already pointed out, the function $x(t) = A\cos(\omega t)$ will *also* be a solution to this equation. The most general solution to the harmonic oscillator equation is in fact

The most general solution to the simple harmonic oscillator equation

$$x(t) = A\cos(\omega t + \theta) \qquad\qquad \text{(N11.6)}$$

Purpose: This equation describes the most general solution to the harmonic oscillator equation.

Symbols: $x(t)$ is the oscillating object's x-position at time t; A is a constant with units of distance called the oscillation's **amplitude;** θ is a constant with units of angle called the oscillation **initial phase;** and ω is a constant with units of angle/time called the oscillation's **phase rate.**

Limitations: This equation assumes that the harmonic oscillator equation adequately describes the motion of the object.

Notes: Again, $\omega \equiv \sqrt{k_s/m}$, where m is the object's mass and k_s is the spring constant appearing in Hooke's law. The values of A and θ are determined by the object's position and velocity at time $t = 0$.

The object's motion is thus described by a cosine wave that cycles back and forth between the limits $x = +A$ and $x = -A$, as shown in figure N11.2. Note that the *total* distance the oscillating object travels from one extreme to the other is $2A$.

Notes about the solution

As shown in figure N11.2, this constant specifies how far along a given cycle the oscillation is at time $t = 0$ (measured from the first peak to the left of the origin). For example, if $\theta = 0$, the object is at $+A$ at $t = 0$; if $\theta = \pi/2$, then the object is at $x = 0$ at $t = 0$; and so on. Different values of θ essentially shift the oscillation back and forth along the ωt axis of figure N11.2.

Note that if $\theta = \pi/2$, then $\cos(\omega t + \theta)$ is equivalent to $-\sin(\omega t)$ (if you shift the peak to the left of the origin back a full quarter-cycle from the origin, you can see that it is like an upside-down sine function). Similarly, if

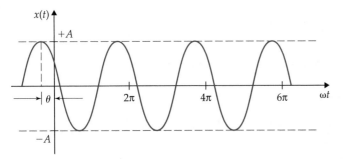

Figure N11.2
The solution to the harmonic oscillator equation is $x(t) = A\cos(\omega t + \theta)$. This function oscillates between the limits $\pm A$ and goes through one complete oscillation every time that ωt increases by 2π. The initial phase θ specifies how far the function is along its cycle at time $t = 0$ (a positive value of θ thus corresponds to shifting the wave crest left from zero by the magnitude of θ). The value of ω is determined by the values of k and m, while the values of A and θ are determined by initial conditions.

$\theta = \pi$, $\cos(\omega t + \theta) = -\cos\theta$; and if $\theta = 3\pi/2$, then $\cos(\omega t + \theta) = +\sin\theta$. The point is that equation N11.6 embraces *both* the sine and cosine solutions to the harmonic oscillator equation and everything in between. Since we can also always change the sign of the function by choosing the right initial phase θ, we conventionally choose θ so that the amplitude A is positive.

The constant ω specifies the rate at which the quantity $\omega t + \theta$ (which is sometimes called the *phase* of the oscillation) increases with time. Since the object goes through one complete cycle every time that $\omega t + \theta$ increases by 2π, the larger ω is, the more cycles the object will complete in a given time.

The **period** T of the oscillation is defined to be the time that it takes the object to go through one complete oscillation of the cosine function. This will happen when the value of ωt (and thus $\omega t + \theta$) increases by 2π rad, that is,

The *period* of an oscillation

$$\omega(t + T) = \omega t + 2\pi \qquad \text{(N11.7)}$$

Subtracting ωt from both sides, we see that

$$\omega T = 2\pi \quad \Rightarrow \quad T = \frac{2\pi}{\omega} = 2\pi\sqrt{\frac{m}{k_s}} \qquad \text{(N11.8)}$$

The **frequency** f of an oscillation is defined to be the number of *cycles* completed per unit time. Since exactly 1 cycle is completed in time T by definition, the frequency f is given by

The *frequency* of an oscillation

$$f = \frac{1\,\text{cycle}}{T} = \frac{\omega}{2\pi}\,\text{cycle} = \frac{\text{cycle}}{2\pi}\sqrt{\frac{k_s}{m}} \qquad \text{(N11.9)}$$

[The Greek letter ν (nu) is also commonly used for frequency, but since this letter is easy to confuse with the letter v used for velocity, I'll avoid it here.] The standard unit for frequency f in the SI system is the **hertz** (abbreviation: Hz), where $1\,\text{Hz} \equiv 1\,\text{cycle/s}$, whereas the standard SI units for phase rate ω are simply s^{-1} (which we can also think of as being radians per second).

Note that an oscillation's frequency f and/or period T is determined by the spring constant k_s and the object's mass m ($\omega = \sqrt{k_s/m}$); but the oscillation's amplitude A and initial phase θ are determined by the oscillator's initial state at time $t = 0$ (as we will see).

Exercise N11X.3

Show by direct substitution that $x(t) = A \cos(\omega t + \theta)$ does indeed solve the harmonic oscillator equation (equation N11.4).

Exercise N11X.4

Verify that $\sqrt{k_s/m}$ has units of s^{-1}.

Exercise N11X.5

An object with a mass of 2.0 kg is attached to a spring with a spring constant of 100 N/m. The object oscillates a distance of 10 cm from one extreme to another. What is the amplitude of the oscillation? What is its period? What is its phase rate?

N11.3   The Oscillator as a Model

The simple harmonic oscillator model has many applications

The simple harmonic oscillator model not only is useful for describing the behavior of objects connected to springs, but also is a good model for an astounding range of physical systems (from ocean waves to electrical oscillations to atomic vibrations), making it one of the most useful models in all physics.

We have already discussed part of the reason why this model turns out to be so useful in section C7.4. There, I argued that anytime the potential energy function for an interaction has a valley, we can approximate the bottom of that valley by the parabolic harmonic oscillator potential energy function. In this section, I want to present a somewhat different way of saying the same thing.

Characteristics of an oscillating system

The most important characteristics of *any* oscillating object are as follows: (1) It has a position or configuration (called its **equilibrium position**) where the force on it is zero; (2) if it is displaced from that position, it experiences a force (called a **restoring force**) that pushes it *back toward* the equilibrium position; and (3) this force (at least for small displacements) grows in magnitude as the object's displacement increases. Any object satisfying these criteria will oscillate about its equilibrium position if it is displaced and then released.

Clearly, an object attached to a spring has these characteristics. The special characteristic of a simple harmonic oscillator is that the magnitude of the restoring force is strictly proportional to the distance that the object is displaced. Many oscillating systems do *not* share this characteristic. For example, the force on an atom in a solid is a complicated function of its position as a result of the complicated electrostatic interactions between the atom and its neighbors.

For small displacements, $F_x = -k_s x$ in many cases

Even so, calculus tells us that we can approximate the curve of almost any physically reasonable (i.e., differentiable) function $F_x(x)$ that is positive for $x < 0$ and negative for $x > 0$ by a straight line $F_x = -k_s x$, where $k_s = -dF_x/dx$ evaluated at $x = 0$. This is a good approximation when x is sufficiently small in magnitude, as shown in figure N11.3.

This means an object responding to almost *any* restoring force $F_x(x)$ will find that for small displacements $F_x \approx -k_s x$. In this *small oscillation limit*, therefore, almost any oscillating object behaves as a simple harmonic oscillator. This is why the harmonic oscillator model is so useful and important.

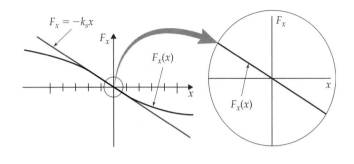

Figure N11.3
Almost *any* function $F_x(x)$ that is negative for $x > 0$ and positive for $x < 0$ can be approximated for small $|x|$ by a straight line $-k_s x$ for an appropriately chosen value of k_s.

Exercise N11X.6

Consider an object moving along the x axis under the influence of a force whose x component is $F_x = a[(b - x)^2 - b^2]$, where a and b are positive constants. Note that this force component is negative when $b > x > 0$ and positive when $x < 0$, so it qualifies as a restoring force. Argue that for small $|x|$, this formula becomes $F_x = -k_s x$, and find k_s in terms of a and/or b.

N11.4   A Mass Hanging from a Spring

The simplest way to construct a practical simple harmonic oscillator is to simply hang an object vertically at the end of a spring. This is much easier than constructing the system illustrated in figure N11.1 (frictionless surfaces are hard to come by). But is a hanging system really the same as that in figure N11.1?

The complication here is that the net force on the object is a sum of the vertical component of the spring force and the force of gravity. If we set up our coordinate system so that the x axis is along the direction of oscillation (vertical in this case) and set $x = 0$ to be the object's position when the spring is relaxed, then the net x-force on the object is

$$F_{\text{net},x} = -k_s x - mg \qquad \text{(N11.10)}$$

The net force in this case includes gravity

This is *not* the same as the harmonic oscillator force law (equation N11.2).

But it turns out that this force law really does work out to be the simple harmonic oscillator force law if we choose our reference frame cleverly. Let's choose a reference frame whose x axis is vertical (and let's choose the $+x$ direction to be upward), but let's define $x = 0$ not to be the position of the hanging object when the spring is relaxed, but rather to be its position at *equilibrium*, where the spring tension force exactly balances the object's weight. (The spring will be somewhat stretched at this position!) Let x_R be the position of the object when the spring is relaxed: the distance s by which the spring is stretched when the object is at any given position x is then

We can get rid of this term by redefining the origin

$$s = x_R - x \qquad \text{(assuming that } x < x_R \text{)} \qquad \text{(N11.11)}$$

(see figure N11.4). Note that since the $+x$ direction is upward, as x becomes more positive, the object is moving *up*, and the distance that the spring is stretched will *decrease:* this is why x is *subtracted* in equation N11.11.

Since the spring will exert an upward force proportional in magnitude to s, the net force acting on the object at position x in this reference frame is

$$F_{\text{net},x} = +k_s s - mg = +k_s(x_R - x) - mg = k_s x_R - k_s x - mg \qquad \text{(N11.12)}$$

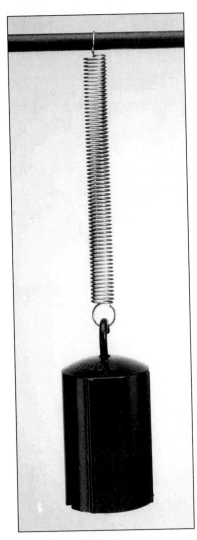

A practical simple harmonic
oscillator.

So a hanging mass behaves
as a harmonic oscillator

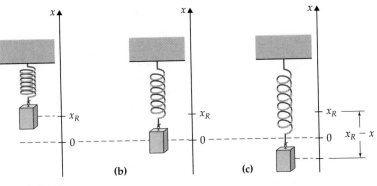

Figure N11.4
A system in which an object hangs from a spring. (a) The object located the
position x_R where the spring is relaxed. (b) The position of the object when
the spring tension exactly balances the object's weight: this position is
defined to be $x = 0$. (c) Here the spring is stretched farther: the distance by
which the spring is stretched when the object is at x is $x_R - x$.

Note that $k_s s$ is *positive* here because the force is upward (in the $+x$ direction)
when the spring is stretched by s. (As a check, note that this formula implies
that as the object gets lower, x decreases and the upward force exerted by the
spring increases, as we would expect.) At $x = 0$, the net force on the movable
object is supposed to be zero, so equation N11.12 tells us that at $x = 0$,

$$0 = F_{net,x} = k_s(x_R - 0) - mg \quad \Rightarrow \quad k_s x_R = mg \qquad \text{(N11.13)}$$

If we substitute this last result into equation N11.12, the two constant terms
cancel, so

$$F_{net,x} = -k_s x \qquad \text{(N11.14)}$$

which is the right force law for a simple harmonic oscillator. Choosing the
right origin thus enables us to cancel the gravitational force with a constant
term in the spring tension force: the resulting net force is directly proportional to x.

The point of all this is that an object hanging from a spring, even though
it is not physically the same thing as a mass oscillating horizontally back and
forth, will *behave* in exactly the same manner. The gravitational force here
simply has the effect of displacing the equilibrium point ($x = 0$ in the reference frame we've been using) downward from the position where the spring
is relaxed. The hanging oscillator even oscillates at the same frequency that it
would if it were horizontal!

Example N11.1: Boys on the Hood

Problem   You see a couple of neighborhood boys bouncing on the hood of
your car. When one jumps on the hood, you see the car's front end oscillate
with a period of about 1.5 s. After you yell at the boys, you get to wondering
about the spring constant k_s of the car's suspension. Estimate k_s from what
you saw.

Model   The weight of a typical car $\approx$ 3000 lb, corresponding to a mass of
roughly 1500 kg. Let's say that the front suspension effectively suspends
about one-half of this mass, or $\approx$ 800 kg when the $\approx$ 50-kg mass of the boy is
included.

Solution   According to equation N11.8, the period of oscillation is $T = 2\pi\sqrt{m/k_s}$, so

$$k_s = \frac{4\pi^2 m}{T^2} = \frac{4\pi^2(800\,\text{kg})}{(1.5\,\text{s})^2}\left(\frac{1\,\text{N}}{1\,\text{kg}\cdot\text{m/s}^2}\right) = 14{,}000\,\text{N/m} \quad \text{(N11.15)}$$

Evaluation   Here is a way to check this estimate. If you were to sit on the hood (assume that your weight $\approx 700\,\text{N} \approx 155\,\text{lb}$), the hood should sink until the compressed springs exert an upward force equal to your weight, that is, by about $|\Delta x| = \Delta F / k_s = (700\,\text{N})/(14{,}000\,\text{N/m}) \approx 0.05\,\text{m} = 5\,\text{cm}$. This seems about right.

N11.5   An Analogy to Circular Motion

According to the discussion in section N11.4, equation N11.6 should accurately describe the oscillatory motion of an object hanging from a spring. Note that the motion described by this equation is mathematically equivalent to the x coordinate of an object attached to a rotating disk (see figure N11.5). This provides a very useful way to visualize the meaning of the oscillation's amplitude A and initial phase θ: the amplitude corresponds to the radius of the circular motion, and θ to the disk's angle at time $t = 0$ relative to its position when the object's x-position is maximum. If we imagine the disk to be rotating counterclockwise, a negative phase angle θ means the disk is oriented clockwise of this maximal position at $t = 0$, while a positive phase angle means that it is oriented counterclockwise of this position.

An oscillating object's x component behaves like that of an object on a rotating disk

So, for example, say that we know that an object hanging from a spring is moving upward through the equilibrium point $x = 0$ at time $t = 0$. If we look at figure N11.5, we can see that this will only happen if the corresponding object on the disk is 90° clockwise of its maximum upward position, because only then will that object be moving *upward* through $x = 0$. Therefore, the initial phase θ for the oscillation must be 90° clockwise or $-\pi/2$. Similarly, we can immediately see that if the object is passing through its lowest point at $t = 0$, the initial phase must be ±180° or $\theta = \pm\pi$ (either sign is acceptable in this case).

Finding A and θ from initial conditions with the help of this analogy

Examples N11.2 and N11.3 illustrate how one can calculate both the oscillation's amplitude A and its initial phase θ from initial conditions. Even when the initial phase is *not* a multiple of $\pi/2$ (see example N11.3), figure N11.5 can help us determine the sign of the initial phase and check its value.

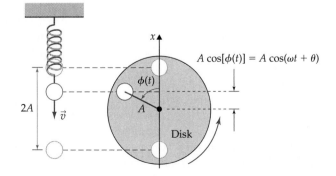

Figure N11.5
The vertical motion of a hanging oscillator is mathematically identical to the projection on the x axis of the motion of an object in uniform circular motion with angular speed ω. Such a diagram helps us visualize the meaning of the initial phase θ.

Example N11.2

Problem An object of mass 0.52 kg hangs from a spring with spring constant $k_s = 130\,\text{N/m}$. The object is measured to pass through its equilibrium point with a speed of 1.0 m/s. Define the time when it passes the equilibrium point going downward to be $t = 0$. (a) What is initial phase θ? (b) What is the amplitude of its oscillation?

(a) *Solution* If we study figure N11.5, we can see that the object will pass downward through the origin only when the corresponding object on the disk is 90° counterclockwise of its maximum upward position. Since this is the situation at time $t = 0$, we must have $\theta = +\pi/2$.

(b) *Model* We can calculate the object's velocity at any time t by calculating the time derivative of equation N11.6:

$$v_x(t) = \frac{dx}{dt} = \frac{d}{dt}A\cos(\omega t + \theta) = -A\omega\sin(\omega t + \theta) \qquad \text{(N11.16)}$$

At time $t = 0$, this becomes

$$v_x(0) = -A\omega\sin\theta \quad \Rightarrow \quad A = \frac{-v_x(0)}{\omega\sin\theta} = \frac{-v_x(0)}{\sin\theta}\sqrt{\frac{m}{k_s}} \qquad \text{(N11.17)}$$

where in the last step, I used $\omega = \sqrt{k_s/m}$. Since we know that $\theta = \pi/2$ from part (a), and all the other quantities are given, we can calculate A.

Solution

$$A = \frac{-(-1.0\text{ m/s})}{\sin\frac{1}{2}\pi}\sqrt{\frac{0.52\text{ kg}}{130\text{ N/m}}\left(\frac{1\text{ N}}{1\text{ kg·m/s}^2}\right)} = 0.062\text{ m} \qquad \text{(N11.18)}$$

Evaluation This is positive (as it should be) and has the right units.

Example N11.3

Problem Imagine that an object hanging from a spring oscillates with a period of $T = 2.0$ s. At $t = 0$, it is 5 cm above the equilibrium point and is moving upward at 10 cm/s. What are A and θ for this oscillation?

Model The equation $x(t) = A\cos(\omega t + \theta)$ gives us the object's position as a function of time, while equation N11.16 gives us its velocity. At time $t = 0$, these equations reduce to

$$x(0) = A\cos\theta \qquad \text{and} \qquad v_x(0) = -A\omega\sin\theta \qquad \text{(N11.19)}$$

We are given T, so equation N11.8 allows us to calculate ω: $\omega = 2\pi/T$. Since we are also given $x(0)$ and $v_x(0)$, we have two equations in our two unknowns A and θ, so we can solve.

Solution We can eliminate the unknown A by dividing the second of equations N11.19 by the first:

$$-\frac{v_x(0)}{\omega x(0)} = \frac{A\sin\theta}{A\cos\theta} = \tan\theta \quad \Rightarrow \quad \theta = \tan^{-1}\left[\frac{-v_x(0)}{\omega x(0)}\right] \qquad \text{(N11.20)}$$

Plugging in the numbers, we get

$$\tan\theta = \frac{-v_x(0)}{\omega x(0)} = \frac{T}{2\pi}\left[\frac{-v_x(0)}{x(0)}\right] = \frac{-(2.0\,\cancel{s})(10\,\cancel{cm}/\cancel{s})}{2\pi(5\,\cancel{cm})} = -\frac{2}{\pi} \quad (\text{N11.21})$$

So $\theta = \tan^{-1}(-2/\pi) = -32°$ or $+148°$ [since $\tan(\theta + \pi) = \tan\theta$, either one of these results would satisfy the equation]. However, we can see from figure N11.5 that if the object is above the equilibrium point and moving upward at $t = 0$, only the first solution makes any sense. We can plug this value for θ back into either of equations N11.19 to solve for A:

$$A = \frac{x(0)}{\cos\theta} = \frac{5.0\text{ cm}}{\cos(-32°)} = +5.9\text{ cm} \quad (\text{N11.22})$$

Evaluation The amplitude is positive, as it should be, has the right units, and is plausibly comparable to the value of $x(0)$.

Note that equation N11.20 in combination with figure N11.5 provides a general way to compute θ from initial conditions when $x(0) \neq 0$.

Yet another way to find the amplitude from initial conditions is to use conservation of energy. Whenever the cosine in $x(t) = A\cos(\omega t + \theta)$ is equal to ± 1 (which will be at either extreme end of the oscillation), then the object's velocity is equal to zero, since $v_x(t) = -\omega A\sin(\omega t + \theta)$ and $\sin(\omega t + \theta) = 0$ whenever $\cos(\omega t + \theta) = 1$ [we also know that an object will change direction at an extreme point, so $v_x(t)$ must be passing through zero at that time]. Therefore, the system's total energy at an extreme point must be

Using conservation of energy to compute A

$$E = K + V(x) = 0 + \tfrac{1}{2}k_s x^2 = \tfrac{1}{2}k_s A^2 \quad (\text{N11.23})$$

Since energy is conserved, this must be the system's energy at time $t = 0$, too, so

$$\tfrac{1}{2}k_s A^2 = E = \tfrac{1}{2}m[v_x(0)]^2 + \tfrac{1}{2}k_s[x(0)]^2 \quad (\text{N11.24})$$

So, if you know $v_x(0)$ and $x(0)$, you can find A in this way.

Exercise N11X.7

A 1.0-kg object hangs from a spring whose spring constant is 100 N/m. You take the mass, pull it down 10 cm, and then give it an initial downward speed of 0.50 m/s. What are the values of A and θ here?

Exercise N11X.8

A 68-kg friend of yours goes bungee jumping, and you watch. You notice that after the jump, the friend oscillates once up and down in about 6.0 s. Estimate the effective spring constant of the bungee cord.

N11.6   The Simple Pendulum

A *simple pendulum* is an example of a system that has nothing to do with a mass connected to a spring, and yet (for small oscillations) it obeys the same mathematics as a simple harmonic oscillator does. A **simple pendulum**

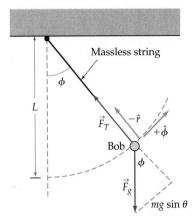

Figure N11.6
The simple pendulum. Note that ϕ is conventionally taken to increase in the counterclockwise direction.

Reexpressing the tangential acceleration term

consists of an object (called a **bob**) that swings back and forth at the end of a massless inextensible string of length L tied to a fixed point (see figure N11.6).

The bob is confined to a circular path by the string, so this is an example of *nonuniform circular motion*. According to the results of section N8.3, the bob's acceleration at any instant in this situation is

$$\vec{a}(t) = \frac{dv}{dt}\hat{v} - \frac{v^2}{L}\hat{r} \qquad (N11.25)$$

where v is the bob's speed, $\hat{v}$ is a unit vector in the direction of its velocity, and $\hat{r}$ points directly away from the center of the object's circular motion.

Since the bob's velocity changes direction at the extreme points, it is awkward to use $\hat{v}$ as a directional in this case. Instead, let's define the directional $\hat{\phi}$ as the direction in which the bob is moving when it moves counterclockwise (the direction in which ϕ increases). Note that $\hat{\phi} = \pm\hat{v}$, where the plus sign applies when the bob moves counterclockwise and the minus sign when it moves clockwise. Now, if the bob's angle changes by $d\phi$ during a short interval dt, the bob travels an arclength $L|d\phi|$ during dt, so its speed is $v = L|d\phi/dt|$, implying that

$$\frac{dv}{dt} = L\frac{d}{dt}\left|\frac{d\phi}{dt}\right| = +L\frac{d}{dt}\left(\pm\frac{d\phi}{dt}\right) = \pm L\frac{d^2\phi}{dt^2} \qquad (N11.26)$$

where the plus sign applies if $d\phi/dt$ is positive, which is again when the bob is moving counterclockwise. Combining this with $\hat{\phi} = \pm\hat{v}$, we see that in *both* cases, equation N11.25 becomes

$$\vec{a}(t) = L\frac{d^2\phi}{dt^2}\hat{\phi} - \frac{v^2}{L}\hat{r} \qquad (N11.27)$$

Now, Newton's second law tells us that $\vec{a} = \vec{F}_{net}/m$, where $\vec{F}_{net}$ is the net force on the bob and m is the bob's mass. The bob is acted on by two forces, a gravitational force due to the earth and the tension force exerted by the string. The latter always acts in the $-\hat{r}$ direction, so only the gravitational force ever has a component in the $\hat{\phi}$ direction. When the string makes an angle ϕ with the vertical, the component of $\vec{F}_g$ in that direction is $F_{g,\hat{\phi}} = -mg\sin\phi$, so

The pendulum equation

$$L\frac{d^2\phi}{dt^2} = \frac{F_{g,\phi}}{m} = -g\sin\phi \qquad \Rightarrow \qquad \frac{d^2\phi}{dt^2} = -\frac{g}{L}\sin\phi \qquad (N11.28)$$

Now for small ϕ in radians (less than a few tenths of a radian), $\sin\phi \approx \phi$. (Try this on a calculator and see for yourself!) Using this, we get

$$\frac{d^2\phi}{dt^2} = -\frac{g}{L}\phi \qquad \text{for small angles} \qquad (N11.29)$$

The low-angle limit is equal to the oscillator equation

Now, compare this equation to equation N11.4: you should see that this equation is mathematically identical to that equation, with ϕ here playing the role of x there and $\sqrt{g/L}$ here playing the role of ω there. We don't need to solve this differential equation a second time: all that we need to do is to use the answers that we found before, substituting ϕ for x and $\sqrt{g/L}$ for ω. So the angle of the pendulum bob depends on time as follows (by analogy to equation N11.6):

So the solution must be analogous as well

$$\phi(t) \approx A\cos(\omega t + \theta) \qquad \text{for small angles} \qquad (N11.30)$$

where $\omega = \sqrt{g/L}$ here, A expresses the absolute value of the angle at the extreme points, and θ is the initial phase. Its period is

$$T = 2\pi/\omega = 2\pi\sqrt{L/g} \qquad \text{(N11.31)}$$

Exercise N11X.9

Imagine that a pendulum hangs vertically at rest. We pull the bob aside to an initial angle of $+15°$ and release the bob from rest. It swings with a period of exactly 1.0 s. What are the values of A, θ and L here?

TWO-MINUTE PROBLEMS

N11T.1 If you double the amplitude of a harmonic oscillator, the oscillator's period
   A. Decreases by a factor of 2.
   B. Decreases by a factor of $\sqrt{2}$.
   C. Does not change.
   D. Increases by a factor of $\sqrt{2}$.
   E. Increases by a factor of 2.
   F. Changes by some other factor (specify).

N11T.2 If you double the amplitude of a harmonic oscillator, the object's maximum speed does what? (Use the answers for problem N11T.1.)

N11T.3 If you double the spring constant of a harmonic oscillator, the oscillation frequency does what? (Use the answers for problem N11T.1.)

N11T.4 A glider on an air track is connected by a spring to the end of the air track. If it takes 0.30 s for the glider to travel the distance of 12 cm from one turning point to the other, its amplitude is
   A. 12 cm
   B. 6 cm
   C. 24 cm
   D. 36 cm
   E. 3.6 cm
   F. We are not given enough information to answer.

N11T.5 Consider the glider described in problem N11T.4. Its phase rate is
   A. $0.30\,\text{s}^{-1}$
   B. $0.15\,\text{s}^{-1}$
   C. $0.60\,\text{s}^{-1}$
   D. $3.77\,\text{s}^{-1}$
   E. $0.096\,\text{s}^{-1}$
   F. Some other result (specify)

N11T.6 A glider on an air track is connected by a spring to the end of the air track. If it is pulled 3.5 cm in the $+x$ direction away from its equilibrium point and then released from rest at $t = 0$, what is the initial phase θ?

   A. 0
   B. $\pi/4$
   C. $\pi/2$
   D. π
   E. $3\pi/2$
   F. Some other result (specify)

N11T.7 A glider on an air track is connected by a spring to the end of the air track. If it is pulled 3.5 cm in the $-x$ direction away from its equilibrium point and then released from rest at $t = 0$, what is the initial phase θ?
   A. 0
   B. $\pi/4$
   C. $\pi/2$
   D. π
   E. $3\pi/2$
   F. Some other result (specify)

N11T.8 A mass hanging from the end of a spring has a phase rate of $\omega = 6.3\,\text{s}^{-1}$ (≈ 1 cycle/s). Let's define $t = 0$ to be when the mass passes $x = 0$ going up. If its speed as it passes is 1.0 m/s, what is its amplitude A?
   A. 0
   B. 0.16 m
   C. 1.0 m
   D. 6.3 m
   E. We are not given enough information to answer.
   F. Some other result (specify).

N11T.9 To double the period of a pendulum, you need to multiply its length by a factor of
   A. $\frac{1}{2}$
   B. 2
   C. $\sqrt{\frac{1}{2}}$
   D. $\sqrt{2}$
   E. 4
   F. Some other result (specify)

HOMEWORK PROBLEMS

Basic Skills

N11B.1 An oscillating object repeats its motion every 3.3 s.
(a) What is the period of this oscillation?
(b) What is its frequency?
(c) What is its phase rate?

N11B.2 An object of mass 0.30 kg hanging from a spring is observed to have an oscillation frequency of 2.2 Hz. What is the spring's spring constant k_s?

N11B.3 An object of mass 0.36 kg hanging at the end of a spring oscillates with an amplitude of 4.8 cm and a frequency of 1.2 Hz. What is the spring's spring constant k_s?

N11B.4 A magnesium atom (mass ≈ 24 proton masses) in a crystal is measured to oscillate with a frequency of roughly 10^{13} Hz. What is the effective spring constant of the forces holding that atom in the crystal?

N11B.5 An object of mass 0.30 kg hanging from a spring is pulled 2.5 cm below its equilibrium position and then is released from rest. What is θ for this oscillation in equation N11.6? Explain your reasoning.

N11B.6 A pendulum is observed to swing with a period of 2.0 s. How long is it?

N11B.7 What will be the natural oscillation period of a 30-kg child on a swing whose seat is 3.2 m below the bar where the chains from the seat are attached?

What is this pendulum's natural frequency of oscillation? (See problem N11B.7.)

Synthetic

N11S.1 An object of mass 0.25 kg extends a spring by a distance of 5.0 cm when it hangs from the spring at rest. If it is then set in vertical motion, what will be its period of oscillation?

N11S.2 An object of mass 0.30 kg hanging from a spring is lifted 2.5 cm above its equilibrium position and is dropped from rest. What is the amplitude of the subsequent oscillation? Explain your reasoning.

N11S.3 Any real spring has mass. Do you think that this mass would make the actual period of a real harmonic oscillator longer or shorter than the period predicted by equation N11.10? Explain your reasoning.

N11S.4 An object of mass 1.0 kg hanging from the end of a spring (whose spring constant is 120 N/m) is observed at $t = 0$ to pass downward through position $x = 2.5$ cm traveling at a speed of 0.60 m/s.
(a) What is the amplitude of oscillation?
(b) What is the initial phase θ for this oscillation? Please explain your reasoning.

N11S.5 An object of mass 0.60 kg hangs from the end of a spring. Imagine that the object is lifted upward and held at rest at the position where the spring is not stretched. The object is then released. It is observed that the lowest point in the object's subsequent oscillation is 12 cm below the point where it was released.
(a) What is the amplitude of the oscillation?
(b) What is the spring constant of the spring?
(c) What is the object's maximum speed? Please explain your reasoning and show your work for each step.

N11S.6 When a 55-kg friend of yours sits on a trampoline, your friend sinks about 45 cm below the trampoline's normal level surface. If your friend were to bounce gently on the trampoline (never leaving its surface), what would be your friend's period of oscillation? (*Hint:* Model the trampoline as if it were a harmonic oscillator. Do you think that this will be a good model?)

What is this person's period of small oscillations? (See problem N11S.6.)

N11S.7 If you stand on a pogo stick, you note that its spring-loaded foot is pushed in about 18 cm. If you bounce *gently* up and down on it (so that the foot never leaves the ground), what is your approximate oscillation frequency?

A pogo stick (see problem N11S.7).

N11S.8 Do you think that a pendulum swinging through a large angle will have a longer or shorter period than the period predicted by equation N11.31? Explain your reasoning carefully. (Possibly helpful hints: consider extreme cases, or compare $\sin\theta$ to θ.)

N11S.9 The net x-force on a 1.0-kg oscillating object moving along the x axis is $F_x = -b(x - a^2 x^3)$, where $b = 120$ and $a = 5$ in appropriate SI units.
(a) What are the units of a and b?
(b) Show that for small oscillations, this force law reduces to the harmonic oscillator law $F_x \approx -k_s x$, where k_s depends on the value of a and/or b.
(c) What is the frequency of such small oscillations?
(d) How small is "small"? For what range of x will $F_x = -k_s x$ to within 1%?

N11S.10 Imagine that the x-force on a 1.0-kg oscillating object in a certain situation is given by $F_x = -a\sin(bx)$, where $a = 100$ and $b = 10$ in appropriate SI units.
(a) What are the SI units of a and b?
(b) Show that for small x, this force law reduces to $F_x \approx -k_s x$, where k_s can be calculated from a and b.
(c) Find the frequency of small oscillations.

N11S.11 We can actually determine the motion of a harmonic oscillator from *conservation of energy* instead of solving the harmonic oscillator equation. Here's how.
(a) Show that a bit of manipulation of the conservation of energy formula for the harmonic oscillator yields

$$A^2 = x^2(t) + \left[\frac{v_x(t)}{\omega}\right]^2 \qquad \text{(N11.32)}$$

where $A = 2E/k_s$ and $\omega = \sqrt{k_s/m}$ as usual.
(b) Argue that this means that at any instant of time, we can find an angle ψ such that

$$x(t) = A\cos\psi \qquad \text{and} \qquad v_x(t) = -\omega A\sin\psi \qquad \text{(N11.33)}$$

Thus these expressions give x and v_x for *all* times, with ψ depending in some unknown way on time. [*Hint:* $|x|$ will always be less than A, so we can *define* $x(t) = A\cos\psi$. Then solve equation N11.32 for the other term.]
(c) Use the definition $v_x = dx/dt$ and the chain rule to show that the most general possible expression for ψ is

$$\psi = \omega t + \theta \qquad \text{(N11.34)}$$

where θ is some constant. Therefore

$$x(t) = A\cos\psi = A\cos(\omega t + \theta) \quad !! \quad \text{(N11.35)}$$

Rich-Context

N11R.1 How do you measure the mass of an astronaut in orbit? (You can't just use a scale!) For the Skylab program, NASA engineers designed a Body Mass Measuring Device (BMMD). This is essentially a chair of mass m mounted on a spring with a carefully measured spring constant $k_s = 605.6\,\text{N/m}$. (The other end of the spring is connected to the Skylab itself, which has a mass much larger than the astronaut or the chair, and so remains essentially fixed as the astronaut oscillates.) The period of oscillation of the empty chair is measured to be 0.90149 s. When an astronaut is sitting in the chair, the period is 2.12151 s. What is the mass of the astronaut? Please describe your reasoning! (Adapted from Halliday and Resnick, *Fundamentals of Physics*, 3/e, New York: Wiley, 1988, p. 324.)

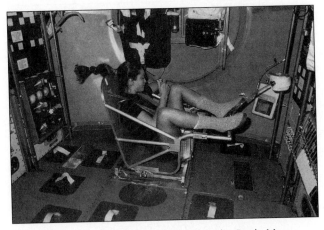

Shuttle astronaut Tamara Jernigan uses the Body Mass Measuring Device (BMMD). (See problem N11R.1.)

N11R.2   Known mobster and gambler Larry the Loser is found dead, hanging at about the 7-story level from a bungee cord tied to the top of the 15-story (45-m) Prudential Building in downtown Chicago. Larry has a mass of 90 kg, and when the bungee cord is cut down, it is found to have a relaxed length of 5 stories. Detective Lestrade from the Chicago Police Department thinks that Larry must have been bungee-jumping from the top of the building and just unluckily hit the ground, but *you* know that Larry was *murdered*. What is your evidence?

(*Note:* A bungee cord will not exert any force on the jumper until the jumper has fallen a distance equal to the bungee cord's relaxed length. Also, the force a bungee cord exerts actually exceeds $k_s|x|$ in magnitude when it is stretched a lot.)

Advanced

N11A.1   Imagine that the force law expressing the interaction between a hydrogen atom and a chlorine atom in an HCl molecule is approximately

$$F_x = -a\left[\left(\frac{b}{r}\right)^2 - \left(\frac{c}{r}\right)^3\right] \tag{N11.36}$$

where F_x is the x-force on the hydrogen atom, a is a constant having units of force and b and c are constants having units of distance, and r is the distance measured from the chlorine atom (which is so massive compared to the hydrogen atom that we will consider it fixed).

(a) What is the equilibrium position r_0 for the hydrogen atom (express your answer in terms of b and c)?

(b) Define $x = r - r_0$, and show that for small x, this force law becomes approximately $F_x = -k_s x$. (One approach is to find the first few terms of a Taylor series expansion of the exact force law. Alternatively, one can use the approximation $(1+q)^n \approx 1 + nq$ if $q \ll 1$.)

(c) What is the frequency of small oscillations of the hydrogen atom in terms of its mass m, and a, b, and c?

ANSWERS TO EXERCISES

N11X.1   The appropriate units for k_s in equation N11.3 are newtons per meter. Since $1\,\text{J} = 1\,\text{N·m}$, this is the same as joules per meter squared.

N11X.2 $k_s = F_{Sp}/(r - r_0) = 3.0\,\text{N}/(0.08\,\text{m} - 0.05\,\text{m}) = 100\,\text{N/m}$.

N11X.3   If we define $u = \omega t + \theta$, then $x(t) = A\cos u$, and

$$\frac{d}{dt}A\cos u = -A\sin u\,\frac{du}{dt} = -\omega A\sin u \tag{N11.37}$$

$$\text{since } \frac{du}{dt} = \frac{d}{dt}(\omega t + \theta) = \omega + 0 = \omega \tag{N11.38}$$

Taking the time derivative again, we get

$$\frac{d^2x}{dt^2} = -\omega A\frac{d}{dt}\sin u = -\omega^2 A\cos u = -\omega^2 x \tag{N11.39}$$

So this solution does indeed solve the SHO equation.

N11X.4   The quantity k_s/m has the units

$$\text{units}\left(\frac{k_s}{m}\right) = \frac{\text{N/m}}{\text{kg}}\left(\frac{1\,\text{kg·m/s}^2}{1\,\text{N}}\right) = \frac{1}{\text{s}^2} \tag{N11.40}$$

So $\sqrt{k_s/m}$ indeed does have units of s^{-1}.

N11X.5   The amplitude is one-half the total distance between extremes, or 5 cm. The period is $T = 2\pi\sqrt{m/k_s} = 0.90\,\text{s}$. Its phase rate is $\sqrt{k_s/m} = 7.1\,\text{s}^{-1}$.

N11X.6   Multiplying out the square, we find that

$$F_x = a(x^2 - 2bx + b^2 - b^2) = ax^2 - 2abx \tag{N11.41}$$

For $x \ll b$, the first term will be very small compared to the second, and we can neglect it, leaving

$$F_x \approx -2abx = -k_s x \qquad \text{where } k_s = 2ab \tag{N11.42}$$

N11X.7 Using equation N11.20, we get

$$\theta = \tan^{-1}\left[\frac{-v_x(0)}{\omega x(0)}\right] = \tan^{-1}\left[\frac{-(-0.5\,\text{m/s})}{(10\,\text{s}^{-1})(-0.1\,\text{m})}\right]$$
$$= \tan^{-1}(-0.5) = -0.46 \qquad \text{(N11.43)}$$

As a fraction of an oscillation, this result corresponds to shifting the cosine wave by $0.46/2\pi = 0.073 = 7.3\%$ of a complete oscillation. But the result for the phase shift θ could *also* be $\pi - 0.46 = 2.68$ (43% of an oscillation), since $\tan\theta = \tan(\theta + \pi)$. It is in fact the latter value that gives the correct positive value for the amplitude:

$$A = \frac{x(0)}{\cos\theta} = \frac{-0.1\,\text{m}}{\cos(2.68)} = +0.11\,\text{m} \qquad \text{(N11.44)}$$

N11X.8 Since $T = 2\pi\sqrt{m/k_s}$, we have

$$k_s = \frac{4\pi^2 m}{T^2} = \frac{4\pi^2(68\,\text{kg})}{(6.0\,\text{s})^2}\left(\frac{1\,\text{N}}{1\,\text{kg·m/s}^2}\right) = 75\,\text{N/m} \qquad \text{(N11.45)}$$

N11X.9 Since $0 = v_x(0) = L\,d\phi/dt = -LA\sin(0+\theta) = -LA\sin\theta$, θ must be either 0 or π. Only the first possibility gives $\cos\theta > 0$, which we need since $\phi(0) = A\cos\theta = +15°$, so we must have $\theta = 0$ and $A = +15°/\cos(0) = +15° = \pi/6$ rad. Since for the pendulum $\omega = \sqrt{g/L}$, we have

$$T = \frac{2\pi}{\omega} = 2\pi\sqrt{\frac{L}{g}} \quad \Rightarrow \quad L = \frac{T^2 g}{4\pi^2} \qquad \text{(N11.46)}$$

Plugging in the numbers, we get $L = 0.25\,\text{m}$.

N12 Introduction to Orbits

Chapter Overview

Introduction
In the previous chapters of this unit, we applied Newton's second law to essentially terrestrial situations. Here we will finally turn to the problem of celestial physics. In this chapter and chapter N13, we will explore how to use the gravitational force law discussed in chapter C8 to predict the motion of the planets and other celestial objects. We will see that Newton's laws work just as well for celestial physics as terrestrial physics, thus providing a *universal* model for (macroscopic) mechanics.

Section N12.1: Kepler's Laws
Johannes Kepler used Tycho Brahe's collection of extremely accurate naked-eye observations of the planets to develop three empirical laws of planetary motion, which we call **Kepler's laws of planetary motion:**

1. All planets move in ellipses, with the sun at one focus.
2. A line from the sun to the planet sweeps out equal areas in equal times.
3. If a is the planet's **semimajor axis** (one-half the ellipse's greatest width), then $T^2 \propto a^3$ where T is the planet's **period.**

    Newton's triumph was to offer an *explanation* of these empirical laws, using a theory that also applied to terrestrial physics.

Section N12.2: Orbits Around a Massive Primary
Consider an isolated (or freely falling) system of two interacting objects with masses M and m. A nonrotating reference frame connected to the system's center of mass (CM) will be inertial in either case (as long as we ignore external gravitational interactions in the freely falling case). If in addition we have $M \gg m$, we call the massive object the **primary** and the other its **satellite,** and the following simplifications apply in the CM frame:

1. The primary is essentially at rest at the origin.
2. The primary's kinetic energy is negligible.
3. The objects' angular momenta are parallel and are separately conserved.
4. The primary's angular momentum is negligible.
5. The objects' separation is essentially equal to the satellite's distance from the origin.

The approximation $M \gg m$ works well for almost all pairs of objects in the solar system.

Section N12.3: Kepler's Second Law
Conservation of angular momentum implies that the orbit of either object around the system's center of mass must lie in a *plane* perpendicular to the fixed angular momentum vector. The same principle also implies Kepler's second law. This is so because as an object moves through an angle $d\theta$ in a tiny time dt, its radius vector from the system's CM sweeps out an area $dA \approx \frac{1}{2}r^2 d\theta$. Since the magnitude of the object's angular momentum is $L = mr^2 d\theta/dt$, it follows that $dA/dt = L/2m =$ a constant. **Kepler's**

second law actually states that $dA/dt = $ constant applies to the line connecting the planet to the *sun* (not the system's CM), but since the sun's mass M is much larger than any planet's mass m, the sun's position is essentially equal to the system's CM.

Section N12.4: Circular Orbits and Kepler's Third Law

From now on, we will assume that $M \gg m$ unless otherwise specified. In this approximation, the primary basically provides a fixed gravitational field to which the satellite responds. **Newton's law of universal gravitation** states that the magnitude of the gravitational force that the primary exerts on the satellite is

$$F_g = \frac{GMm}{r^2}$$ (N12.10)

> **Purpose:** This equation specifies the magnitude of the gravitational force $\vec{F}_g$ that an object of mass M exerts on an object of mass m (or vice versa) when their centers of mass are separated by a distance r.
> **Symbols:** $G = 6.67 \times 10^{-11}$ N·m^2/kg^2 is the **universal gravitational constant.**
> **Limitations:** This equation applies to point masses or spheres, but not to irregularly shaped objects. Also the equation does not apply to *extremely* strong gravitational fields (*much* stronger than any fields in our solar system) or to objects moving at close to the speed of light.

If we apply Newton's second law, the law of universal gravitation, and the equation for the acceleration of an object in uniform circular motion, we find that for a satellite in a circular orbit

and

$$v = \sqrt{\frac{GM}{R}}$$ (N12.12)

$$T^2 = \frac{4\pi^2}{GM}R^3$$ (N12.14)

> **Purpose:** These equations specify the orbital speed v and **period** T of a satellite in a circular orbit of radius R around a primary with mass M.
> **Symbols:** G is the universal gravitational constant.
> **Limitations:** The orbit must be circular, and the primary must be much more massive than the satellite. The limitations on equation N12.10 also apply here.

Equation N12.14 is **Kepler's third law** for circular orbits.

Section N12.5: Circular Orbit Problems

Circular orbit problems are very much like the constrained-motion problems in chapter N8, except that we use equation N12.12 or N12.14 or the *magnitude* of Newton's second law as the master equation. This means that setting up coordinate axes is not very important. Moreover, the net force on any object of interest is always only a gravitational force, so a force diagram is not needed either. In the conceptual model section, you really only need to (1) describe the interacting objects, (2) check that one is much more massive than the other, and (3) check that the satellite's orbit is essentially circular.

Section N12.6: Black Holes and Dark Matter

This section discusses how astrophysicists have used Kepler's third law to show that black holes exist in the centers of galaxies and to discover the existence of dark matter in the universe.

N12.1   Kepler's Laws

In the year 1600, Johannes Kepler came to Prague to join the research staff at an observatory operated by Tycho Brahe, an astronomer who was both the official "imperial mathematician" of the Holy Roman Empire and a friend of Galileo. The following year, Brahe died, and Kepler succeeded him as imperial mathematician and as director of the observatory. His new position gave him complete access to Tycho Brahe's extraordinary collection of careful astronomical observations of the planets, the result of a lifetime of work by perhaps the greatest naked-eye astronomer who ever lived. (The telescope was not invented until about 1610.)

Kepler's laws

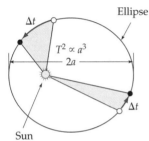

Figure N12.1
Kepler's laws illustrated. The colored regions show the area swept out by a line connecting the planet and the sun during equal time intervals Δt during different parts of the orbit: their areas are equal by Kepler's second law.

Kepler's laws are empirical

Newton offered an explanation of these laws

In 1609, Kepler published a work entitled *Astronomia Nova* ("New Astronomy") in which he stated two empirical laws that seemed to be consistent with Brahe's planetary observations. In modern language, these laws state that

1.   The orbits of the planets are *ellipses*, with the sun at one focus.
2.   The line from the sun to a planet sweeps out *equal areas in equal times.*

(I'll define the *focus* of an ellipse in chapter N13.) Kepler offered no theoretical explanation for these laws: he simply presented them as being *descriptive* of planetary orbits, according to Brahe's observational data. These laws represented a rather radical departure from the accepted wisdom of the time: up to then, it was assumed by most astronomers that the motions of the planets could be described in terms of combinations of uniform circular motions (although it was becoming clear that such schemes had to be extraordinarily complex to fit the best observational data available).

Ten years later, Kepler published in his *Harmonice Mundi* ("Harmonics of the World") a third empirical law:

3.   The square of a planet's **period** T (the time that it takes to complete one orbit) is proportional to the cube of the **semimajor axis** a of its orbit.

(An ellipse's *semimajor axis* is defined to be one-half of the distance measured across its widest part.) The three numbered laws above are known as **Kepler's three laws of planetary motion,** and they are illustrated in figure N12.1.

Again let me emphasize that these laws are entirely empirical: they do not so much explain as *describe* the motion of the planets. However, because Kepler supported them so carefully with observational data of extraordinary quality, these laws became widely known and accepted in spite of their radical character.

For more than six decades after the publication of the last of these laws, the scientific community was unable to say anything about why these laws should be true. Isaac Newton's incredible triumph was to show that each of these laws follows directly from his second law and the assumption that the force of gravity between two objects depends on the inverse square of the distance separating them. In other words, Newton offered an *explanation* of Kepler's laws in terms of physical principles that applied equally well to terrestrial motion: one simply had to accept that the planets were endlessly falling around the sun.

Let me emphasize how radical *this* suggestion was at the time! Before Newton, scholars had believed that the laws of physics pertaining to the motion of heavenly bodies (where unceasing motion in approximate circles seemed to be the rule) were completely distinct from the laws pertaining to terrestrial motion (where objects generally come to rest rather quickly). It took Newton's genius not only to see that this very credible division between celestial and terrestrial physics was in fact not necessary at all (which, granted, was *beginning* to occur to others as well), but also to provide a complete theoretical perspective that unified terrestrial and celestial physics in a manner that demonstrably *worked*. The simplicity, beauty, and extraordinary predictive power of Newton's ideas were so compelling that it brought the physics community to its first real consensus on a grand theoretical structure for physics. This first consensus (as discussed in chapter C1) in some sense marks the birth of physics as a scientific discipline.

Our goal in this chapter and chapter N13 is to prove that Kepler's laws are a consequence of the laws of mechanics that we have been studying. We will not quite follow the same path that Newton did in proving this: the laws of conservation of energy and angular momentum give us more powerful tools than even Newton had at his disposal. (Using these laws will enable us to do in a few pages what it took Newton scores of pages to show in the *Principia*.)

Our goal: prove that Kepler's laws follow from principles we have studied

N12.2   Orbits Around a Massive Primary

The first step toward understanding what Newton's laws say about planetary motion is to take advantage of the simplifications that result when (1) our system of interest consists of a very massive object interacting with a much lighter object and (2) we choose the origin of our reference frame to be the system's center of mass.

Consider an isolated system consisting of an object of mass M interacting with a smaller object of mass m (see figure N12.2). If the system is *really* isolated, then its center of mass will move at a constant velocity and thus can be used as the origin of an inertial reference frame. In practical situations, it actually is more likely that the system is *freely falling* in some external gravitational field (e.g., the earth and moon falling around the sun), but we can *still* treat the system's center of mass as the origin of an inertial reference frame if we ignore the external gravitational field (as we saw in chapter N9).

General situation: an isolated pair of interacting objects

If we define the center of mass of the system to be the origin, then the definition of the center of mass means that

Results in a frame based on the system's CM

$$0 = \frac{m\vec{r} + M\vec{R}}{M + m} \quad \Rightarrow \quad M\vec{R} = -m\vec{r} \quad \Rightarrow \quad \vec{R} = -\frac{m}{M}\vec{r} \quad \text{(N12.1)}$$

This implies that in *all* circumstances (no matter how the two objects move and/or interact with each other) the positions $\vec{R}$ and $\vec{r}$ of the objects relative

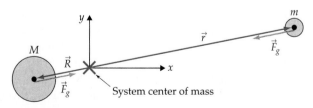

Figure N12.2

An isolated system of two objects interacting gravitationally. Note that if we define the origin of our reference frame to be attached to the system's center of mass, the positions of the objects are opposite. The gravitational force exerted on each object by the interaction points directly toward the other.

to the center of mass are opposite and have magnitudes that are strictly proportional to each other: if r gets bigger or smaller, then so does R (proportionally).

Taking the time derivative of both sides of equation N12.1, we get

$$\vec{V} = -\frac{m}{M}\vec{v} \quad \Rightarrow \quad V = \frac{m}{M}v \qquad \text{(N12.2)}$$

This means that

$$\frac{1}{2}MV^2 = \frac{1}{2}M\left(\frac{m}{M}v\right)^2 = \left(\frac{m}{M}\right)\left(\frac{1}{2}mv^2\right) \quad \Rightarrow \quad K_M = \frac{m}{M}K_m \quad \text{(N12.3)}$$

where K_M and K_m are the kinetic energies of the massive and light objects, respectively. Thus (as we've seen before), the kinetic energy of the more massive object in this frame is smaller than that of the lighter object by the factor m/M.

We can also express the angular momentum $\vec{L}_M$ of the larger object around the system's center of mass in terms of $\vec{L}_m$ as follows:

$$\vec{L}_M \equiv \vec{R} \times M\vec{V} = \left(-\frac{m}{M}\right)\vec{r} \times M\left(-\frac{m}{M}\right)\vec{v} = +\frac{m}{M}(\vec{r} \times m\vec{v}) = \frac{m}{M}\vec{L}_m \quad \text{(N12.4)}$$

The system's total angular momentum around the center of mass is thus

$$\vec{L}_{\text{tot}} = \vec{L}_M + \vec{L}_m = \left(\frac{m}{M} + 1\right)\vec{L}_m \qquad \text{(N12.5)}$$

The system's total angular momentum around its center of mass will be conserved *if and only if* the light object's angular momentum is conserved around the center of mass.

Implications when $M \gg m$

Equations N12.1 through N12.5 apply to *any* isolated (or freely falling) system of two interacting objects described in a reference frame whose origin is the system's center of mass. If, in addition, we have $M \gg m$, then

1.   The position of the massive object is $\vec{R} \approx 0$ (by N12.1).
2.   This object is essentially at *rest* at the origin: $\vec{V} \approx 0$ (by N12.2).
3.   Its kinetic energy is negligible: $K_M \approx 0$ (by N12.3).
4.   Its angular momentum is negligible: $\vec{L}_M \approx 0$ (by N12.4).
5.   The objects' separation $\approx$ distance of lighter object from the origin.

In such a case, the massive object (which we call the **primary** of this system under these circumstances) is essentially at rest at the origin, and the lighter object (which we call a **satellite**) orbits it. The primary then provides an essentially fixed origin for the gravitational force exerted on the satellite.

This approximation is very useful in the solar system

This approximation holds very well in the solar system. Even Jupiter's mass is more than 1000 times smaller than the sun's mass, and the earth's mass is more like 330,000 times smaller. Similarly, the moons that orbit the major planets typically have masses much smaller than their primary: even our own moon, which is the second largest moon in the solar system compared to its primary (after the Pluto/Charon system), has 81 times less mass than the earth.

The fact that planetary masses are so small compared to the sun has another important implication. When computing the orbit of one planet, we can ignore the gravitational effects of the others (to an excellent degree of approximation): the sun is so much more massive than anything else that the gravitational force that it exerts is by far the greatest influence on each planet's motion. Therefore, as an excellent approximation, we can pretend that each planet orbits the sun as if it were alone.

Exercise N12X.1

The radius of the earth's orbit is about 1.5×10^{11} m on the average. What is the distance between the sun and the center of mass of the earth/sun system? How does this compare to the sun's radius (700,000 km)?

N12.3   Kepler's Second Law

Equations N12.4 and N12.5 imply that the angular momentum of *either* object in an isolated interacting pair is *separately* conserved. This has two important consequences: (1) the object's orbit lies in a plane, and (2) its position vector sweeps out equal areas in equal times.

The first of these consequences follows from the definition of the angular momentum, which says that $\vec{L}$ for either object is defined to be $\vec{L} \equiv \vec{r} \times m\vec{v}$. By definition of the cross product, this means that $\vec{L}$ is always perpendicular to $\vec{r}$. But if $\vec{L}$ has a fixed orientation in space (because it is conserved), then the object's position vector $\vec{r}$ must always lie in the fixed plane perpendicular to $\vec{L}$. Therefore the object's orbit lies in a certain fixed plane.

The orbit of either object must lie in fixed plane in space

We can see that conservation of angular momentum also implies Kepler's second law as follows. Imagine that in an infinitesimal time dt, the object moves a certain infinitesimal angle $d\theta$ (shown greatly exaggerated in figure N12.3) as it moves in its orbit around the system's center of mass. The area swept out by the line between the object and the system's center of mass is the colored pie slice in figure N12.3. If $d\theta$ is very small, the shape of the slice is very nearly triangular, so its area is $dA \approx \frac{1}{2}$(base)(height). Now, as shown in the drawing, the height h of the triangle is very nearly equal to the arclength $r\, d\theta$, and this approximation gets better and better as $d\theta \rightarrow 0$. So if $d\theta$ is small,

The position vector of either object sweeps out equal areas in equal times

$$dA \approx \tfrac{1}{2}r(r\, d\theta) = \tfrac{1}{2}r^2\, d\theta \qquad (N12.6)$$

If we divide both sides by dt and take the limit as dt and $d\theta$ go to zero, we get

$$\frac{dA}{dt} = \frac{1}{2}r^2\frac{d\theta}{dt} \qquad (N12.7)$$

But according to equation C13.8, $L = \text{mag}(\vec{L})$ for the object is

$$L = mr^2\omega = mr^2\frac{d\theta}{dt} \qquad (N12.8)$$

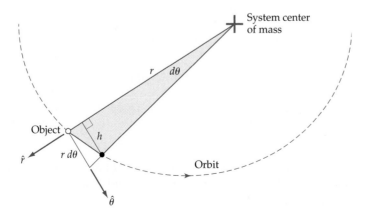

Figure N12.3
The colored region shows the area swept out by the line between the object and the system's center of mass as the object moves a tiny angle $d\theta$ around the center of mass (the object's initial position is the white dot and its final position is the black dot). This area is almost equal to $\frac{1}{2}rh$, and h in turn is approximately equal to the arclength $r\, d\theta$.

This means that

$$\frac{dA}{dt} = \frac{1}{2}r^2\frac{d\theta}{dt} = \frac{1}{2m}\left(mr^2\frac{d\theta}{dt}\right) = \frac{L}{2m} \qquad \text{(N12.9)}$$

Therefore, since L is conserved, dA/dt is constant: *the object's radius vector sweeps out equal areas in equal times.* Note that this applies to *either* object, independent of the objects' relative masses or the nature of their interaction.

Application to Kepler's second law

Now, Kepler's second law actually says that the line between the planet and the *sun* sweeps out equal areas in equal times, not the line between the planet and the system's *center of mass*. But if the sun is much more massive than the planet, it essentially *is* located at the system's center of mass, and Kepler's second law is essentially correct.

N12.4   Circular Orbits and Kepler's Third Law

We will assume $M \gg m$ from now to the end of unit

From now on through the rest of the unit, we will assume the primary-satellite approximation, in which the mass of the primary is much greater than the mass of the satellite. To simplify notation, when we refer to the satellite's kinetic energy and/or angular momentum from now on, we will drop the subscripts on K_m and $\vec{L}_m$ and simply use K and $\vec{L}$.

Circular orbits at constant speed are possible

If the primary is much more massive than its satellite, then it is possible for the satellite to follow an essentially circular orbit at constant speed *around the primary*. Let us see whether such an orbit is consistent with what we know about uniform circular motion and the gravitational interaction.

We know that an object moving in a circular trajectory at a constant speed is accelerating toward the center of its circular path, and that the magnitude of its acceleration is $a = v^2/R = $ a *constant*, where R is the radius of the object's orbit and v is its constant orbital speed. Newton's second law tells us that this acceleration must be caused by some force that is directed toward the center of the satellite orbit and which has a *constant* magnitude.

The gravitational force $\vec{F}_g$ exerted on the satellite by its gravitational interaction with its primary satisfies these criteria. It is directed toward the center of the satellite's trajectory (to the extent that we can consider the massive object to be at rest). According to equation C8.24, its magnitude is given by Newton's **law of universal gravitation:**

$$F_g = \frac{GMm}{r^2} \qquad \text{(N12.10)}$$

Purpose: This equation specifies the magnitude of the gravitational force $\vec{F}_g$ that an object of mass M exerts on an object of mass m (or vice versa) when their centers of mass are separated by a distance r.

Symbols: $G = 6.67 \times 10^{-11}$ N·m^2/kg^2 is the **universal gravitational constant.**

Limitations: This equation applies to point masses or spheres, but not to irregularly shaped objects. Also it does not apply to *extremely* strong gravitational fields (*much* stronger than any fields in our solar system) or to objects moving at close to the speed of light.

(We derived this in section C8 from the gravitational potential energy formula.) Therefore, in the case of a truly circular orbit (where $r = R = $ constant) the magnitude of the force will be GMm/R^2, which is a constant, as needed.

This means that it is at least *plausible* that the gravitational force exerted on the satellite due to its interaction with the primary can hold the satellite in a circular orbit. This does not mean that orbits *always* have to be circular: indeed, orbits generally are *not* circular (as we will see in chapter N13). But it does mean that a circular orbit is a possibility. In fact the orbital radii of most major objects in the solar system are constant to within a few percent. Most artificial satellites orbit the earth with approximately circular orbits as well. Therefore, this "special case" of all the kinds of orbits possible is in fact approximately applicable to a wide variety of realistic situations.

Let's see what Newton's second law can tell us quantitatively about such orbits. Assuming that the gravitational force is the *only* force acting on the satellite, then $F_{net} = F_g = GMm/R^2$. Therefore, the magnitude of Newton's second law tells us that

Implications of Newton's second law for circular orbits

$$ma = F_{net} = \frac{GMm}{R^2} \qquad \text{(N12.11)}$$

Dividing both sides by m and plugging $a = v^2/R$ into this equation, we find that

$$\frac{v^2}{R} = \frac{GM}{R^2} \quad \Rightarrow \quad v = \sqrt{\frac{GM}{R}} \qquad \text{(N12.12)}$$

Purpose: This equation specifies the orbital speed v of a satellite in a circular orbit of radius R around a primary with mass M.
Symbols: G is the universal gravitational constant.
Limitations: The orbit must be circular, and the primary must be much more massive than the satellite. The limitations on equation N12.10 also apply here.

(Note that v is appropriately constant.)

We can use this information to determine how long it will take the satellite to go once around its circular orbit. The orbit's period T in this case is the time it takes the satellite to travel a distance $2\pi R$ at a constant speed v:

$$T = \frac{2\pi R}{v} \qquad \text{(N12.13)}$$

If we square equation N12.12 and plug in equation N12.11, we can show that

$$T^2 = \frac{4\pi^2}{GM} R^3 \qquad \text{(N12.14)}$$

Purpose: This equation specifies the orbital period T of a satellite in a circular orbit of radius R around a primary with mass M.
Symbols: G is the universal gravitational constant.
Limitations: The orbit must be circular, and the primary must be much more massive than the satellite. The limitations on equation N12.10 also apply here.

Exercise N12X.2

Verify equation N12.14.

Kepler's third law for circular orbits

Kepler's third law states that *the square of a planet's period is proportional to the cube of its semimajor axis.* When an orbit is circular, its semimajor axis (one-half of the distance measured across the widest part of the orbit) *is* its radius R. So we see here that in a few lines we have *derived* Kepler's third law (for the special case of circular orbits anyway). Moreover, this derivation even gives us the constant of proportionality between T^2 and R^3!

Exercise N12X.3

The earth orbits the sun in an approximately circular orbit once each year. If it were 4 times as far from the sun, how long would it take to orbit once?

N12.5   Circular Orbit Problems

We can use equations N12.12 and N12.14 to answer many questions about orbiting objects and their primaries. In this section, we'll explore some examples of how the equations derived in section N12.4 can be applied to problems involving nearly circular orbital motion in the solar system. Since these examples involve circularly "constrained" motion (although only because the orbits *happen* to be roughly circular, not because they are required to be), we can easily adapt the framework we used for constrained-motion problems for use with these problems.

Differences between circular orbit and general constrained-motion problems

Doing circular orbit problems is much like doing the constrained-motion problems in chapter N8, with some important modifications. For example, we do not need to use Newton's laws in component form (equation N12.12 or N12.14 will probably be more helpful), and this means that we don't really need to define reference frame axes. (We will assume that the origin of our reference frame is at the center of the massive primary.) A force diagram of an orbiting object is also optional (since it would almost always show only one force vector). The only important issues to be addressed in a conceptual model section are the validity of the circular orbit approximation and the assumption that the primary is indeed much more massive than the satellite.

A plan for solving circular orbit problems

So a terse outline of our adaptation of the problem-solving framework for orbital motion problems might look as follows:

A. *Translation*
    1. Draw a sketch of the situation.
    2. Label it with appropriate symbols.
    3. List all known symbols.
B. *Conceptual model*
    1. Describe the two interacting objects.
    2. Make sure that one is very massive.
    3. Make sure that the satellite's orbit is circular (or state this as an assumption).
C. *Solution:* Apply equation N12.12 or N12.14.
D. *Evaluation:* Check the result for correct sign, correct units, and a plausible magnitude (as usual).

The examples that follow illustrate solutions following this format as well as some of the many applications of equations N12.12 through N12.14.

Example N12.1: Duration of a Shuttle Orbit

Problem   Imagine that the space shuttle *Atlantis* is in a circular orbit at an altitude of 250 km above the earth's surface. What is its orbital speed? What is its orbital period?

The space shuttle *Atlantis* in orbit around the earth.

Translation

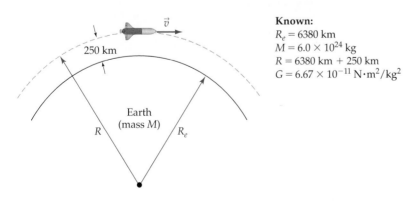

Known:
$R_e = 6380$ km
$M = 6.0 \times 10^{24}$ kg
$R = 6380$ km $+ 250$ km
$G = 6.67 \times 10^{-11}$ N·m²/kg²

Model   According to the problem statement, the *Atlantis* is following a circular orbit around the earth, and the earth is much more massive than the *Atlantis*, so equations N12.12 and N12.14 should apply.

Solution   Equation N12.12 tells us that the speed of a circularly orbiting object is

$$v = \sqrt{\frac{GM}{R}} = \sqrt{\frac{(6.67 \times 10^{-11} \text{ N·m}^2/\text{kg}^2)(6.0 \times 10^{24} \text{ kg})}{6{,}380{,}000 \text{ m} + 250{,}000 \text{ m}} \left(\frac{1 \text{ kg·m/s}^2}{1 \text{ N}}\right)}$$

$$= 7770 \text{ m/s} = 7.8 \text{ km/s} \tag{N12.15}$$

Once we have the speed, it is easier to use equation N12.13 rather than equation N12.14. The orbit's period is thus

$$T = \frac{2\pi R}{v} = \frac{2\pi(6380 \text{ km} + 250 \text{ km})}{7.77 \text{ km/s}} \left(\frac{1 \text{ min}}{60 \text{ s}}\right) = 89.4 \text{ min} \tag{N12.16}$$

Evaluation   Note that this is a little bit longer than the result we found in example N8.4. This is plausible: since the shuttle orbits some distance above the earth's surface, the force of gravity on the orbiting shuttle will be a bit smaller than it would be on the earth's surface. This means that the shuttle must move a bit more slowly if its acceleration is to match the reduced force.

Example N12.2: How to Weigh Jupiter

Problem   Ganymede, the largest moon of Jupiter, has a nearly circular orbit whose radius is 1.07 Gm. Measurements show that the moon goes around Jupiter once every 7 d, 3 h, and 43 min. What is Jupiter's mass?

Jupiter and its moon Ganymede.

Translation

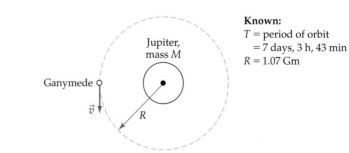

Known:
T = period of orbit
   = 7 days, 3 h, 43 min
R = 1.07 Gm

Model   The interacting objects here are Ganymede and Jupiter. We are told that Ganymede's orbit is circular and given its radius R and period T, and we can safely assume that Ganymede is much less massive than Jupiter. Since we know R and T, we can solve equation N12.14 for the mass M of Jupiter.

Solution   Doing this, we get

$$T^2 = \frac{4\pi^2}{GM}R^3 \quad \Rightarrow \quad M = \frac{4\pi^2 R^3}{GT^2} \qquad \text{(N12.17)}$$

The period T, expressed in seconds, is

$$T = 7\,\cancel{d}\left(\frac{24\,\cancel{h}}{1\,\cancel{d}}\right)\left(\frac{3600\text{ s}}{1\,\cancel{h}}\right) + 3\,\cancel{h}\left(\frac{3600\text{ s}}{1\,\cancel{h}}\right) + 43\,\cancel{min}\left(\frac{60\text{ s}}{1\,\cancel{min}}\right)$$

$$= 6.18 \times 10^5 \text{ s} \qquad \text{(N12.18)}$$

Plugging this into equation N12.17, we get

$$M = \frac{4\pi^2(1.07 \times 10^9\,\cancel{m})^3}{(6.67 \times 10^{-11}\,\cancel{N \cdot m^2}/kg^2)(6.18 \times 10^5\,\cancel{s})^2}\left(\frac{1\,\cancel{N}}{1\,\cancel{kg \cdot m/s^2}}\right)$$

$$= 1.9 \times 10^{27} \text{ kg} \qquad \text{(N12.19)}$$

Evaluation   Note that we can determine the mass of an object by observing the motion of a smaller object orbiting it (this is in fact the standard method of determining the mass of astronomical objects).

Example N12.3

Problem   Astronomical measurements show that Jupiter orbits the sun in a nearly circular orbit once every 11.86 years. How does the radius of Jupiter's orbit compare with that of the earth?

Translation

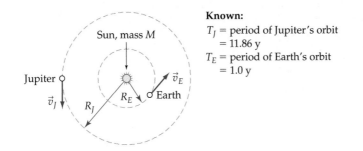

Known:
T_J = period of Jupiter's orbit
    = 11.86 y
T_E = period of Earth's orbit
    = 1.0 y

Model   In this problem, Jupiter and earth are both orbiting the sun, which is much more massive than either planet. We are ignoring the comparatively tiny interaction between earth and Jupiter (J). To do this problem, we have to *assume* that both the earth's orbit and Jupiter's orbit are circular (which is pretty closely true). If this is so, then equation N12.14 applies to both orbits, so

$$T_J^2 = \frac{4\pi^2}{GM}R_J^3 \quad \text{and} \quad T_E^2 = \frac{4\pi^2}{GM}R_E^3 \qquad \text{(N12.20)}$$

Solution   Since we are really looking for the ratio R_J/R_E, the fastest way to find that ratio is to divide the first expression by the second:

$$\frac{T_J^2}{T_E^2} = \frac{(4\pi^2/GM)R_J^3}{(4\pi^2/GM)R_E^3} = \frac{R_J^3}{R_E^3} \quad \Rightarrow \quad \frac{R_J}{R_E} = \left(\frac{T_J}{T_E}\right)^{2/3} = \left(\frac{11.86\,\cancel{y}}{1.0\,\cancel{y}}\right)^{2/3}$$

(N12.21)

So $R_J = (11.86)^{2/3}R_E = 5.20R_E$.

Evaluation   This result is consistent with the result given in the inside front cover of the book.

Example N12.4: The Rutherford Atom

In 1911, Ernest Rutherford proposed (on the basis of certain scattering experiments that he and his graduate students performed) that all atoms consist of a tiny positively charged nucleus surrounded by comparatively lightweight orbiting electrons. Each electron was assumed to be held in its orbit by the force of electrostatic attraction between the electron's negative charge $-e$ (where $e = 1.6 \times 10^{-19}$ C) and the nucleus's positive charge. We now know that this model is a tremendous oversimplification, but it was an important step toward understanding the structure of atoms.

The potential energy formula that describes the electrostatic interaction between two particles with charges q_1 and q_2 is $V = kq_1q_2/r$, where k is the Coulomb constant $= 9.0 \times 10^9$ J·m/C² (see section C7.1). Since this is the same formula as the one for gravitational potential energy $V = -GMm/r$ except that the constants $-GMm$ are replaced by kq_1q_2, it follows that the magnitude of the electrostatic force must be $F_e = k|q_1q_2|/r^2$ by analogy to the gravitational force law $F_g = +GMm/r^2$.

Problem   So, let us consider hydrogen, the simplest possible atom. In the Rutherford model a hydrogen atom would consist of a single electron orbiting a single proton. The proton has a mass about 1836 times greater than that of the electron and has a charge of $+e$. For the sake of argument, let's assume that the electron's orbit is approximately circular and has a radius equal to the hydrogen atom's measured radius of about 0.54 Å, or 5.4×10^{-11} m. What is the approximate orbital speed of such an electron according to this model?

Translation

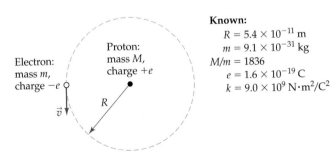

Known:
$R = 5.4 \times 10^{-11}$ m
$m = 9.1 \times 10^{-31}$ kg
$M/m = 1836$
$e = 1.6 \times 10^{-19}$ C
$k = 9.0 \times 10^9$ N·m²/C²

Model Since the proton is so much more massive than the electron, we can consider the proton to be at rest at the system's center of mass. In this case, the force that causes the electron's acceleration in its circular orbit is not a gravitational force but an electrostatic attraction force whose magnitude is $F_e = ke^2/R^2$, where $R = 5.4 \times 10^{-11}$ m. Newton's second law for this electron thus implies that (since the electron's acceleration in its circular orbit is $a = v^2/R$)

$$m\frac{v^2}{R} = F_{net} = F_e = \frac{ke^2}{R^2} \qquad \text{(N12.22)}$$

Since we know (or can look up) m, R, k, and e, we can solve for our unknown v.

Solution Solving for v yields

$$v = \sqrt{\frac{ke^2}{mR}} = \sqrt{\frac{(9.0 \times 10^9 \ \text{N·m}^2/\text{C}^2)(1.6 \times 10^{-19} \ \text{C})^2}{(9.1 \times 10^{-31} \ \text{kg})(5.4 \times 10^{-11} \ \text{m})} \left(\frac{1 \ \text{kg·m/s}^2}{1 \ \text{N}}\right)}$$

$$= 2.2 \times 10^6 \ \text{m/s} \left(\frac{c}{3.0 \times 10^8 \ \text{m/s}}\right) = 0.0073c = \frac{c}{140} \qquad \text{(N12.23)}$$

Evaluation Interestingly, independent measures of the electron's orbital speed (having to do with the magnetic field produced by the circulating charge and various relativistic effects) yield approximately the same result. So even though this model may be oversimplified, it must be onto something!

The point of this example is that circular orbit problems don't have to be limited to astronomical situations!

N12.6   Black Holes and Dark Matter

In 1994, H. C. Ford and R. J. Harms announced that they had discovered a giant black hole at the center of the galaxy known as M87. This was the first time that anyone had presented an apparently iron-clad argument for the existence of a giant black hole at the center of any galaxy.

A recent Hubble photograph of galaxy M87 clearly shows a jet of gas emerging from the elliptical galaxy's central black hole.

A black hole is an object so dense that light cannot escape it. So how can we hope to "discover" an object that by definition doesn't emit any light? If you can find something orbiting the black hole and you can measure the orbit's period and radius, you can compute the black hole's mass by using Kepler's third law!

Ford and Harms used the Hubble Telescope to take spectra from very small regions of a gas cloud near the center of M87. By measuring the Doppler shifts of the spectra obtained, they were able to determine that gas that was a radius $R = 60$ light-years (ly) ($\approx 5.7 \times 10^{17}$ m) from the center of the cloud was orbiting the center with a speed of 450 km/s. If we assume that the orbit is roughly circular, $T = 2\pi R/v$. Plugging the latter equation into Newton's version of Kepler's third law and solving for M, we get

$$\frac{4\pi^2 R^3}{GM} = T^2 = \left(\frac{2\pi R}{v}\right)^2 \quad \Rightarrow \quad M = \frac{4\pi^2 R^3 v^2}{G(2\pi R)^2} = \frac{Rv^2}{G} \quad \text{(N12.24)}$$

Plugging in $R = 5.7 \times 10^{17}$ m and $v = 450$ km/s, we get 1.7×10^{39} kg, which is about 10^9 solar masses (1 solar mass $= 2.0 \times 10^{30}$ kg). This is just one of many bits of evidence cited by Ford and Harms.

How do we know that this is a black hole? No other explanation works! A star cluster containing a billion stars within a radius of 60 ly would emit lots of light that is not seen. No plausible model *other* than the black hole model explains how a billion solar masses could fit within a sphere 60 ly in radius and yet be consistent with all the other available data taken by Ford and Harms.

For more than a decade, astronomers also have been collecting evidence that 90% or more of the mass of the universe is *dark matter*, that is, matter that is *not* in stars that emit light or dust clouds that emit radio signals, reflect or obstruct light. Consistent evidence for this dark matter comes from a variety of studies of the orbital motion of stars within galaxies, satellite galaxies around large galaxies, and galaxies in galactic clusters. Many of these studies use Newton's version of Kepler's third law to determine the mass of the unseen matter by its gravitational effects.

In one recent study, D. N. C. Lin, B. F. Jones, and A. R. Klemona carefully measured the transverse movement of the *Large Magellanic Cloud* (LMC) relative to background objects. The LMC is a small galaxy that is a companion to and presumably orbits our own much larger galaxy. Lin and his collaborators measured the transverse velocity of the LMC to be about 200 km/s. Since the LMC is about 170,000 ly (1.6×10^{21} m) from our galaxy, if it were in a circular orbit, that would imply that the mass of our galaxy is

$$M = \frac{Rv^2}{G} = \frac{(1.6 \times 10^{21}\ \text{m})(200{,}000\ \text{m/s})^2}{6.67 \times 10^{-11}\ \text{N·m}^2/\text{kg}^2} \left(\frac{1\ \text{N}}{1\ \text{kg·m/s}^2}\right) = 1.0 \times 10^{42}\ \text{kg}$$

$$\text{(N12.25)}$$

which is about 500 billion solar masses (a more accurate calculation based on the LMC's actual trajectory puts the estimate at more like 600 billion solar masses). Visible matter in our galaxy amounts to about 100 billion solar masses.

What is this dark matter? No one knows! Very recent experiments have strongly suggested that it *cannot* be dwarf stars, large planets, or black holes; mounting evidence suggests that this matter is *not* even ordinary matter made of protons, neutrons, and electrons. The mystery is still in the process of being solved, but part of the point here is that we would not even know that this major fraction of the universe existed if it weren't for Newton's form of Kepler's third law!

TWO-MINUTE PROBLEMS

N12T.1 Kepler's second law implies that as a planet's distance from the sun *increases* in an elliptical orbit, its orbital speed
   A. Increases.
   B. Decreases.
   C. Remains the same.
   D. Changes in a way we cannot determine.

N12T.2 The sun's mass is about 1000 times that of Jupiter, and the radius of Jupiter's orbit is about 1100 times the sun's radius. The center of mass of the sun/Jupiter system is inside the sun, true (T) or false (F)?

N12T.3 Two stars, one with radius r and the other with radius $3r$, orbit each other so that their centers of mass are $25r$ apart. Assume that the stars have the same uniform density. The center of mass of this system is inside the larger star, T or F?

N12T.4 The speed of a satellite in a circular orbit of radius R around the earth is 3.0 km/s. The speed of another satellite in a different circular orbit around the earth is one-half this value. What is the radius of that satellite's orbit?
   A. $4R$
   B. $2R$
   C. $\sqrt{2}R$
   D. $R/\sqrt{2}$
   E. $R/2$
   F. $R/4$
   T. Some other multiple of R (specify)

N12T.5 A satellite orbits the earth once every 2.0 h. What is the orbital period of another satellite whose orbital radius is 4.0 times larger?
   A. 4.0 h
   B. 8.0 h
   C. 16 h
   D. 64 h
   E. Some other period (specify)

N12T.6 The radius of Saturn's orbit is 9.53 times that of the earth. What is the period of Saturn's orbit (assuming that it is nearly circular)?
   A. 9.53 y
   B. 29 y
   C. 91 y
   D. 866 y
   E. Some other period (specify)

N12T.7 The radius of the earth's (almost circular) orbit around the sun is 150,000,000 km, and it takes 1 y for the earth to go around the sun. Imagine that a certain satellite goes in an almost circular orbit of radius 15,000 km around the earth (this radius is 10,000 times smaller than the earth's orbital radius around the sun). What is the period of this orbit?
   A. 10^{-6} y
   B. 10^{-4} y
   C. 10^{-3} y
   D. 10^{6} y
   E. These periods are not related in any simple way.

HOMEWORK PROBLEMS

Basic Skills

N12B.1 A satellite in a circular orbit around the earth has a speed of 3.00 km/s. What is the radius of this orbit?

N12B.2 The circular orbit of a satellite going around the earth has a radius of 10,000 km. What is the satellite's orbital speed?

N12B.3 *Geostationary* satellites are placed in a circular orbit around the earth at such a distance that their orbital period is exactly equal to 24 h (this means that the satellite seems to hover over a certain point on the earth's surface). What is the radius of such a geocentric orbit?

N12B.4 What is the orbital speed of the moon? The radius of the moon's orbit is roughly 384 Mm.

N12B.5 What is the earth's speed as it orbits the sun?

N12B.6 Triton is the largest moon of Neptune (roughly the same size as the earth's moon). It orbits Neptune once every 5.877 days at a distance of roughly 354 Mm from Neptune's center. What is the mass of Neptune?

A photograph of Neptune and its moon Triton (see problem N12B.6).

N12B.7 The radius of Neptune's nearly circular orbit is about 30 times larger than that of earth's orbit. What is the period of Neptune's orbit?

N12B.8 The radius of Venus's nearly circular orbit is about 0.723 times that of earth's orbit. What is the period of Venus's orbit?

N12B.9 A neutron star is an astrophysical object having a mass of roughly 2.8×10^{30} kg (about 1.4 times the mass of the sun) but a radius of only about 12 km. If you were in a circular orbit of radius 320 km (about 200 mi), how long would it take you to go once around the star?

Synthetic

N12S.1 Is the center of mass of the earth/moon system inside the earth?

The earth-moon system (see problem N12S.1).

N12S.2 Imagine that the sun, earth, and Jupiter are aligned so that all three are in a line. What is the magnitude of the gravitational force exerted by Jupiter on the earth compared to that exerted by the sun on the earth?

N12S.3 What is the magnitude of the gravitational force exerted on the earth by the moon compared to that exerted on the earth by the sun?

N12S.4 During a certain 5-day time period, the line connecting a comet with the sun changes angle by about 3.2°. Assume that the comet's distance from the sun is roughly 130 million km during this time. During a 5-day time period sometime later, the angle of this line changes by 0.80°. How far is the comet from the sun now? Explain.

N12S.5 Consider a light object and a massive object connected by a spring. Both objects are floating in deep space.

(a) Will the path of the light object around the massive object lie in a plane?

(b) Will it obey Kepler's second law? Explain your reasoning in both cases.

N12S.6 What would be the orbital speed of the electron in a hydrogen atom if the interaction between the proton and electron were gravitational instead of electrostatic? (Assume that the electron's orbit is still circular and its radius is still 5.4×10^{-11} m.)

N12S.7 Imagine that two objects with masses M and m are connected by a spring with zero relaxed length, so that the attractive force that each exerts on the other is $F = k_s r$. Assume that $M \gg m$, and that the satellite orbits in a circular orbit of radius R around its primary. Find an expression (analogous to Kepler's third law) that gives the period of the orbit T as a function of the orbital radius R, the spring constant k_s, and whatever else you need.

N12S.8 Imagine that in a different universe the magnitude of the force of gravity exerted by one object on another were given by

$$F_g = \frac{B M m}{r^3} \qquad (N12.26)$$

where r is the separation between the two objects and B is some constant. What would Kepler's third law for circular orbits be in this universe?

N12S.9 The force law for the magnitude of the force between two quarks separated by a distance r turns out to be very roughly $F = br$, where b is some constant. If we pretend that Newton's laws apply to quarks, and imagine that one quark is very much lighter than the other and is in a circular orbit around it, what would be the equation for the lighter quark's orbital period T as a function of its orbital radius R?

N12S.10 In Newton's time, the distance between the earth and the moon was known to be about 60 times the radius of the earth. According to the law of universal gravitation, about how many times smaller is the earth's gravitational field strength at the radius of the moon's orbit than it is on the earth's surface? Use this information (rather than the mass of the earth) to estimate the period of the moon's orbit in days. How does this compare with the moon's actual orbital period? Please show your work. (This was one of the ways that Newton supported his inverse-square law of gravitation.)

N12S.11 In a certain binary star system, a small red star with a mass ≈ 0.22 solar masses orbits a bright white-hot star with a mass ≈ 4.2 solar masses. (These masses are estimated from the stars' color and luminosity.)

The red star is observed to eclipse the other every 482 days.

(a) Assume that the white star's mass is large enough that we can use the $M \gg m$ approximation. What is the approximate distance D between the centers of these stars? What other assumptions (if any) do you have to make to solve the problem?

(b) Problem N12A.1 discusses a more exact version of Kepler's third law that applies even when M and m are comparable. Use this result to calculate the distance between the stars' centers. How does this compare to the result of part (a)?

Rich-Context

N12R.1 Consider a spherical asteroid made mostly of iron (whose density is $7.9 \, \text{g/cm}^3$) whose radius is 22 km. Could you run fast enough to put yourself in orbit around this asteroid?

N12R.2 Imagine that you wanted to put a satellite in such a circular orbit that it appeared on the western horizon every Monday morning at 6 a.m. and was never at any other time on the western horizon. What radius should the orbit have? (Don't forget to account for the earth's rotation!)

N12R.3 Astrophysicists believe that a collision between galaxies might lead to the resulting combined galaxy having a *pair* of supermassive black holes at its center. If this is so, then interactions between the black-hole binary and the stars and gas in the galactic center will quickly circularize the orbits of the black holes around their common center of mass.

(a) Imagine that the two black holes have *equal* masses M and that their orbits are circular and have a radius R_0 around the system's center of mass. Show that the speed of each black hole in its orbit is given by

$$v = \frac{1}{2}\sqrt{\frac{GM}{R_0}} \qquad (N12.27)$$

(*Hint:* Equation N12.12 does not apply in this case because the black holes have comparable masses.)

(b) Interactions with the surrounding galactic matter will cause the distance between the black holes to shrink very gradually with time. Imagine that over a period of 60 My (60 million years) the orbital radius shrinks by one-half to $R = \frac{1}{2}R_0$. Compute the change in the black-hole binary system's total energy (kinetic + potential) during this time (express your answer in terms of G, M, and R_0).

(c) Has the system lost energy to or gained energy from the surrounding matter? Does your answer make intuitive sense?

(d) If $M = 2.0 \times 10^{36}$ kg and the initial separation of the black holes is $R_0 \approx 10 \, \text{ly} \approx 10^{17}$ m, roughly what is the average rate at which energy is transferred, in watts? (For comparison, the rate at which the sun radiates energy is about 3.9×10^{26} W.)

Advanced

N12A.1 Consider an isolated two-object system interacting gravitationally such that M is not necessarily much greater than m. Assume that the smaller object's orbit around the system's center of mass is circular and has radius r. Argue that both objects have the same orbital period T around the system's center of mass and that

$$T^2 = \frac{4\pi^2 D^3}{G(M+m)} \qquad (N12.28)$$

where D is the separation of the two objects.

ANSWERS TO EXERCISES

N12X.1 According to equation N12.1, if the distance between the earth and system's center of mass is r (which is then also the radius of the earth's orbit) and the distance between the sun and the system's center of mass is R, then

$$R = \frac{m}{M}r = \frac{6.0 \times 10^{24} \, \text{kg}}{2.0 \times 10^{30} \, \text{kg}}(1.5 \times 10^{11} \, \text{m})$$

$$= 450{,}000 \, \text{m} = 450 \, \text{km} \qquad (N12.29)$$

(Note that you can look up the earth's and sun's masses in the inside front cover.) This is nearly 2000 times smaller than the sun's radius.

N12X.2 Plugging equation N12.12 into equation N12.13, we get

$$T = \frac{2\pi R}{\sqrt{GM/R}} = 2\pi R\sqrt{\frac{R}{GM}} = 2\pi\sqrt{\frac{R^3}{GM}} \qquad (N12.30)$$

Squaring both sides of this gives us equation N12.14.

N12X.3 According to equation N12.14, $T \propto R^{3/2}$. If we increase R by a factor of 4, then $R^{3/2}$ increases by a factor of $(\sqrt{4})^3 = 2^3 = 8$. So the orbit would last 8 y.

N13 Planetary Motion

Chapter Overview

Introduction

In this final chapter of the unit, we will bring many of the tools that we have developed in this unit (and in unit C) to bear on the problem that represents the crowning triumph of the newtonian synthesis. We will see here how Newton's laws predict that the orbit of a satellite moving near a massive primary will be either an ellipse or a hyperbola, and that a simple generalization of Kepler's third law exists when the orbit is an ellipse. We will also see how we can use conservation of energy and angular momentum as tools for determining orbit characteristics.

Section N13.1: Ellipses and Hyperbolas

An **ellipse** is the curve comprised of all points P on a plane whose distances r_1 and r_2 to two given **foci** F_1 and F_2 add up to a constant $2a$, where $2a$ is the distance across the ellipse at its widest point (a is called the ellipse's **semimajor axis**). An ellipse's **eccentricity** ε is defined to be c/a, where $2c$ is the distance between the two foci. When $\varepsilon \to 0$, the ellipse becomes a circle; and when $\varepsilon \to 1$, the ellipse becomes very long and thin. The **vertices** are the points where the ellipse crosses the axis connecting the foci. For a given focus, the distance to the closest vertex is $r_c = a(1 - \varepsilon)$, and the distance to the farthest vertex is $r_f = a(1 + \varepsilon)$.

A **hyperbola** is the curve comprised of all points P on a plane whose distances r_1 and r_2 to two given foci F_1 and F_2 have a *difference* that is a constant $2a$. As r_2 grows large, the hyperbola approaches two **asymptotes** that make the same angle $\theta = \cos^{-1}(a/c)$ with respect to the axis connecting the foci, where $2c$ is the distance between the two foci. We define a hyperbola's eccentricity to be $\varepsilon \equiv c/a$ (note that $\varepsilon > 1$ in the case of a hyperbola). The hyperbola has a single vertex whose distance from the closest focus is $r_c = a(\varepsilon - 1)$.

The section discusses some other interesting things about these curves, which are examples of **conic sections.**

Section N13.2: Trajectory Diagrams for Orbits

This section discusses how we can use the computer program Newton to model planetary orbits. *You will get the most out of this section if you read it near a computer running the program Newton and if you do the exercises.* Using this program, you can show that Newton's second law and the law of universal gravitation do indeed imply that the orbit of a planet or satellite around a massive primary must be either an ellipse or a hyperbola and that the following equation is true for ellipses:

$$T^2 = \frac{4\pi^2}{GM}a^3 \qquad\qquad \text{(N13.8)}$$

Purpose: This equation generalizes Kepler's third law for elliptical orbits.

Symbols: T is the period of the orbit, G is the universal gravitational constant $= 6.67 \times 10^{-11}$ N·m^2/kg^2, M is the mass of the primary, and a is the semimajor axis of the elliptical orbit.

The section also derives an equation (see equation N13.10) that is useful for locating the second focus of a hyperbola when you know the location of only one. This equation is particularly useful for locating the second focus on a hyperbola graph drawn by the program Newton.

Section N13.3: Conservation Laws and Orbits

Both the energy E and the angular momentum L of a primary/satellite system will be conserved throughout the orbit. This section discusses how one can use this fact to link the values of E and L for a given system to the values of a and ε that determine the shape of the orbit. Table N13.1 summarizes the results of this section. One of the most important basic consequences of the equations listed there is that

$$\text{If}\quad E < 0 \quad\text{then}\quad \varepsilon < 1 \text{ (the orbit is } \textit{elliptical}\text{)}$$

$$\text{If}\quad E > 0 \quad\text{then}\quad \varepsilon > 1 \text{ (the orbit is } \textit{hyperbolic}\text{)}$$

Section N13.4: Solving Orbit Problems

Given the powerful set of equations summarized in table N13.1, one can solve an astonishing range of orbit problems given only two pieces of information about the orbit (such as the satellite's distance from the primary and its velocity at a given time, or the points of the orbit that are nearest or farthest from the primary). This section contains examples that illustrate approaches to solving such problems.

Table N13.1 Table of useful equations for solving orbit problems

Item	Equation		Symbols
Definitions of E and L	$\dfrac{2E}{m} = v^2 - \dfrac{2GM}{r}$ and	$\dfrac{L}{m} \equiv rv\sin\phi$	E = total system energy L = system angular momentum
Calculation of eccentricity	$\varepsilon^2 = 1 + \dfrac{1}{(GM)^2}\left(\dfrac{L}{m}\right)^2\left(\dfrac{2E}{m}\right)$		G = gravitational constant m = satellite mass M = primary mass

Item	Elliptical Case	Hyperbolic case	
Connection between a and E	$\dfrac{2E}{m} = -\dfrac{GM}{a}$	$\dfrac{2E}{m} = +\dfrac{GM}{a}$	r = distance from primary v = satellite speed at that point ϕ = angle between $\vec{r}$ and $\vec{v}$ a = semimajor axis for ellipse, analogous for hyperbola
Location of extremes	$r_c = a(1-\varepsilon)$ $\qquad\qquad r_c + r_f = 2a$ $r_f = a(1+\varepsilon)$	$r_c = a(\varepsilon - 1)$	r_c = distance from primary to closest point on the orbit r_f = distance from primary to farthest point on the orbit
Other useful relations	$T^2 = \dfrac{4\pi^2}{GM}a^3$ (Kepler's third law)	$\tan\theta = \sqrt{\varepsilon^2 - 1}$ $\qquad = \dfrac{1}{GM}\left(\dfrac{1}{m}\right)\sqrt{\dfrac{2E}{m}}$	T = orbital period θ = angle of asymptotes

N13.1   Ellipses and Hyperbolas

The core of the newtonian synthesis was Newton's proof that his three laws of mechanics and the law of universal gravitation imply that planetary orbits must be *ellipses*, consistent with Kepler's surprising but empirical first law. Newton's proof was even more general than this: he showed that the path of any object moving around a massive primary under the influence of gravity alone must be either an *ellipse* or a *hyperbola*, with the massive object's center of mass located at one focus of the curve. In this section, we will discuss the mathematical definitions and properties of ellipses and hyperbolas so as to better understand the meaning of this statement.

What is an ellipse?

An **ellipse** is the curve comprised of all points P on a plane whose distances r_1 and r_2 to two given points F_1 and F_2 add up to a constant greater than the distance between F_1 and F_2 (see figure N13.1a). The points F_1 and F_2 are called the **foci** (and each point individually is a **focus**) of the ellipse. The ellipse's **center** is halfway between the two foci, a distance c from each focus.

The semimajor axis a

The ellipse is widest along the axis going through the foci. The ellipse's **vertices** P_1 and P_2 are the points where it crosses this axis. The distance between either vertex and the ellipse's center is called the ellipse's **semimajor axis** a. Note that at the vertices, $r_1 + r_2$ is clearly equal to $2a$ (at P_2, for example, $r_1 = a + c$ and $r_2 = a - c$, so $r_1 + r_2 = 2a$). Since $r_1 + r_2$ is a constant for all points on the ellipse by definition, *all* points on the ellipse satisfy $r_1 + r_2 = 2a$.

The ellipse is narrowest along the axis through the center that is perpendicular to the wide axis (see figure N13.1b). The half-width of the ellipse across this axis is the ellipse's **semiminor axis** b. Since the distances from point P_3 on that axis to both F_1 and F_2 are the same and must add up to $2a$, the distance to each must be a. The pythagorean theorem implies that $b = (a^2 - c^2)^{1/2}$.

The **eccentricity** ε of an ellipse is defined to be

The eccentricity ε

$$\varepsilon = \frac{c}{a} \qquad \text{(note that } 0 \leq \varepsilon < 1) \qquad \text{(N13.1)}$$

When ε approaches zero, the separation of the foci becomes negligible compared to the semimajor axis a, and the ellipse basically becomes a circle with radius a. When ε approaches 1, $b = (a^2 - c^2)^{1/2} = [a^2 - (\varepsilon a)^2]^{1/2} = a(1 - \varepsilon^2)^{1/2}$

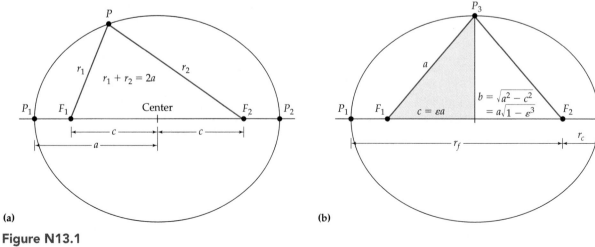

(a)   **(b)**

Figure N13.1
The geometric properties of an ellipse.

becomes small compared to a, so the ellipse becomes very skinny compared to its length.

Note in figure N13.1b that the point P_1 on the widest axis of the ellipse is the farthest point on the ellipse from the focus F_2, while P_2 is the point closest to that focus. If we define the distances to the close and far points to be r_c and r_f, respectively, then you can easily derive these handy relations:

Useful formulas for calculating the nearest and farthest points from a given focus

$$r_c = a(1 - \varepsilon) \qquad r_f = a(1 + \varepsilon) \qquad \text{(N13.2a)}$$

$$a = \tfrac{1}{2}(r_c + r_f) \qquad a\varepsilon = \tfrac{1}{2}(r_f - r_c) \qquad \text{(N13.2b)}$$

Exercise N13X.1

Derive equations N13.2.

A **hyperbola** is the curve comprised of all points P on a plane whose distances r_1 and r_2 to two given points F_1 and F_2 have a *difference* is a constant less than the distance between F_1 and F_2 (see figure N13.2a). The two points F_1 and F_2 are the *foci* of the hyperbola. We will call the point halfway between the two foci the hyperbola's *center,* in analogy with the center of an ellipse.

What is a hyperbola?

The curve of a hyperbola is closest to this center at the point where the curve crosses the axis connecting the two foci: we call this point the hyperbola's *vertex.* Let us define the distance between the hyperbola's vertex and center to be a (in analogy with the ellipse's semimajor axis). Note that at the vertex, $r_1 = c - a$ and $r_2 = c + a$, so $r_2 - r_1 = 2a$. Since $r_2 - r_1$ is the same for all points on the hyperbola by definition, *all* points on the hyperbola must satisfy $r_2 - r_1 = 2a$.

Definition of a for a hyperbola

An interesting feature of a hyperbola is that as r_2 grows large, the hyperbola asymptotically approaches two lines (the dashed lines in figure N13.2) that go through the hyperbola's center and make the same angle θ with respect to the axis connecting the foci. Indeed, if we consider a point P on the hyperbola that is extremely far away, the lines that connect that point to the foci become essentially parallel to each other and to the **asymptote** (the dashed line). Since r_2 must be $2a$ longer than r_1 according to the relation

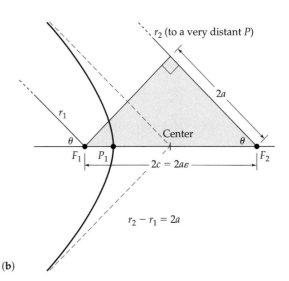

(a)   (b)

Figure N13.2
The geometric properties of a hyperbola.

$r_2 - r_1 = 2a$, the upper right leg of the colored triangle shown in figure N13.2 must have a length of $2a$. This implies that the angle θ that the asymptotes make with the line connecting the foci must be such that $\cos\theta = a/c$.

We define a hyperbola's eccentricity to be $\varepsilon \equiv c/a$, just as in the case of the ellipse (note that $\varepsilon > 1$ in the case of a hyperbola). This means that

A hyperbola's eccentricity

$$\cos\theta = \frac{1}{\varepsilon} \qquad (N13.3a)$$

Exercise N13X.2

Show that $\cos\theta = 1/\varepsilon$ implies that

$$\sin\theta = \frac{\sqrt{\varepsilon^2 - 1}}{\varepsilon} \qquad \text{and} \qquad \tan\theta = \sqrt{\varepsilon^2 - 1} \qquad (N13.3b)$$

Therefore, as the eccentricity of a hyperbola approaches 1, $\sin\theta \to 0$, implying that $\theta \to 0$. In this limit, therefore, the hyperbola becomes very thin vertically, almost lying parallel to the axis through the foci. On the other hand, as ε becomes large, $\cos\theta \to 0$, implying that $\theta \to 90°$. In this limit, therefore, the hyperbola essentially becomes a straight vertical line.

Note also that the vertex is the point on the hyperbola that is closest to the near focus F_1: if we define the distance between this focus and the vertex to be r_c in analogy with the ellipse, then we have

A useful expression for the point of closest approach

$$r_c = c - a = a\varepsilon - a = a(\varepsilon - 1) \qquad (N13.4)$$

We will find equations N13.3 and N13.4 useful in section N13.5.

It is worth noting in passing that as $\varepsilon \to 1$ from either above or below, the hyperbola or ellipse in question approaches a **parabola,** which is defined to be the set of all points on a plane that are the same distance d from both a focus point F and a straight line (see figure N13.3). If you think about it, you can see that this curve is what one would get from the definition of either an ellipse or a hyperbola in the limit that the curve's other focus is essentially an infinite distance away (in either case, d at the vertex represents the absolute value of the tiny difference between a and c). So we can think of a parabola as the limit of an extremely long and skinny ellipse (whose far vertex is essentially at infinity) or an extremely long and thin hyperbola (whose asymptotic angle $\theta \approx 0$).

Additional (less crucial) tidbits about these curves (for your edification)

Ellipses, parabolas, and hyperbolas are collectively called **conic sections,** because they happen to represent the possible shapes that one could get if one were to slice a cone at different angles. This is not really relevant to us here, but you can probably find a proof of this in a good text on analytic geometry.

You also may have previously learned about ellipses and hyperbolas in terms of the analytical equations that define these curves. If we use a coordinate system whose origin is located at the center of the curve and whose x axis goes through both foci, then the x, y coordinates of points on the curve satisfy

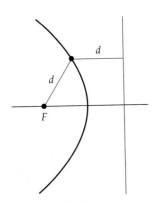

Figure N13.3
The definition of a parabola.

$$\text{Ellipse:} \quad 1 = \frac{x^2}{a^2} + \frac{y^2}{b^2} \qquad \text{where } b^2 = a^2 - c^2 = a^2(1 - \varepsilon^2) \quad (N13.5a)$$

$$\text{Hyperbola:} \quad 1 = \frac{x^2}{a^2} - \frac{y^2}{b^2} \qquad \text{where } b^2 = c^2 - a^2 = a^2(\varepsilon^2 - 1) \quad (N13.5b)$$

Again, this is really not relevant for our purposes here, but you can show fairly easily that these equations follow from the definitions given above.

Exercise N13X.3

Show that equations N13.5 follow from the definitions of the ellipse and hyperbola given above. [*Hint:* For the ellipse, $r_1^2 = (x+c)^2 + y^2$ and $r_2^2 = (x-c)^2 + y^2$ by the pythagorean theorem. Plug these into the definition $r_1 + r_2 = 2a$ and manipulate the resulting expression to get equation N13.5a.]

N13.2   Trajectory Diagrams for Orbits

In the more than three centuries since the publication of Newton's *Principia*, people have invented many mathematical proofs that Kepler's first law follows from Newton's second law and the law of universal gravitation. Unhappily, none are easy, and most require mathematics quite beyond our level. However, *we* have the program Newton, which automates the trajectory construction process discussed in chapter N4, which in turn is a completely general method for finding an object's trajectory whenever we know all the forces acting on the object.

To use Newton, one must do three things: (1) specify the object's initial position and velocity, (2) describe how each force acting on the object contributes to its total acceleration, and (3) specify an appropriate time step. Let's see how we might do these things intelligently in the case at hand.

Using the program Newton for orbit problems

For orbit problems, it turns out to be convenient to define the origin of our coordinate system to be the center of mass of the massive primary, and to position the satellite somewhere along the x axis some distance r_0 from the origin. If we also define the satellite's initial velocity to be $\vec{v}_0 = [0, +v_0, 0]$ (i.e., totally in the $+y$ direction), it turns out that the initial position will then correspond to a vertex of the resulting ellipse or hyperbola.

Defining the initial position and velocity

We also generally assume in orbit problems that the only force acting on the object is the gravitational force due to its interaction with the primary. The magnitude of this force is $F_g = GMm/r^2$, so the magnitude of the acceleration this force causes is $a = F_g/m = GM/r^2$. In the Newton program, one has the option of entering an acceleration term that is proportional to a constant times r raised to some power. One can thus set up Newton for this situation by entering -2 for the power of r in this term, using the pull-down menu at the end of the term to select the direction of the force to be the "r direction," and then typing the value of $-GM$ as the term's constant (the minus sign is necessary because the force in this case is actually in the $-r$ direction, i.e., toward the origin).

Defining the acceleration due to gravity

One can use SI units for r_0, v_0, and $-GM$, but if one is considering orbits of planetlike objects in the solar system, it is more convenient to measure distances in astronomical units (where 1 AU is the mean distance between the earth and the sun) and times in years. One can easily determine the value of GM for the sun in these units by applying Kepler's third law to the essentially circular orbit of the earth:

Using units of astronomical units and years for solar system problems

$$T^2 = \frac{4\pi^2}{GM}R^3 \quad \Rightarrow \quad GM = 4\pi^2 \frac{R^3}{T^2} = 4\pi^2 \frac{(1\ \text{AU})^3}{(1\ \text{y})^2} = 4\pi^2 \frac{\text{AU}^3}{\text{y}^2} \quad \text{(N13.6)}$$

Since GM has the same value for all objects orbiting the sun, once we have computed it using data from the earth's orbit, we know it for all objects orbiting the sun. Since $4\pi^2 = 39.48$, we therefore enter $-39.48\ \text{AU}^3/\text{y}^2$ for the constant required by the program Newton. (You should also be sure to enter "AU" as the distance unit and "y" as the time unit in the appropriate slots to

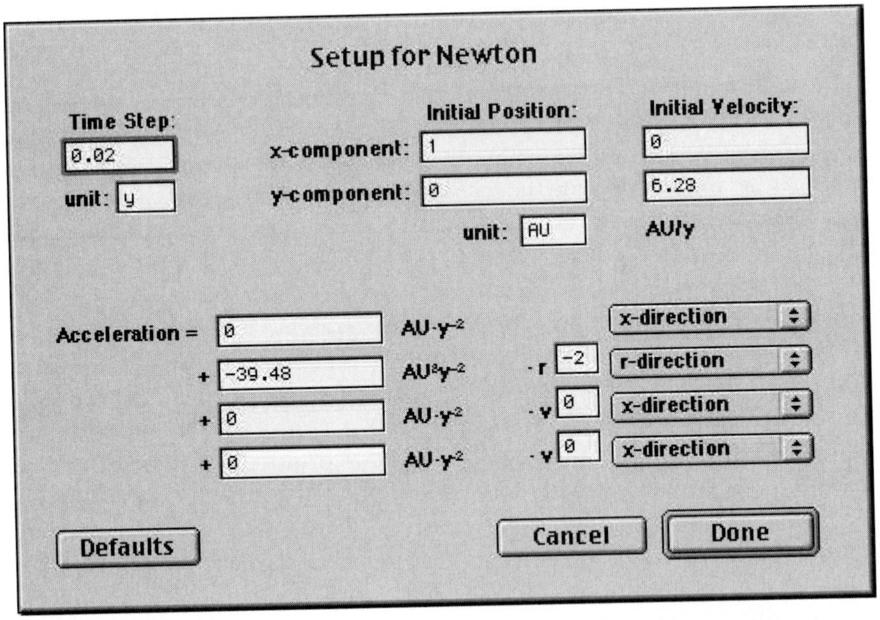

Figure N13.4
Setting up the Newton program to model the earth's orbit around the sun.

Choosing an appropriate time step

Reproducing the earth's orbit as a check

"Proving" Kepler's first law

ensure that the graphs that Newton produces are labeled correctly. If you do this, the units of the constant will be specified automatically by the program.)

As for choosing a time step, one often does this by trial and error. If the acceleration changes dramatically during a time step, the approximations assumed in the trajectory construction method will be poor. Conversely, if the time step is "small enough," then making it smaller does not change the answer much. In general, I find that in orbit problems, a complete orbit should be divided into at least 30 to 100 steps for good results (maybe more if the eccentricity is near to 1).

It is good to check that the program is operating correctly by showing that it correctly reproduces the earth's orbit, given the correct initial conditions. To duplicate the earth's orbit, choose $r_0 = 1$ AU and $v_0 = 2\pi$ AU/y, since the earth covers 2π AU of distance in the 1 y of its orbit and its speed is approximately constant. If you enter these quantities and choose $\Delta t \approx 0.02$ y (as shown in figure N13.4), you should find that the program Newton indeed produces a circular orbit.

Now let's consider a situation in which the orbit will not be circular. Imagine that we have an object whose initial distance from the sun is 1 AU (same as the earth) but has an initial speed v_0 that is too high for a circular orbit, let's say, 7.695 AU/y instead of 2π AU/y = 6.28 AU/y. If you enter this into Newton, you should get an elliptical orbit whose closest point to the sun is $r_c = 1$ AU but whose farthest point from the sun is $r_f = 3$ AU.

Exercise N13X.4

Run Newton and show that this is true.

How could we verify that this is indeed an ellipse? We can use the basic definition! The sun corresponds to one of the ellipse's foci. Since the two foci are supposed to be symmetrically placed along the wide axis of the ellipse, you should be able to see that the other focus must be at $x = -2$ AU. Now pick an arbitrary point on the ellipse and measure the distances r_1 and r_2 to the two foci with a ruler. You should find that these add up to a distance that on the graph that is equivalent to $2a = 4$ AU. Check another point and see that r_1 and r_2 add up to the same number. Continue the process until you are convinced that the orbit Newton has drawn is indeed an ellipse.

Exercise N13X.5

Carry out the process described in the previous paragraph.

If you count the time steps, you should find that the complete orbit in this situation takes about 2.82 y. In chapter N12, we proved Kepler's third law for a *circular* orbit, showing in fact that the period T for a circular orbit of radius R is given by $T^2 = 4\pi^2 R^3 / GM$. Just for the fun of it, let's compute what the radius R would be for a *circular* orbit having a period of 2.82 y. Solving for R and using the value of GM expressed in AU^3y^{-2} found previously, we get

A generalization of Kepler's third law

$$R = \left(\frac{GM}{4\pi^2} T^2 \right)^{1/3} = \left[\frac{4\pi^2 \, \text{AU}^3 \text{y}^{-2}}{4\pi^2} (2.82 \, y)^2 \right]^{1/3} = 2.0 \text{ AU} \quad \text{(N13.7)}$$

But this corresponds exactly to the semimajor axis a of our elliptical orbit! Since we just pulled this orbit out of a hat, it makes sense to hypothesize that for *all* elliptical orbits

$$T^2 = \frac{4\pi^2}{GM} a^3 \qquad\qquad \text{(N13.8)}$$

Purpose: This equation generalizes Kepler's third law for elliptical orbits.

Symbols: T is the period of the orbit, G is the universal gravitational constant $= 6.67 \times 10^{-11}$ N·m^2/kg^2, M is the mass of the primary, and a is the semimajor axis of the elliptical orbit.

Limitations: The equation assumes that the primary's mass is much larger than the satellite mass.

Notes: For planets orbiting our sun, $GM = 39.48 \text{ AU}^3/\text{y}^2$. Testing this formula shows that it does indeed correctly describe other orbits.

Exercise N13X.6

Change the object's initial speed to something arbitrary between 6.28 AU/y and 8.8 AU/y, and use Newton to check that our generalized Kepler's third law correctly describes the new orbit as well.

If we try increasing higher initial speeds, eventually we reach a point where the object's orbit is no longer elliptical. For example, if we set the initial speed to $v_0 = 10$ AU/y, the object moves continually away from the sun, eventually approaching a straight line that makes an angle of $\approx 50°$ with respect to the horizontal. This orbit exhibits the characteristic behavior of a hyperbola.

A hyperbolic orbit

To check that this curve is indeed a hyperbola with the sun at one focus, we have to find the other focus. This is much harder than in the elliptical case, but here is one method for doing this. Assume that we set up an xy coordinate system centered on the focus F_1 that we know, and we let $r_1 = h$ be the vertical distance between that focus and the point P where the hyperbola crosses

How to locate the second focus of the hyperbola

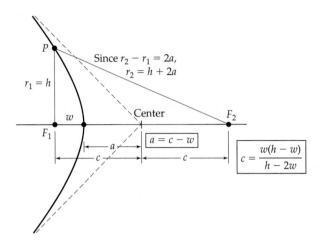

Figure N13.5
How to locate the second focus of a hyperbola, given *w* and *h*.

the y axis, as shown in figure N13.5. Let w be the distance between that focus and the hyperbola's vertex (w is the same as r_c in equation N13.4). Assuming that the curve really is a hyperbola, the distance r_2 from P to the other focus must satisfy $r_2 - r_1 = 2a$, so $r_2 = h + 2a$. But $F_2 F_1 P$ is a right triangle, so we know that

$$h^2 + 4c^2 = (h + 2a)^2 = h^2 + 4ah + 4a^2 \qquad \text{(N13.9}a\text{)}$$

We can also see from the diagram that

$$w = c - a \qquad \text{(N13.9}b\text{)}$$

Now, given a graph created by the program Newton, we can easily measure both h and w, and equations N13.9a and N13.9b give us two equations that can be solved for the remaining unknowns a and c. Solving equation N13.9b for a and plugging the result into equation N13.9a, we get (after a bit of algebra)

$$c = \frac{w(h - w)}{h - 2w} \qquad \text{(N13.10)}$$

Exercise N13X.7

Verify equation N13.10.

So if we measure w and h for our Newton-generated curve, then we can use equation N13.10 to calculate c and draw our hypothetical second focus F_2 a distance $2c$ from focus F_1 along the horizontal axis. Once we have this point, we can measure r_1 and r_2 for other points along the curve, and if $r_2 - r_1 = 2a = 2(c - w)$ consistently, then the curve really is a hyperbola.

Exercise N13X.8

Check that the curve generated by the program Newton for the case where $v_0 = 10$ AU/y really is a hyperbola.

Summary of what we have found

We have therefore seen that the Newton program predicts that (1) the orbit of a satellite around a very massive primary is either an ellipse or a hyperbola with the primary at one focus (this is a generalization of Kepler's first law!) and (2) the period of an elliptical orbit is given by $T^2 = 4\pi^2 a^3/GM$ (a generalization of Kepler's third law!). While we have not *mathematically* proved these things, it is worth emphasizing that Newton is simply a calculation tool that (at least if Δt is appropriately small) computes the numerical consequences of Newton's second law and the law of universal gravitation directly from these laws and the definitions of velocity and acceleration. Therefore, if we can believe that we are not horribly deluded by the few specific examples we have examined, we can confidently accept that these statements do indeed follow from these laws.

N13.3   Conservation Laws and Orbits

Once we know that orbits around a massive primary are either elliptical or hyperbolic, we can use conservation of energy and angular momentum to calculate many things about an orbit without having to solve Newton's second law directly. In this section we will explore what these conservation laws can tell us about elliptical orbits.

Consider first an object in an elliptical orbit around a massive object with mass M. At the closest and farthest points of the orbit, the orbiting object's velocity is entirely perpendicular to the radial direction (see figure N13.6), since at these extreme points, the object's velocity is switching from being outward to inward (or vice versa) so its radial component is passing through zero. This implies that the magnitude of the orbiting object's angular momentum around the primary is simply $L = \mathrm{mag}(\vec{r} \times m\vec{v}) = rmv \sin 90° = rmv$ at these points. Therefore, conservation of angular momentum and conservation of energy for these extreme points imply that

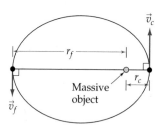

Figure N13.6
An orbiting object's velocity vectors are perpendicular to the radial direction at the orbit's extreme points.

$$r_c v_c = \frac{L}{m} = r_f v_f \qquad (\text{N13.11}a)$$

$$v_c^2 - \frac{2GM}{r_c} = \frac{2E}{m} = v_f^2 - \frac{2GM}{r_f} \qquad (\text{N13.11}b)$$

Conservation of energy and angular momentum for an elliptical orbit

The ratios L/m and $2E/m$ will appear in many of the equations that follow. I recommend that you think about them less as ratios and more as if they were single numbers whose symbols happen to be L/m and $2E/m$ respectively. I will place these ratios in parentheses in most of the equations that follow so that you can see them more easily.

Note that according to equation N13.2a, $r_c = a(1 - \varepsilon)$ and $r_f = a(1 + \varepsilon)$. If one multiplies both sides of the left equality in equation N13.11b by r_c^2 and substitutes the information from equations N13.2a and N13.11a, one gets

$$\left(\frac{L}{m}\right)^2 - 2GMa(1 - \varepsilon) = \frac{2E}{m}a^2(1 - \varepsilon)^2 = \frac{2E}{m}a^2(1 - 2\varepsilon + \varepsilon^2) \quad (\text{N13.12})$$

Similarly, the right equality in equation N13.11b yields

$$\left(\frac{L}{m}\right)^2 - 2GMa(1 + \varepsilon) = \frac{2E}{m}a^2(1 + 2\varepsilon + \varepsilon^2) \qquad (\text{N13.13})$$

If we subtract equation N13.13 from equation N13.12 and rearrange terms, we find that

$$4GMa\varepsilon = -\frac{2E}{m}a^24\varepsilon \quad \Rightarrow \quad E = -\frac{GMm}{2a} \qquad (\text{N13.14})$$

The link between energy and semimajor axis

Exercise N13X.9

Fill in the missing steps between equations N13.13 and equation N13.14.

This surprising result tells us that *the total conserved energy of an object in an elliptical orbit depends on the orbit's semimajor axis alone* (i.e., *not* on the object's angular momentum or the orbit's eccentricity), and that the energy gets larger (less negative) as the semimajor axis increases. This very powerful equation is useful in a number of contexts.

If we add equations N13.12 and N13.13, we get

$$2\left(\frac{L}{m}\right)^2 - 4GMa = \frac{2E}{m}a^2 2(1+\varepsilon^2) \qquad \text{(N13.15)}$$

If we divide this equation through by 2 and plug in $a = -GM/(2E/m)$ (from equation N13.14) and rearrange things, we find that

The link between energy, angular momentum, and eccentricity

$$\varepsilon^2 = 1 + \frac{1}{(GM)^2}\left(\frac{L}{m}\right)^2\left(\frac{2E}{m}\right) \qquad \text{(N13.16)}$$

This equation gives us the relationship between the orbit's eccentricity and the conserved energy and angular momentum of the orbiting object.

Exercise N13X.10

Verify equation N13.16.

The analogous results for hyperbolic orbits

We can do the same kind of analysis for hyperbolic orbits as well. This is somewhat trickier because a hyperbolic orbit does not have an easily analyzed far point. However, we can use the limiting behavior as the object goes to very large r as the equivalent of the far point in the elliptical orbit calculation.

The calculations are involved, and I don't think they give enough illumination to be worth the effort, so I will leave them as a problem for those interested (see problem N13A.1). The results, however, are *very* interesting. We find that

$$E = +\frac{GMm}{2a} \qquad \text{(N13.17)}$$

$$\varepsilon^2 = 1 + \frac{1}{(GM)^2}\left(\frac{L}{m}\right)^2\left(\frac{2E}{m}\right) \qquad \text{(N13.18)}$$

$$\tan\theta = \frac{1}{GM}\left(\frac{L}{m}\right)\sqrt{\frac{2E}{m}} \qquad \text{(N13.19)}$$

where a here is defined as in figure N13.2 and θ is the angle of the hyperbola's asymptotes. Note that equation N13.17 is the same as equation N13.14 except for the sign, and equation N13.18 is exactly the same as equation N13.16!

Note that equations N13.14, N13.16, N13.17, and N13.18 are consistent in implying that a negative system energy implies an elliptical orbit and a positive system energy implies a hyperbolic orbit. We can summarize this by the following table:

An important result linking an orbit's energy to its shape

$$\text{If} \quad E < 0 \quad \text{then} \quad \varepsilon < 1 \text{ (the orbit is } elliptical\text{)}$$
$$\text{If} \quad E > 0 \quad \text{then} \quad \varepsilon > 1 \text{ (the orbit is } hyperbolic\text{)}$$

The case of $E = 0$ corresponds to the crossover between the two cases. This is of mathematical interest only (no real orbit will have its total energy *exactly* equal to zero), but the orbit in this case would be a parabola (see the discussion near the end of section N13.2).

Table N13.1   Table of useful equations for solving orbit problems

Item	Equation		Symbols
Definitions of E and L	$\dfrac{2E}{m} = v^2 - \dfrac{2GM}{r}$ and $\dfrac{L}{m} \equiv rv\sin\phi$		E = total system energy L = system angular momentum G = gravitational constant m = satellite mass
Calculation of eccentricity	$\varepsilon^2 = 1 + \dfrac{1}{(GM)^2}\left(\dfrac{L}{m}\right)^2\left(\dfrac{2E}{m}\right)$		M = primary mass r = distance from primary

Item	Elliptical Case	Hyperbolic case	v = satellite speed at that point
Connection between a and E	$\dfrac{2E}{m} = -\dfrac{GM}{a}$	$\dfrac{2E}{m} = +\dfrac{GM}{a}$	ϕ = angle between $\vec{r}$ and $\vec{v}$ a = semimajor axis for ellipse,
Location of extremes	$r_c = a(1-\varepsilon)$ $r_f = a(1+\varepsilon)$ $r_c + r_f = 2a$	$r_c = a(\varepsilon - 1)$	analogous for hyperbola r_c = distance from primary to closest point on the orbit
Other useful relations	$T^2 = \dfrac{4\pi^2}{GM}a^3$ (Kepler's third law)	$\tan\theta = \sqrt{\varepsilon^2 - 1}$ $= \dfrac{1}{GM}\left(\dfrac{1}{m}\right)\sqrt{\dfrac{2E}{m}}$	r_f = distance from primary to farthest point on the orbit T = orbital period θ = angle of asymptotes

N13.4   Solving Orbit Problems

The point of the last section is that knowing the values of $2E/m$ and L/m for a system involving a satellite orbiting a massive primary allows us to calculate a and ε, which in turn completely specify the shape of the satellite's orbit. Table N13.1 summarizes the crucial equations regarding orbits. Note that since E and L are conserved quantities for the orbit, we can calculate these quantities using position and velocity information at *any* point on the orbit.

We can use the equations in table N13.1 to solve a variety of orbit problems

Applying these equations allows us to solve an astonishing variety of interesting orbit problems. Examples N13.1 through N13.4 are just a sample.

Example N13.1

Problem   Imagine that a strange object is discovered 22 AU (3.3 Tm) from the sun. Measurements show its velocity to be 11.2 km/s in a direction 169.7° from the line between the object and the sun (meaning that the velocity vector points mostly toward the sun). Is this a previously undiscovered member of the solar system, or is it an interloper from outside? How close will it get to the sun?

Translation and Model   Figure N13.7 shows the situation. We can answer both questions if we can calculate the quantities $2E/m$ and L/m from the information given. Since we are given the object's distance r from the sun and its speed v, and since we can easily calculate GM for the sun, we can easily calculate $2E/m = v^2 - GM/r$. We also know the angle $\theta = 169.7°$ that the object's velocity makes with the object's position vector $\vec{r}$, so we should be able to calculate $L/m = rv\sin\theta$ as well.

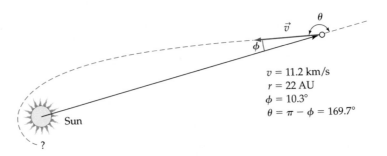

Figure N13.7
The situation described
in example N13.1.

$v = 11.2$ km/s
$r = 22$ AU
$\phi = 10.3°$
$\theta = \pi - \phi = 169.7°$

Solution   The value of GM for the sun is

$$GM = \left(6.67 \times 10^{-11}\ \frac{\text{N·m}^2}{\text{kg}^2}\right)(2.0 \times 10^{30}\ \text{kg})\left(\frac{1\ \text{kg·m/s}^2}{1\ \text{N}}\right)$$

$$= 1.33 \times 10^{20}\ \text{m}^3/\text{s}^2 \tag{N13.20}$$

The value of $2E/m = v^2 - GM/r$ for this object is thus

$$\frac{2E}{m} = \left(11{,}200\ \frac{\text{m}}{\text{s}}\right)^2 - \frac{2(1.33 \times 10^{20}\ \text{m}^3/\text{s}^2)}{3.3 \times 10^{12}\ \text{m}} = +4.4 \times 10^7\ \text{m}^2/\text{s}^2 \tag{N13.21}$$

Since this is positive, the orbit is hyperbolic, so the object must be from *outside* the solar system (assuming that it is unpowered). Ominous! The system's angular momentum per unit mass will be

$$\frac{L}{m} = rv \sin\theta = (3.3 \times 10^{12}\ \text{m})(11{,}200\ \text{m/s})(\sin 10.3°)$$

$$= 6.6 \times 10^{15}\ \text{m}^2/\text{s} \tag{N13.22}$$

Therefore,

$$a = \frac{+GM}{(2E/m)} = \frac{1.33 \times 10^{20}\ \text{m}^3/\text{s}^2}{4.4 \times 10^7\ \text{m}^2/\text{s}^2} = 3.0 \times 10^{12}\ \text{m} \tag{N13.23a}$$

$$\varepsilon = \left[1 + \frac{(4.4 \times 10^8\ \text{m}^2/\text{s}^2)(6.6 \times 10^{15}\ \text{m}^2/\text{s})^2}{(1.33 \times 10^{20}\ \text{m}^3/\text{s}^2)^2}\right]^{1/2} = 1.44 \tag{N13.23b}$$

and the object's distance r_c from the sun at the orbit's vertex will be

$$r_c = a(\varepsilon - 1) = (3.0 \times 10^{12}\ \text{m})(0.44) = 1.33 \times 10^{11}\ \text{m} \tag{N13.23c}$$

Note that r_c is just inside the earth's orbital radius of 1.5×10^{11} m.

Exercise N13X.11

You can see that it is handy to know the value of GM. Show that GM for the earth is 3.99×10^{14} m³/s². (Similarly, GM for the moon is about 4.9×10^{12} m³/s², and for Jupiter it is about 1.3×10^{17} m³/s².)

Example N13.2

Problem   A certain asteroid in an elliptical orbit is a distance $r_c = 3.5$ AU from the sun at the closest point in its orbit and a distance $r_f = 4.5$ AU at the farthest point. What is the period of its orbit in years?

Solution   The semimajor axis of the orbit is $a = \frac{1}{2}(r_c + r_f) = 4.0$ AU. The semimajor axis of the earth's orbit $a_e \approx 1$ AU (by definition). Since $T \propto a^{3/2}$,

$$\frac{T}{T_e} = \left(\frac{a}{a_e}\right)^{3/2} = \left(\frac{4.0 \text{ AU}}{1.0 \text{ AU}}\right)^{3/2} = 8.0 \qquad \text{(N13.24)}$$

Since the period of the earth's orbit $T_e = 1$ y by definition, the period of the asteroid's orbit must be 8.0 y.

Example N13.3

Problem   At the point of its orbit closest to the earth, a satellite is 200 km above the earth's surface and is traveling at a speed of 10.0 km/s. Find the farthest distance that this satellite gets from the earth, and determine its period.

Translation   See figure N13.8 for a sketch of the orbit. We are given that at the closest point in its orbit, the satellite is 200 km from the earth's surface, and thus 200 km + 6380 km = 6580 km = r_c from the earth's center, and that its speed $v_c = 10.0$ km/s at this point. We want to find r_f.

Model   The earth is extremely massive compared to any human-made satellite, so the massive-primary model we have developed in this chapter should work. We can calculate $2E/m$ and L/m using data using the values of r_c and v_c given for the close point, and we calculate anything we want from that.

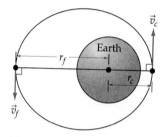

Figure N13.8
The situation described in example N13.3.

Solution   Using the given data and the result from exercise N13X.11, we find that

$$\frac{2E}{m} = \left(10{,}000 \, \frac{\text{m}}{\text{s}}\right)^2 - \frac{2(3.99 \times 10^{14} \text{ m}^3/\text{s}^2)}{6580 \times 10^3 \text{ m}} = +2.13 \times 10^7 \text{ m}^2/\text{s}^2 \quad \text{(N13.25)}$$

$$a = -\frac{GM}{(2E/m)} = -\frac{3.99 \times 10^{14} \text{ m}^3/\text{s}^2}{-2.16 \times 10^7 \text{ m}^2/\text{s}^2} = 1.88 \times 10^7 \text{ m} \quad \text{(N13.26)}$$

which is the same as 18,500 km. Since $r_f + r_c = 2a$,

$$r_f = 2a - r_c = 37{,}500 \text{ km} - 6580 \text{ km} = 30{,}900 \text{ km} \qquad \text{(N13.27)}$$

and according to the generalized version of Kepler's third law,

$$T = \sqrt{\frac{4\pi^2 (1.88 \times 10^7 \text{ m})^3}{3.99 \times 10^{14} \text{ m}^3/\text{s}^2}} = 25{,}600 \text{ s} \left(\frac{1 \text{ h}}{3600 \text{ s}}\right) = 7.1 \text{ h} \quad \text{(N13.28)}$$

Example N13.4

Problem   Imagine that we have a space probe in a circular orbit 200 km above the earth and we want to put it into an elliptical orbit whose farthest point coincides with the moon. How much do the probe's rocket engines have to add to its speed to put it into this orbit? Assume that the rocket engines fire for so brief a time that the object's distance from the earth does not change much during the time the engines are firing (i.e., during the *burn* in astronautical language).

Translation   Figure N13.7 still provides a qualitative sketch of the elliptical orbit (although solving the orbit in this case is much longer and thinner).

Model The earth is extremely massive compared to any probe, so the massive-primary model should be accurate. If the rocket engines fire only very briefly, the probe will be essentially at the same radius $r_c = 6580$ km both before and just after the engines fire. We also know $r_f = 384,000$ km from the inside front cover. We can use this information to calculate a, use that to calculate $2E/m$, and use that to calculate the speed v_c the probe must have just after the burn. We can calculate the probe's original speed v_0 in its circular orbit by using equation N12.12.

Solution Since $2a = r_f + r_c$, we have

$$a = \frac{1}{2}(r_f + r_c) \quad \Rightarrow \quad \frac{2GM}{r_f + r_c} = \frac{2E}{m} = v_c^2 - \frac{2GM}{r_c} \qquad \text{(N13.29)}$$

$$\Rightarrow \quad v_c = \sqrt{\frac{2GM}{r_f + r_c} + \frac{2GM}{r_c}}$$

$$= \sqrt{\frac{2(3.99 \times 10^{14} \text{ m}^3/\text{s}^2)}{391,000,000 \text{ m}} - \frac{2(3.99 \times 10^{14} \text{ m}^3/\text{s}^2)}{6,580,000 \text{ m}}}$$

$$= 22,200 \text{ m/s} \qquad \text{(N13.30)}$$

Since the probe's original circular orbit speed was

$$v_0 = \sqrt{\frac{GM}{r_c}} = \sqrt{\frac{3.99 \times 10^{14} \text{ m}^3/\text{s}^2}{6,580,000 \text{ m}}} = 7800 \text{ m/s} \qquad \text{(N13.31)}$$

the engines need to boost the probe's speed by 14,400 m/s.

TWO-MINUTE PROBLEMS

N13T.1 According to the text, an orbiting object having energy $E < 0$ follows an elliptical orbit. When we specify the numerical value for an object's energy, we are implicitly comparing it to the energy the object has in some reference situation. So, when we say that $E < 0$ in this case, we are comparing the energy of the orbiting object to the energy of an object with the same mass that is
A.   At rest at the primary's center.
B.   At rest on the primary's surface.
C.   In a circular orbit just above the primary's surface.
D.   At rest at $r = \infty$.
E.   Some other situation (specify).

N13T.2 An object is discovered near the earth with values of $2E/m = 2.12 \times 10^7 \text{ m}^3/\text{s}^2$ and $L/m = 7.8 \times 10^{10} \text{ m}^2/\text{s}$. This object is in what kind of orbit around the earth?
A.   Elliptical
B.   Parabolic
C.   Hyperbolic

N13T.3 The eccentricity of this object's orbit is
A.   1.08
B.   1.81
C.   0.81
D.   0.90
E.   1.35
F.   Other (specify)

N13T.4 A comet is discovered in an elliptical orbit around the sun. Its closest distance from the sun is 1.0 AU, and measurements of its speed at this distance imply that its greatest distance from the sun is 7 AU. About how many years will pass between the comet's closest approaches to the sun?
A.   8 y
B.   16 y
C.   22.6 y
D.   64 y
E.   226 y
F.   Other (specify)

N13T.5 For an elliptical orbit, the relationship $r_c v_c = r_f v_f$ follows from the fact that the satellite's velocity is perpendicular to its position vector at the two extreme points and from
A. Conservation of angular momentum.
B. Conservation of energy.
C. Conservation of momentum.
D. Newton's second law.
E. Some other properties of ellipses.
F. The given relationship is false.

N13T.6 Imagine that a satellite orbits the earth so closely that it experiences some drag due to the earth's upper atmosphere. This will drain away some of orbital energy of this system, converting it to thermal energy. If this happens fairly slowly, the satellite's orbit will remain nearly circular. What happens to the radius of this satellite's orbit as time passes in this case?
A. It slowly decreases.
B. It remains the same: just the satellite's speed decreases.
C. It slowly increases.

N13T.7 In the situation described in problem N13T.6, what do you think will happen to the satellite's speed as time passes?
A. It slowly decreases.
B. It remains roughly the same.
C. It slowly increases.

HOMEWORK PROBLEMS

Basic Skills

N13B.1 An object is discovered near the earth with values of E, L, and m such that $2E/m = -8.2 \times 10^6$ m^3/s^2 and $L/m = 7.8 \times 10^{10}$ m^2/s. Find the eccentricity of this orbit and the radius of the closest point of the orbit to the earth; and classify the orbit as being elliptical, parabolic, or hyperbolic.

N13B.2 A space probe near the earth has values of E, L, and m such that $L/m = 7.8 \times 10^{10}$ m^2/s and $2E/m = 0$. Find the eccentricity of this orbit and the radius of the closest point of the orbit to the earth, and classify the orbit as elliptical, parabolic, or hyperbolic.

N13B.3 An asteroid orbiting the sun has a semimajor axis of 3.0 AU. What is the period of its orbit in years?

N13B.4 A space probe is put in an orbit around the sun that is 0.5 AU from the sun at the closest and 2.0 AU at the farthest. What is its period in years?

N13B.5 Satellite A travels around the earth in a circular orbit of radius R. Another satellite orbits in an elliptical orbit that is R from the earth at its closest point and $3R$ from the earth at its farthest point. How does the period of satellite B compare to that of satellite A? Show your work.

N13B.6 Imagine that you are in an orbit around the earth whose most distant point from the earth is 5 times farther from the earth's center than the closest point. If your speed is 10.0 km/s at the closest point, what is your speed at the farthest point?

N13B.7 A satellite orbits the earth in such a way that its speed at its point of closest approach is roughly 3 times its speed at the most distant point. How many times more distant from the earth's center is the farthest point than the closest point?

Synthetic

N13S.1 A small asteroid is discovered 14,000 km from the earth's center moving at a speed of 9.2 km/s. Can you tell from the information provided whether this asteroid is in an elliptical or hyperbolic orbit around the earth? Is the *direction* of its velocity vector important in determining this? Please explain.

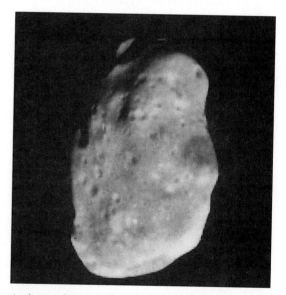

A photo of the small asteroid 243 Ida taken by the Galileo spacecraft.

N13S.2 A mysterious object moving at a speed of 21 km/s at a distance of 3.8 AU from the sun is sighted by astronomers (1 AU = earth's orbital radius = 1.5×10^{11} m). Can you tell from the information provided whether this object is in an elliptical or hyperbolic orbit around the sun? Is the *direction* of its velocity vector important in determining this? Please explain.

N13S.3 Imagine that you are in a circular orbit of radius $R = 7500$ km around the earth. You'd like to get to a geostationary space station whose circular orbit has a radius of $3R$, so you'd like to put yourself in an elliptical orbit whose closest point to the earth has radius R and whose most distant point has radius $3R$ (such an orbit is called a *Hohmann transfer* orbit, and it represents the lowest-energy way to get from one circular orbit to another). By what percentage would you have to increase your speed while at radius R to put yourself in this transfer orbit? (Assume that your rocket engines fire for such a brief time that your distance from the earth remains essentially R throughout the time the engines are on.)

N13S.4 Imagine that you are an astronaut in a circular orbit of $R = 6500$ km around the earth.
(a) What is your orbital speed?
(b) Say that you fire a rocket pack so that in a very short time, you increase your speed in the direction of your motion by 20%. What are the characteristics of your new orbit? Calculate a, ε, T, r_c, and r_f for the new orbit.

N13S.5 A certain satellite is in an orbit around the earth whose nearest and farthest points from the earth's center are 7000 km and 42,000 km, respectively. Find the satellite's orbital speed at its point of closest approach.

N13S.6 An asteroid is in an orbit around the sun whose closest point to the sun has a radius R and whose most distant point has a radius $9R$, where R is equal to the radius of the earth's orbit $= 1$ AU $= 1.5 \times 10^{11}$ m.
(a) What is the asteroid's speed when it is closest to the sun?
(b) How does this compare with the earth's orbital speed?
(c) What is the period of this orbit, in years?

N13S.7 A new comet is discovered 6.6 AU from the sun (1 AU $=$ earth's orbital radius $= 1.5 \times 10^{11}$ m) moving with a speed of 17 km/s. At that time, its velocity vector $\vec{v}$ makes an angle of 174.3° with respect to its position vector $\vec{r}$ (meaning that the comet's velocity points almost toward the sun).
(a) Is this comet in a hyperbolic orbit?
(b) What will be its distance from the sun at the point of closest approach?

N13S.8 Some recent space probes have made several hyperbolic passes past the earth to help give them the right direction and speed to go to their final destination. Imagine that one such probe has a speed of 12 km/s at a distance of 650,000 km from the earth. If the angle that the probe's velocity vector $\vec{v}$ makes with its position vector $\vec{r}$ at that time is 177°, how near will it pass by the earth at its point of closest approach?

N13S.9 Argue that for a given value of angular momentum L, the smallest (most negative) energy that an orbiting object can have is

$$E = -\frac{G^2 M^2 m^3}{2L^2} \tag{N13.32}$$

and that the orbit in this case will be circular.

N13S.10 (a) Do exercises N13X.4, N13X.5, N13X.6, and N13X.8. Hand in a printout of the graph that program Newton generates for each exercise, and do any additional work required on the printout.
(b) For the hyperbolic orbit, determine the value of ε from appropriate measurements you can make directly on the graph (once you have located the other focus). Then calculate the asymptote angle θ, using equation N13.3. Check that this angle is consistent with the asymptote angle displayed on your printout.

Rich-Context

N13R.1 You are the commander of the starship *Execrable*. You are currently in a standard orbit around a class-M planet whose mass is 4.4×10^{24} kg and whose radius is 6100 km. Your current circular orbit around the planet has a radius of $R = 50{,}000$ km. Your exobiologist wants to get in closer (say to an orbital radius of 10,000 km, or $R/5$) to look for signs of life. Your planetary geologist wants to stay at the current radius so that the entire face of the planet can be scanned with the sensors at once. Because you are tired of the bickering, you decide to put the *Execrable* into an elliptical orbit whose minimum distance from the planetary center is $R/5$ and whose farthest point is R from the same. Your navigational computers are down again (of course), so you have to compute by hand how to insert yourself in this new orbit. Your impulse engines are capable of causing the ship to accelerate at a rate of $1\ g = 9.8$ m/s^2. In what direction should you fire your engines, relative to your current direction of motion (forward or backward)? For how many seconds should you fire them?

N13R.2 You'd like to put your spaceship into a Hohmann transfer orbit between earth and Mars (i.e., an elliptical orbit whose closest point to the sun is the same as the earth's orbital radius and whose farthest point is the same as Mars's orbital radius: see problem N13.4).
(a) If your spaceship is initially traveling around the sun with the same orbital speed as the earth

and at the same distance from the sun as the earth is, by what factor will you need to increase your speed to put you into the transfer orbit? Assume that you fire your rocket engines for a short enough time that your distance from the sun is still essentially 1.0 AU when the engines cease firing.

(b) How long will it take you to get to Mars along this orbit?

(c) Check your work, using the Newton computer program: set up the probe as having an initial position of 1.0 AU along the x axis, and give it an initial velocity parallel to the y axis whose magnitude is the value of v_c you calculated in part (b). I recommend a time step of about 0.02 y. Show that running some time steps does indeed produce an elliptical orbit having the desired value of $r_f = R$; and by counting time steps, verify the flight time that you predicted in part (a).

N13R.3 Imagine that you are the orbital engineer for the first NASA space shot to Ceres, the largest known asteroid. Ceres's nearly circular orbit around the sun has a radius of $R = 2.77$ AU. After being launched from the earth, the probe will initially be in a circular orbit around the sun with the same radius ($r_e = 1.0$ AU) and the same orbital speed ($v_e = 6.28$ AU/y) as the earth's orbit. The probe's rocket engines will then fire briefly to increase the probe's speed to that speed v_c needed to put the probe into an elliptical orbit whose initial (and minimum) distance from the sun is $r_e = 1.0$ AU and whose final (and largest) distance from the sun is $R = 2.77$ AU. (Such an orbit is called a *Hohmann transfer orbit* between the earth and Ceres: this happens to be the orbit between the two that requires the least energy expenditure.)

(a) How long will it take the probe to get from the earth to Ceres in such an orbit?

(b) What is the speed v_c that the probe has to have just after firing its engines to be inserted into this orbit? (Assume that the duration of the boost is short enough that its distance from the sun is still $\approx r_e$ just after the engines have been fired.)

(c) Check your work, using the Newton computer program: set up the probe as having an initial position of 1.0 AU along the x axis, and give it an initial velocity parallel to the y axis whose magnitude is the value of v_c that you calculated in part (b). I recommend a time step of about 0.02 y. Show that running some time steps does indeed produce an elliptical orbit having the desired value of $r_f = R$; and by counting time steps, verify the flight time that you predicted in part (a).

Advanced

N13A.1 Consider an object in a hyperbolic orbit around a massive primary.

(a) Argue that the object's speed v and angular momentum per unit mass at a point a very long distance from the primary are

$$v \approx \sqrt{\frac{2E}{m}} \quad \text{and} \quad \frac{L}{m} = va \qquad \text{(N13.33)}$$

(*Hint:* Look at figure N13.2b, and recall that the magnitude of the cross product for $\vec{L}$ can be written $L = r_\perp mv$.)

(b) Use this information to derive equations N13.17 and N13.18. (*Hint:* Use the calculation of the corresponding equations for the ellipse as a template.)

ANSWERS TO EXERCISES

N13X.1 Note that from figure N13.1, $r_f = a + c$ and $r_c = a - c$. But $c \equiv a\varepsilon$, so $r_f = a + a\varepsilon = a(1 + \varepsilon)$ and $r_c = a - a\varepsilon = a(1 - \varepsilon)$. If we add $r_f = a + c$ and $r_c = a - c$, we get $r_f + r_c = a + c + a - c = 2a \Rightarrow a = \frac{1}{2}(r_f + r_c)$. On the other hand, if we subtract r_c from r_f, we get $r_f - r_c = a + c - a + c = 2c = 2a\varepsilon$. Dividing by 2 yields the second equation in equation N13.2b.

N13X.2 Since $\sin^2\theta + \cos^2\theta = 1$, $\sin\theta = (1 - \cos^2\theta)^{1/2} = [1 - (1/\varepsilon)^2]^{1/2} = \varepsilon^{-1}(\varepsilon^2 - 1)^{1/2}$. Also

$$\tan\theta = \frac{\sin\theta}{\cos\theta} = \frac{\varepsilon^{-1}\sqrt{\varepsilon^2 - 1}}{\varepsilon^{-1}}$$

$$= \sqrt{\varepsilon^2 - 1} \qquad \text{(N13.34)}$$

N13X.3 In the case of the ellipse, we have

$$2a = r_1 + r_2 = \sqrt{(x + c)^2 + y^2} + \sqrt{(x - c)^2 + y^2}$$

$$\Rightarrow \quad 2a - \sqrt{(x + c)^2 + y^2} = \sqrt{(x - c)^2 + y^2}$$

$$\Rightarrow \quad 4a^2 - 4a\sqrt{(x + c)^2 + y^2} + (x + c)^2 + y^2$$
$$= (x - c)^2 + y^2$$

$$\Rightarrow \quad 4a^2 - 4a\sqrt{(x + c)^2 + y^2} + x^2 + 2xc + c^2 + y^2$$
$$= x^2 - 2xc + c^2 + y^2$$

$$\Rightarrow \quad 4a^2 - 4a\sqrt{(x + c)^2 + y^2} + 4xc = 0$$

$$\Rightarrow \quad a^2 + xc = a\sqrt{(x + c)^2 + y^2}$$

$\Rightarrow \quad a^4 + 2a^2xc + x^2c^2 = a^2(x^2 + 2xc + c^2 + y^2)$

$\Rightarrow \quad a^4 + 2a^2xc + x^2c^2$
$$= a^2x^2 + 2a^2xc + a^2c^2 + a^2y^2$$

$\Rightarrow \quad a^2(a^2 - c^2) = (a^2 - c^2)x^2 + a^2y^2 \qquad \text{(N13.35)}$

Dividing through by $a^2(a^2 - c^2)$ gives the desired result. The hyperbolic case is completely analogous.

N13X.4 You have to do this one yourself.

N13X.5 The answer is given.

N13X.6 The answer is given.

N13X.7 According to equation N13.9b, $a = c - w$. Plugging this into equation N13.9a, we get

$$h^2 + 4c^2 = h^2 + 4(c - w)h + 4(c^2 - 2cw + w^2)$$
$$= h^2 + 4ch - 4wh + 4c^2 - 8cw + 4w^2$$
$$\text{(N13.36)}$$

Canceling h^2 and $4c^2$ from both sides and dividing both sides by 4, we get

$$0 = ch - wh - 2cw + w^2 = c(h - 2w) - wh + w^2$$
$$\text{(N13.37)}$$

If we now add $wh - w^2$ to both sides and divide both sides by $h - 2w$, we get the desired result.

N13X.8 *Hints:* For this curve, you should find that $c = 2.89$ AU, so the other focus is at $x = 2c = 5.78$ AU. The value of $2a$ is $2(c - w) = 3.77$ AU. You can find the distance in centimeters that is equivalent to $2a$ on your graph, measure r_1 and r_2 for various points

on your curve in centimeters, and then show that $r_2 - r_1$ is equal to $2a$ cm.

N13X.9 Performing the specified subtraction, we get

$$\left(\frac{L}{m}\right)^2 - \left(\frac{L}{m}\right)^2 + 2GMa(-1 + \varepsilon + 1 + \varepsilon)$$
$$= \frac{2E}{m}a^2(1 - 2\varepsilon + \varepsilon^2 - 1 - 2\varepsilon - \varepsilon^2)$$

$$\Rightarrow \quad 0 + 2GMa(2\varepsilon) = \frac{2E}{m}a^2(-4\varepsilon)$$

$$\Rightarrow \quad 4GMa\varepsilon = -\frac{2E}{m}4a\varepsilon \qquad \text{(N13.38)}$$

Dividing both sides by $8a\varepsilon/m$ yields the desired result.

N13X.10 After dividing both sides of equation N13.15 by 2 and substituting $a = -GM/(2E/m)$, we get

$$\left(\frac{L}{m}\right)^2 - 2GM\left[-\frac{GM}{(2E/m)}\right]$$
$$= \frac{2E}{m}\left[-\frac{GM}{(2E/m)}\right]^2(1 + \varepsilon^2) \qquad \text{(N13.39)}$$

Multiplying both sides of this by $(2E/m)/(GM)^2$ yields

$$\frac{1}{(GM)^2}\left(\frac{2E}{m}\right)\left(\frac{L}{m}\right)^2 + 2 = 1 + \varepsilon^2$$

$$\Rightarrow \quad \frac{1}{(GM)^2}\left(\frac{2E}{m}\right)\left(\frac{L}{m}\right)^2 + 1 = \varepsilon^2 \qquad \text{(N13.40)}$$

N13X.11 This is just a matter of plugging in numbers.

Appendix **NA**
Differential Calculus

NA.1   Derivatives

The purpose of this section is to review (in the relatively informal language used by physicists) some basic principles of differential calculus, principles that we will use repeatedly throughout this unit and in later units.

Consider a function of $f(t)$, where t is some *arbitrary* variable (not *necessarily* time, although functions of time will be our main concern in this unit). The t derivative of $f(t)$ evaluated at t is defined to be

$$\frac{df}{dt} \equiv \lim_{\Delta t \to 0} \frac{f(t + \Delta t) - f(t)}{\Delta t} \qquad \text{(NA.1)}$$

Definition of the derivative

That is, the derivative is the limit as Δt goes to zero of the change in f during the interval Δt starting at t, divided by the value of Δt.

The best way to show what we mean by "the limit as Δt goes to zero" here is to do an example. Imagine that $f(t) = t^2$. The derivative of $f(t)$ in this case is

$$\frac{df}{dt} \equiv \frac{d}{dt}[t^2] \equiv \lim_{\Delta t \to 0} \frac{(t + \Delta t)^2 - t^2}{\Delta t} \qquad \text{(NA.2)}$$

Examples of evaluating the limit as Δt goes to zero

Multiplying out the binomial in the numerator, we find that

$$\frac{(t + \Delta t)^2 - t^2}{\Delta t} = \frac{t^2 + 2t\,\Delta t + \Delta t^2 - t^2}{\Delta t} = \frac{2t\,\Delta t + \Delta t^2}{\Delta t} = 2t + \Delta t \quad \text{(NA.3)}$$

As Δt becomes smaller, this clearly gets closer and closer to the value $2t$. Thus

$$\text{If} \quad f(t) = t^2 \quad \text{then} \quad \frac{df}{dt} \equiv \lim_{\Delta t \to 0} (2t + \Delta t) = 2t \qquad \text{(NA.4)}$$

Here is a second example. Imagine that $f(t) = 1/t$. Its derivative is

$$\frac{df}{dt} \equiv \frac{d}{dt}\left(\frac{1}{t}\right) \equiv \lim_{\Delta t \to 0} \frac{1}{\Delta t}\left(\frac{1}{t + \Delta t} - \frac{1}{t}\right) \qquad \text{(NA.5)}$$

Putting the expression in parentheses over a common denominator, we find that

$$\frac{1}{\Delta t}\left[\frac{1}{t + \Delta t} - \frac{1}{t}\right] = \frac{1}{\Delta t}\left[\frac{t - (t + \Delta t)}{t(t + \Delta t)}\right] = \frac{1}{\Delta t}\left[\frac{-\Delta t}{t(t + \Delta t)}\right] = \frac{-1}{t(t + \Delta t)} \qquad \text{(NA.6)}$$

In the limit that Δt becomes very small, the value of $t + \Delta t$ approaches t, so

$$\text{If} \quad f(t) = \frac{1}{t} \quad \text{then} \quad \frac{df}{dt} = \lim_{\Delta t \to 0} \frac{-1}{t(t + \Delta t)} = -\frac{1}{t^2} \qquad \text{(NA.7)}$$

NA.2   Some Useful Rules

In general, one can show, using the same general approach, that

$$\text{If} \quad f(t) = t^n \quad \text{then} \quad \frac{df}{dt} = nt^{n-1} \qquad \text{(NA.8)}$$

The derivative of $f(t) = t^n$

for any integer $n \neq 0$. (Although the proof is more complicated, this equation applies to *noninteger* values of n as well.)

If $f(t) = c$ (a constant), taking the limit is easy:

The derivative of a constant

$$\text{If} \quad f(t) = c \quad \text{then} \quad \frac{df}{dt} = \lim_{\Delta t \to 0} \frac{c - c}{\Delta t} = \lim_{\Delta t \to 0} \frac{0}{\Delta t} = 0 \quad \text{(NA.9)}$$

This case makes it abundantly clear that the *limit* of $0/\Delta t$ as Δt approaches zero (which is zero here) is *not* the same as the *value* of $0/\Delta t$ when $\Delta t = 0$ (which is undefined, as are the ratios in equations NA.2 and NA.5). The limit is the value that the function *approaches* as Δt gets small: since $0/\Delta t$ is zero for all $\Delta t \neq 0$, we have to say that $0/\Delta t$ approaches (indeed *is exactly*) zero as Δt gets small.

The following useful theorems follow from the definition of the derivative: if $f(t)$, $g(t)$, and $h(t)$ are arbitrary functions of t, then

The sum rule

$$\text{If} \quad f(t) = g(t) + h(t) \quad \text{then} \quad \frac{df}{dt} = \frac{dg}{dt} + \frac{dh}{dt} \quad \text{(NA.10)}$$

The product rule

$$\text{If} \quad f(t) = g(t)h(t) \quad \text{then} \quad \frac{df}{dt} = \frac{dg}{dt}h + g\frac{dh}{dt} \quad \text{(NA.11)}$$

The inverse rule

$$\text{If} \quad f(t) = \frac{1}{g(t)} \quad \text{then} \quad \frac{df}{dt} = -\frac{1}{[g(t)]^2}\frac{dg}{dt} \quad \text{(NA.12)}$$

Equations NA.9 and NA.11 together imply that if c is a constant

The constant rule

$$\text{If} \quad f(t) = ch(t) \quad \text{then} \quad \frac{df}{dt} = c\frac{dh}{dt} \quad \text{(NA.13)}$$

A physicist's proof of the product rule

The sum, product, and inverse rules are fairly easy to prove. As an example, consider the product rule. If $f(t) = g(t)h(t)$, then

$$\frac{df}{dt} \equiv \lim_{\Delta t \to 0} \frac{g(t + \Delta t)h(t + \Delta t) - g(t)h(t)}{\Delta t} \quad \text{(NA.14)}$$

Adding zero to the numerator in the form $-g(t + \Delta t)h(t) + g(t + \Delta t)h(t)$ yields

$$\frac{df}{dt} \equiv \lim_{\Delta t \to 0} \frac{g(t + \Delta t)h(t + \Delta t) - g(t + \Delta t)h(t) + g(t + \Delta t)h(t) - g(t)h(t)}{\Delta t}$$

$$= \lim_{\Delta t \to 0} \left[g(t + \Delta t)\frac{h(t + \Delta t) - h(t)}{\Delta t} + \frac{g(t + \Delta t) - g(t)}{\Delta t}h(t) \right] \quad \text{(NA.15)}$$

Since the limit of a sum is the sum of the limits and the limit of a product is the product of the limits, this reduces to

$$\frac{df}{dt} = \lim_{\Delta t \to 0} [g(t + \Delta t)] \lim_{\Delta t \to 0} \left[\frac{h(t + \Delta t) - h(t)}{\Delta t} \right]$$

$$+ \lim_{\Delta t \to 0} \left[\frac{g(t + \Delta t) - g(t)}{\Delta t} \right] \lim_{\Delta t \to 0} [h(t)]$$

$$\equiv g(t)\frac{dh}{dt} + \frac{dg}{dt}h(t) \quad \text{Q.E.D.} \quad \text{(NA.16)}$$

The other rules can be proved similarly.

An example application

Equations NA.8 through NA.13 make it possible to evaluate the derivative of any polynomial in t. For example, if $f(t) = at^3 + bt + c$, then

$$\frac{df}{dt} = \frac{d}{dt}(at^3) + \frac{d}{dt}(bt) + \frac{d}{dt}(c) \qquad \text{by the sum rule}$$

$$= a\frac{d}{dt}(t^3) + b\frac{d}{dt}(t) + 0 \qquad \text{by the constant rule and NA.9}$$

$$= a(3t^2) + b(1) = 3at^2 + b \qquad \text{by equation NA.8} \quad \text{(NA.17)}$$

Exercise NAX.1

Use the rules described above to calculate the t derivative of the function $f(t) = (at + b)^{-1}$, where a and b are constants.

NA.3   Derivatives and Slopes

The meaning of the *derivative* of a function $f(t)$ can be more fully understood in terms of graphs of $f(t)$. This section reviews the graphical interpretation of the derivative.

Recall from equation NA.1 that the derivative of the function $f(t)$ with respect to its variable t is defined to be

$$\frac{df}{dt} \equiv \lim_{\Delta t \to 0} \frac{\Delta f}{\Delta t} \equiv \lim_{\Delta t \to 0} \frac{f(t + \Delta t) - f(t)}{\Delta t} \qquad \text{(NA.18)}$$

Figure NA.1 illustrates that on a graph of $f(t)$ plotted versus t, the ratio $\Delta f / \Delta t$ for nonzero Δt is equal to the slope of a straight line drawn between the points on the curve at t and $t + \Delta t$. As Δt approaches zero, the value of this slope approaches a specific value that equals the slope of a line drawn tangent to the graph at t. We call this the **slope of $f(t)$ at that point.**

NA.4   The Chain Rule

The **chain rule** is one of the most important and useful theorems of differential calculus. Assume that $f(u)$ is a function of u, which in turn is a function $u(t)$ of t. This means that f is also a function of t and thus has a meaningful t derivative. The chain rule claims that

Statement of the chain rule

$$\frac{df}{dt} = \frac{df}{du}\frac{du}{dt} \qquad \text{(NA.19)}$$

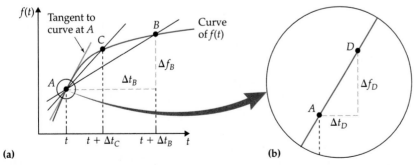

(a)   **(b)**

Figure NA.1
(a) The ratio $\Delta f_B / \Delta t_B$ specifies the slope of the line going through points A and B. The ratio $\Delta f_C / \Delta t_C$ specifies the slope of the line going through points A and C. The latter is closer to the slope of the line tangent to the curve of $f(t)$ at point A (light colored straight line). (b) As Δt gets *very* small, the curve and the tangent line become indistinguishable, and $\Delta f / \Delta t$ specifies the slope of *both* in the neighborhood of A.

This theorem looks relatively pedestrian, but in fact is *very* useful. For example, imagine that we want to find the derivative of the function

$$f(t) = (at^2 + bt + c)^5 \qquad (NA.20)$$

We *could* evaluate this derivative by multiplying out the polynomial and then taking the derivative of each term, using the constant rule and the fact that the derivative of t^n is nt^{n-1} (see equation NA.8). This would be very tedious. It is much simpler to define $u \equiv at^2 + bt + c$, which means that $f(u) = u^5$. The chain rule then tells us that

$$\frac{df}{dt} = \frac{df}{du}\frac{du}{dt} = \left(\frac{d}{du}u^5\right)\left[\frac{d}{dt}(at^2 + bt + c)\right]$$

$$= (5u^4)(2at + b) = 5(at^2 + bt + c)^4(2at + b) \qquad (NA.21)$$

Exercise NAX.2

Use this technique to find the derivative of $f(t) = 1/(t^2 - b)^3$.

The "physicist's proof" as a mnemonic device

The physicist's "proof" of this theorem, by the way, is that we simply multiply the top and bottom of df/dx by du and rearrange things a bit (physicists tend to think of df, dt, and du simply as small numbers). Real mathematicians worry more about the definitions of the derivatives, the limit-taking process, and all the things that might go wrong with this cross-multiplication, but the basic result is the same. At the very least, you can think of "multiplying top and bottom by du" as a way of *remembering* the chain-rule formula.

NA.5　Derivatives of Other Functions

We will sometimes make use of the following derivatives:

Derivatives of some other useful functions

$$\frac{d}{d\theta}\sin\theta = \cos\theta \qquad \frac{d}{d\theta}\cos\theta = -\sin\theta \qquad (NA.22)$$

$$\frac{d}{du}e^u = e^u \qquad \frac{d}{du}\ln u = \frac{1}{u} \qquad (NA.23)$$

We will often see such functions in contexts where the angle θ or the quantity u depends on time t: for example, $\theta(t) = \omega t$ or $u(t) = -bt$, where ω and b are constants. In such cases, we can use the chain rule to evaluate the derivatives of such functions with respect to time. For example, if $\theta(t) = \omega t$:

$$\frac{d}{dt}\sin[\theta(t)] = \frac{d\sin\theta}{d\theta}\frac{d\theta}{dt} = \cos\theta\frac{d}{dt}\omega t = \omega\cos\omega t \qquad (NA.24)$$

Similarly, if $u(t) = -bt$, then

$$\frac{d}{dt}e^{u(t)} = \frac{de^u}{du}\frac{du}{dt} = e^u\frac{d}{dt}(-bt) = -be^{-bt} \qquad (NA.25)$$

Exercise NAX.3

Use this technique to find the time derivative of $f(t) = A\cos(\omega t + \theta_0)$, where A, ω, and θ_0 are constants.

HOMEWORK PROBLEMS

Basic Skills

NAB.1 Calculate the derivatives of the following functions (where a, b, and c are constants).
(a) $at^2 + b$
(b) $1/ct^3$
(c) $b/(1 - at^2)$

NAB.2 Use the chain rule to evaluate the time derivatives of the following functions (where b is a constant).
(a) $3(t^3 + b^3)^4$
(b) $\sqrt{t^2 - b^2}$
(c) $\cos(bt^2)$

NAB.3 Use the chain rule to evaluate the time derivatives of the following functions (where b is a constant).
(a) $5b/(bt - 1)^4$
(b) $(bt + 1)^{3/2}$
(c) $\sin(bt + \pi)$

Synthetic

NAS.1 Using the basic approach illustrated by equations NA.2 through NA.7, *prove* that if $f(t) = at$ (where a is a constant), then $df/dt = a$. (Do *not* use equation NA.8.)

NAS.2 Using the basic approach illustrated by equations NA.2 through NA.7, *prove* that if $f(t) = t^{-2}$, then $df/dt = -2t^{-3}$. (Do *not* use equation NA.8.)

NAS.3 Prove that the *sum rule* (equation NA.10) follows from the definition of the derivative. (You may assume without proof that the limit as $\Delta t \to 0$ of a sum of two quantities is the same as the sum of the limits of each quantity separately.)

NAS.4 Prove that the *inverse rule* (equation NA.12) follows from the definition of the derivative. (You may assume without proof that the limit as $\Delta t \to 0$ of the product of two quantities is the same as the products of the limits of each quantity separately.)

NAS.5 Use the chain rule to prove the inverse rule.

NAS.6 Use the chain rule to prove this useful result:

$$\frac{d}{dt}(t + c)^n = n(t + c)^{n-1} \qquad \text{(NA.26)}$$

ANSWERS TO EXERCISES

NAX.1 Using the inverse rule, we have

$$\frac{df}{dt} = -\frac{1}{(at + b)^2}\frac{d}{dt}[at + b] = -\frac{a}{(at + b)^2} \quad \text{(NA.27)}$$

NAX.2 Define $u(t) \equiv t^2 - b$. Then

$$\frac{df}{dt} = \frac{d}{du}u^{-3}\frac{du}{dt} = -3u^{-4}\frac{d}{dt}(t^2 + b)$$

$$= -\frac{3}{(t^2 - b)^4}(2t + 0) = -\frac{6t}{(t^2 - b)^4} \quad \text{(NA.28)}$$

NAX.3 If we define $\theta(t) \equiv \omega t + \theta_0$, then the derivative is

$$\frac{df}{dt} = A\frac{d}{dt}\cos[\theta(t)] = A\frac{d\cos\theta}{d\theta}\frac{d\theta}{dt}$$

$$= A(-\sin\theta)\frac{d}{dt}(\omega t + \theta_0) = -A\omega\sin(\omega t + \theta_0)$$

$$\text{(NA.29)}$$

Appendix NB
Integral Calculus

NB.1   Antiderivatives

Definition of an antiderivative

The **antiderivative** $F(t)$ of a function $f(t)$ is defined to be any function whose derivative is $f(t)$. For example, one possible antiderivative for $f(t) = at$ (where a is a constant) is $F(t) = \frac{1}{2}at^2$ because

$$\frac{dF}{dt} = \frac{d}{dt}\left(\frac{1}{2}at^2\right) = \frac{1}{2}a\frac{d}{dt}(t^2) = \frac{1}{2}a(2t) = at \qquad \text{(NB.1)}$$

Note that there is not a single *unique* antiderivative for a given function: if $F(t)$ is an antiderivative of $f(t)$, then so is $F(t) + C$ (where C is a constant), since

$$\frac{d}{dt}[F(t) + C] = \frac{dF}{dt} + \frac{dC}{dt} = \frac{dF}{dt} + 0 = \frac{dF}{dt} = f(t) \qquad \text{(NB.2)}$$

Thus the antiderivative $f(t)$ is really a whole *family* of functions $F(t)$ that differ by an additive constant. We usually write the antiderivative of $f(t)$ as $F(t) + C$, where C is an *unspecified* constant called a **constant of integration.**

NB.2   Definite Integrals

Now, the **definite integral** of a function $f(t)$ from a point t_A to a point t_B on the horizontal axis is defined as follows:

$$\int_{t_A}^{t_B} f(t)\,dt \equiv \text{total area under curve of } f(t) \text{ from } t_A \text{ to } t_B \qquad \text{(NB.3}a\text{)}$$

(with area above the t axis being considered positive and area below the t axis negative). Figure NB.1 shows that we can approximate this area by a set of bars, the ith bar of which (where i is an arbitrary integer between 1 and the number of bars N) has a width Δt (the same for all bars) and a height equal to the value of $f(t_i)$, where $t_i = t_A + (i - \frac{1}{2})\Delta t$ is the value of t halfway across

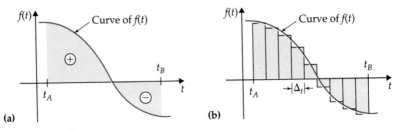

(a)   **(b)**

Figure NB.1

(a) The definite integral of a function $f(t)$ from t_A to t_B is defined to be the area under a graph of $f(t)$ between t_A and t_B (that is, the shaded area shown). Area above the horizontal axis is treated as positive and area below as negative. (b) The area is approximately equal to the total area of the bars shown, where each bar is Δt wide and as tall as the function's value halfway across the bar.

the width of the bar. Perhaps you can convince yourself that as the bar width Δt goes to zero, the area under the curve of the graph can be written

$$\int_{t_A}^{t_B} f(t)\, dt \equiv \lim_{\Delta t \to 0} \sum_{i=1}^{N} f(t_i)\,\Delta t \qquad \text{(NB.3b)}$$

where $N = (t_B - t_A)/\Delta t$ is the number of bars in the set. Indeed, the symbol $\int$ used for the integral is simply a stylized S standing for "sum."

The following properties of the integral follow directly from its definition:

$$\int_{t_A}^{t_B} [f(t) + g(t)]\, dt = \int_{t_A}^{t_B} f(t)\, dt + \int_{t_A}^{t_B} g(t)\, dt \qquad \text{(NB.4a)}$$

The sum rule

$$\int_{t_A}^{t_B} cf(t)\, dt = c \int_{t_A}^{t_B} f(t)\, dt \qquad \text{(NB.4b)}$$

The constant rule

$$\int_{t_A}^{t_C} f(t)\, dt = \int_{t_A}^{t_B} f(t)\, dt + \int_{t_B}^{t_C} f(t)\, dt \quad \text{if} \quad t_C \geq t_B \geq t_A \quad \text{(NB.4c)}$$

Equation NB.4c says that the total area under the curve between t_A and t_C is the sum of the areas between t_A and t_B, and t_B and t_C (something that is not really very surprising). The first two equations are easiest to prove if we remember that the integral is the limit of a sum (see equation NB.3b).

In this course, we will almost always use *definite* integrals instead of antiderivatives (for a variety of reasons).

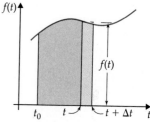

Figure NB.2
The *difference* between the area from t_0 to t and the area from t_0 to $t + \Delta t$ will be *approximately* equal to the area of the white bar, which has width Δt and height $f(t)$. This approximation becomes exact as $\Delta t \to 0$.

NB.3   The Fundamental Theorem

At first glance, it may not seem that the *antiderivative* of $f(t)$ [which is a function whose derivative is $f(t)$] should have anything to do with the *integral* of $f(t)$ [which is the area under the curve of $f(t)$]. However, the **fundamental theorem of calculus** asserts that if $F(t)$ is *any* antiderivative of $f(t)$, then

$$F(t) - F(t_0) = \int_{t_0}^{t} f(t)\, dt \qquad \text{(NB.5)}$$

The fundamental theorem of calculus

where t_0 is some arbitrary fixed value of the variable t. Note that when we use the variable t as the upper limit of the integration, the value of the integral (i.e., the area from t_0 to t) is essentially a function of t.

We can prove the fundamental theorem as follows. The definition of the derivative implies that the time derivative of the right side of equation NB.5 is

Proof of the fundamental theorem of calculus

$$\frac{d}{dt} \int_{t_0}^{t} f(t)\, dt = \lim_{\Delta t \to 0} \frac{1}{\Delta t} \left[\int_{t_0}^{t+\Delta t} f(t)\, dt - \int_{t_0}^{t} f(t)\, dt \right] \qquad \text{(NB.6)}$$

However, as figure NB.2 shows, the difference between the area under the curve between t_0 and t and the area between t_0 and $t + \Delta t$ (as Δt becomes very small) becomes essentially the area of a bar of width Δt and height $f(t)$. Therefore

$$\frac{d}{dt} \int_{t_0}^{t} f(t)\, dt = \lim_{\Delta t \to 0} \frac{1}{\Delta t} [f(t)\,\Delta t] = f(t) \qquad \text{(NB.7)}$$

So we see that the integral is indeed an antiderivative of $f(t)$. Therefore, the *general* antiderivative of $f(t)$ can be written as

$$F(t) = \int_{t_0}^{t} f(t)\, dt + C \qquad C \text{ is an undetermined constant} \qquad \text{(NB.8)}$$

Therefore, for *any* antiderivative $F(t)$ of the function $f(t)$, we have

$$F(t) - F(t_0) = \int_{t_0}^{t} f(t)\,dt + C - \left[\int_{t_0}^{t_0} f(t)\,dt + C\right] = \int_{t_0}^{t} f(t)\,dt \quad \text{(NB.9)}$$

since the C's cancel and the area under the curve from t_0 to t_0 is clearly zero. This is equation NB.5, so we have completed the proof.

NB.4   Indefinite Integrals

Definition of an indefinite integral

The **indefinite integral** $\int f(t)\,dt$ is *any* one of the possible antiderivatives of the function $f(t)$ (usually the antiderivative with the simplest algebraic form). Some useful indefinite integrals to know are the following:

Useful integrals

$$\int \frac{df}{dt}\,dt = f(t) \quad \text{for any function } f(t) \quad \text{(NB.10}a)$$

$$\int c\,dt = ct \quad \text{where } c \text{ is any constant} \quad \text{(NB.10}b)$$

$$\int t^n\,dt = \frac{t^{n+1}}{n+1} \quad \text{for } n \neq -1 \quad \text{(NB.10}c)$$

The last follows from equation NA.8. One can show in fact that

$$\frac{d}{dt}\left[\frac{(t+c)^{n+1}}{n+1}\right] = (t+c)^n \quad \Rightarrow \quad \int (t+c)^n\,dt = \frac{(t+c)^{n+1}}{n+1} \quad \text{(NB.10}d)$$

where c is any constant. This is a useful extension of equation NB.10c. Other useful indefinite integrals include these:

$$\int \sin(\omega t + b)\,dt = -\frac{1}{\omega}\cos(\omega t + b) \quad \text{where } \omega \text{ and } b \text{ are constants}$$

$$\text{(NB.11}a)$$

$$\int \cos(\omega t + b)\,dt = \frac{1}{\omega}\sin(\omega t + b) \quad \text{where } \omega \text{ and } b \text{ are constants}$$

$$\text{(NB.11}b)$$

$$\int e^{bt}\,dt = \frac{1}{b}e^{bt} \quad \text{where } b \text{ is a constant} \quad \text{(NB.12}a)$$

$$\int \frac{dx}{x} = \ln x \quad \text{(NB.12}b)$$

One can prove any of the integrals in equations NB.10 through NB.12 by taking the derivative of both sides.

Tables of integrals

A **table of integrals** provides a list of indefinite integrals for a wider variety of functions. Feel free to look up indefinite integrals in such a table if you have access to one. Remember that a given indefinite integral is only one of a family of possible antiderivatives: all others can be generated by adding some constant C to the one you have.

Once you know an indefinite integral for a function, you can find its definite integral from t_0 to t by simply evaluating the indefinite integral at these two points and subtracting. For example, $\int_{t_0}^{t} c\,dt = ct - ct_0$. This works no matter which member of the family of possible antiderivatives the indefinite integral you have chosen is: equation NB.9 applies to *any* antiderivative of a given function $f(t)$ (note how the arbitrary constant C cancels out).

NB.5   Substitution of Variables

Indefinite integrals for a variety of more complicated functions can be evaluated using the technique of substitution of variables. To illustrate the process, consider the following integral (where b is a constant):

$$\int \frac{t\,dt}{\sqrt{b^2 t^2 + 1}} \qquad\qquad \text{(NB.13)}$$

This looks like an appallingly difficult function to integrate. However, imagine that we define a new variable $u \equiv b^2 t^2 + 1$. Note that this means that

$$\frac{du}{dt} = b^2(2t) + 0 = 2b^2 t \quad \Rightarrow \quad \frac{du}{2b^2} = t\,dt \qquad \text{(NB.14)}$$

If we substitute these results into equation NB.13, we find that

$$\int \frac{t\,dt}{\sqrt{b^2 t^2 + 1}} = \int \frac{du}{2b^2 \sqrt{u}} = \frac{1}{2b^2} \int u^{-1/2}\,du \qquad \text{(NB.15)}$$

where I have used the constant rule in the last step. We can evaluate this final integral by using equation NB.10c:

$$\int \frac{t\,dt}{\sqrt{b^2 t^2 + 1}} = \frac{1}{2b^2}\frac{u^{+1/2}}{+\frac{1}{2}} = \frac{1}{b^2}u^{1/2} = \frac{1}{b^2}\sqrt{b^2 t^2 + 1} \qquad \text{(NB.16)}$$

An illustration of the technique of substitution of variables

---

Exercise NB.1

Use this technique to evaluate the integral

$$\int \sqrt{at + b}\,dt \qquad\qquad \text{(NB.17)}$$

and check your work by taking the derivative of the result and showing that you recover the original function.

It is virtually essential to master this technique to get full use of a table of integrals, as you are unlikely to find the *exact* indefinite integral you want in such a table. If you know this technique, however, you can often massage your integral into a form that you can find in the table.

However, it can be tricky (and in some cases impossible) to find the perfect substitution that makes your integral simple. It is worth pointing out that while one can calculate the derivative of essentially any function by diligent but essentially mechanical application of the chain rule and other rules, there is no foolproof mechanical process for evaluating integrals. Some very simple-looking integrals cannot be evaluated at all in symbolic form, and others require very advanced techniques. Be appropriately appreciative for all the poor mathematicians who ground out all the integrals in a good integral table!

Please note that if you substitute variables in a *definite* integral, you must also change the limits of integration to reflect the new variable. For example, consider a definite integral involving the case just discussed

There is no foolproof way to evaluate an integral

Substitution of variables in definite integrals

$$\int_{t_A}^{t_B} \frac{t\,dt}{\sqrt{b^2 t^2 + 1}} = \frac{1}{2b^2} \int_{u_A}^{u_B} \frac{du}{\sqrt{u}} = \frac{1}{b^2}\big[\sqrt{u}\big]_{u_A}^{u_B} = \frac{1}{b^2}\big[\sqrt{u_B} - \sqrt{u_A}\big] \quad \text{(NB.18)}$$

where $u_A \equiv u$ evaluated at t_A and u_B is defined similarly. For example, if we are evaluating the integral on the left from $t = 0$ to $t = 2$ s and $b = 0.5\,\text{s}^{-1}$, then we would evaluate the integral on the right from $u(0) = b^2 0^2 + 1 = 1$ to $u(2\,\text{s}) = (0.5\,\text{s}^{-1})^2 (2.0\,\text{s})^2 + 1 = 2$.

Equivalently, one can substitute the definition of u in terms of t back into the final expression for the evaluated integral and then use the original limits:

$$\int_{t_A}^{t_B} \frac{t\,dt}{\sqrt{b^2 t^2 + 1}} = \frac{1}{2b^2} \int_{u_A}^{u_B} \frac{du}{\sqrt{u}} = \frac{1}{b^2}\left[\sqrt{u}\right]_{u_A}^{u_B} = \frac{1}{b^2}\left[\sqrt{b^2 t^2 + 1}\right]_{t_A}^{t_B} \quad \text{(NB.19)}$$

Either approach is correct: the point is that you should *think carefully* about what happens to the limits in definite integrals when you change variables.

HOMEWORK PROBLEMS

BASIC SKILLS

NBB.1   Integrate the following functions from 0 to t (b is a constant).
(a) $f(t) = b^2 t^2$
(b) $f(t) = 1/\sqrt{bt + 1}$

NBB.2   Integrate the following functions from 0 to t (b and T are constants).
(a) $f(t) = e^{-bt} + 1$
(b) $f(t) = b(t - T)^{3/2}$

NBB.3   Show that equations NB.10, NB.11, and NB.12 are correct by taking the derivative of both sides.

SYNTHETIC

NBS.1   Using the technique of substitution of variables, evaluate the integral of $f(t) = (1 - bt)^3$.

NBS.2   Use the technique of substitution of variables to evaluate the definite integral

$$\int_0^{2\,\text{s}} \frac{\sin \omega t\, dt}{(1 + \cos \omega t)^2} \quad \text{when} \quad \omega = \frac{1}{2}\pi\,\text{s}^{-1} \quad \text{(NB.20)}$$

ANSWERS TO EXERCISES

NBX.1   Define the function $u \equiv at + b$, and note that

$$\frac{du}{dt} = a + 0 = a \quad \Rightarrow \quad \frac{du}{a} = dt \quad \text{(NB.21)}$$

Substituting these into the integral, we get

$$\int \sqrt{at + b}\, dt = \int \sqrt{u}\, \frac{du}{a} = \frac{1}{a} \int u^{1/2}\, du$$

$$= \frac{1}{a} \frac{u^{3/2}}{\frac{3}{2}} = \frac{2u}{3a} = \frac{2}{3a}\sqrt{at + b}$$

$$\text{(NB.22)}$$

Glossary

acceleration $\vec{a}$: a vector quantity that describes how fast and in what direction an object's velocity vector is changing at an instant:

$$\vec{a} \equiv \lim_{\Delta t \to 0} \frac{\Delta \vec{v}}{\Delta t} \equiv \lim_{\Delta t \to 0} \frac{\vec{v}(t + \Delta t) - \vec{v}(t)}{\Delta t} \equiv \frac{d\vec{v}}{dt} \quad \text{(N2.1)}$$

This quantity is sometimes called the **instantaneous acceleration** to distinguish it from **average acceleration**. (Section N2.3)

acceleration of gravity $\vec{g}$: see **gravitational field vector**. (Sections N4.4, N10.1)

amplitude: the constant A in $x(t) = A \cos(\omega t + \theta)$, equal to one-half the distance between the limits of an oscillation. (Section N11.2)

antiderivative: the antiderivative $F(t)$ of a function $f(t)$ is that function whose derivative is $f(t)$: $dF/dt = f(t)$. The antiderivative $F(t)$ is determined only up to an overall additive constant: if $F(t)$ is an antiderivative of $f(t)$, then so is $F(t) + C$, where C is any constant. Thus the antiderivative of $f(t)$ is usually written $F(t) + C$, where $F(t)$ is any *specific* antiderivative of $f(t)$ and C is an unspecified constant of integration. (Sections N4.1, NB.1)

area method: a method of drawing a graph of a function from a graph of its derivative that involves plotting a value on the former graph at a given t that corresponds to the accumulated area under the derivative graph up to that t. (Section N4.2)

astronomical unit (AU): a unit of distance that astronomers often use to describe distances in the solar system. 1 AU $\equiv$ the average distance between the earth and the sun = 1.5×10^{11} m = 150 million kilometers. (Section N13.2)

asymptote (of a hyperbola): the straight line through the hyperbola's center that the curve of the hyperbola approaches at large distances from the center. (Section N13.1)

average acceleration (during an interval) $\vec{a}_{\Delta t}$: the change in an object's velocity during an interval of time Δt, divided by that interval: $\vec{a}_{\Delta t} \equiv \Delta \vec{v}/\Delta t$. This quantity is most closely equal to the object's instantaneous acceleration at the *middle* of the interval. This approximation becomes better as Δt gets smaller. (Section N2.3)

average velocity (of an object during an interval Δt): the object's displacement during that interval, divided by the duration Δt of the interval. An object's average velocity during an interval is approximately equal to its instantaneous velocity at any instant during the interval: the approximation improves as Δt gets smaller. For any finite Δt, the average velocity during an interval generally best approximates the instantaneous velocity at the *middle* of the interval. (Section N2.2)

banking (into a turn): action taken when an airplane or bicycle rotates about an axis parallel to its motion as it goes around a curve. Both will lean toward the center of the curve (for different physical reasons). A roadbed also may be banked into a curve in the same way. If the speed of a car is just right, the banking will allow it to negotiate the curve without the road exerting a sideways static friction force. (Section N8.4)

bob: the weight at the end of a **simple pendulum**. (Section N11.6)

buoyant force $\vec{F}_B$: an upward force exerted on an object at least partially immersed in a fluid, where F_B = the magnitude of the weight of the fluid displaced by the object. (This force is usually ignored if fluid is air.) (Section N1.5)

center (of a conic section): the midpoint along the line connecting the **conic section's foci**. (Section N13.1)

centrifugal force: the fictitious force that in a rotating reference frame appears to pull objects away from the axis of rotation. (This force, while sometimes useful as a frame-correction force, is entirely fictitious and is usually confusing, so I only rarely use this term.) (Section N9.5)

chain rule: a mathematical theorem that says that if f is a function of u which is a function of t, then

$$\frac{df}{dt} = \frac{df}{du}\frac{du}{dt} \quad \text{(NA.19)}$$

(Section NA.4)

circle: a conic section with $\varepsilon = 0$. Such a curve has constant radius $r = R$. (Section N13.1)

coefficient of kinetic friction μ_k: the constant of proportionality between the magnitude of the steady kinetic friction force exerted by a contact interaction between sliding surfaces and the magnitude of the normal force exerted by the same interaction (which expresses how tightly the interacting surfaces are pressed together): $\text{mag}(\vec{F}_{KF}) = \mu_k \, \text{mag}(\vec{F}_N)$. The unitless values of μ_k are generally somewhat smaller than values of μ_s for the same type of surfaces. (Section N6.3)

coefficient of static friction μ_s: the constant of proportionality between the magnitude of the maximum possible static friction force that a contact interaction can provide and the magnitude of the normal force exerted by the same contact interaction (F_N essentially expresses how tightly the interacting surfaces are pressed together): $\mathrm{mag}(\vec{F}_{SF,\max}) = \mu_s \, \mathrm{mag}(\vec{F}_N)$. The unitless values of μ_s range from 0.04 for very slippery surfaces to 1.1 or more for very sticky surfaces, and they are typically about 0.5. (Section N6.3)

compression force: a force arising from a contact interaction that seeks to keep the interacting objects from becoming closer. (Section N1.5)

conic section: the name describing a family of curves such that either the sum or the difference of the distances r_2 and r_1 between a point on the curve and two fixed points (called the curve's **foci**) is a positive constant $2a$. The curve's **center** is the midpoint between the foci. The curve's **eccentricity** ε is $\varepsilon \equiv c/a$, where c is the distance between the curve's center and either of its foci. (These curves happen to be the intersection between an arbitrarily oriented infinite plane and the surface of an infinite cone.) (Section N13.1)

constant of integration: the arbitrary constant that one can add to an antiderivative of a certain function to get another valid antiderivative of that same function. In physics, one usually determines the constant of integration by insisting that the antiderivative satisfy some **initial condition.** (Sections N4.3, NB.1)

constant rule: if $f(t) = ch(t)$, then

$$\frac{df}{dt} = c \, \frac{dh}{dt} \qquad \text{(NA.13)}$$

(Section NA.2)

constrained-motion problem: a physics problem in which we use an observed constraint on an object's motion to determine the forces acting on that object. (Section N6.6)

contact interactions: interactions that arise when two objects come into physical contact. All known macroscopic interactions are either **long-range interactions** or contact interactions. (Section N1.5)

coupled objects: a set of interacting objects that are constrained to move together in such a way that their acceleration magnitudes are linked (in almost all cases, the acceleration magnitudes are actually *equal*). (Section N7.1)

definite integral: the definite integral from t_0 to t of a function $f(t)$ is defined to be the area under a graph of $f(t)$ between t_0 and t. (Sections N4.3, NB.2)

differential equation: an equation that expresses a relationship between a function and its derivatives. (Section N11.2)

drag coefficient C: the quantity appearing in the drag force equation $\mathrm{mag}(\vec{F}_D) = \frac{1}{2} C \rho A v^2$ for an object moving through air. This empirical coefficient depends on the shape of the object (it has a value ≈ 0.5 for a sphere). (Section N6.4)

drag force $\vec{F}_D$: a force arising from the contact interaction between a fluid and an object moving relative to the fluid that *opposes* the object's motion. (Section N1.5)

eccentricity ε: a quantity that defines the shape of a **conic section.** Technically $\varepsilon = c/a$, where c is one-half of the distance between the section's two foci and a is the section's **semimajor axis.** If $\varepsilon = 0$, the conic section is a **circle;** if ε is less than 1, it is an **ellipse;** if $\varepsilon = 1$, it is a **parabola;** and if $\varepsilon > 1$, the conic section is a **hyperbola.** (Section N13.1)

electrostatic force $\vec{F}_e$: a force arising from an interaction between an electrically charged object and something else (a subcategory of an *electromagnetic* force). (Section N1.5)

ellipse: a **conic section** with $0 < \varepsilon < 1$. An ellipse is a closed curve, having a well-defined radius for all angles θ around one of its **foci.** (Section N13.1)

equilibrium position: the position of an oscillating object at which it experiences zero net force. (Section N11.3)

equivalence principle: the statement that a freely falling frame is physically equivalent to an inertial frame floating in deep space, and a frame in a uniform gravitational field is physically equivalent to a constantly accelerating frame in deep space. This principle (with some suitable qualifiers) is one of the founding principles of the theory of general relativity. (Section N9.7)

fictitious force: a force that we infer to exist in a noninertial reference frame when we try (either intuitively or consciously) to apply Newton's second law in that frame (even though it *doesn't* apply in a noninertial frame). (Section N9.1)

first-law detector: any device that examines the paths of isolated objects for deviations from nonaccelerated motion. Such devices include the *floating-ball* and *floating-puck* detectors described in section N9.3.

focus: one of two mathematical points used to define a **conic section.** The plural is *foci.* (Section N13.1)

force law: an expression that tells how the force exerted on an object by its interaction with another object depends on the separation of the two objects. We can find a force law for any interaction that can be described by a potential energy function that depends on the objects' separation. (Section C8.6)

frame-correction force: A **fictitious force** that we deliberately use so that we can pretend to apply Newton's second law in a **noninertial frame.** This force corrects for the acceleration of the reference frame. (Section N9.6)

free-body diagram: a diagram of an object that displays all the external forces acting on it (as labeled arrows) attached to the points on the object where these forces act. (Section N1.6)

freely falling: an object is *freely falling* if the *only* (significant) force acting on it is its weight. (Sections N4.4, N10.1)

free-particle diagram: a diagram of an object that displays all the external forces acting on it as labeled arrows radiating from its center of mass. This diagram makes it much easier to write out Newton's second law when we are only interested in how the object's center of mass moves. (Contrast with **free-body diagram.**) (Section N6.1)

frequency f: the number of cycles that an oscillating object completes in a unit time. Measured in **hertz** (where 1 Hz = 1 cycle/s). (Section N11.2)

friction force: the component of the force arising from a contact interaction between two solid objects that is parallel to the interface between the objects. (Section N1.5)

fundamental theorem of calculus: if $F(t)$ is any antiderivative of $f(t)$, then the definite integral of $f(t)$ from t_0 to t is

$$\int_{t_0}^{t} f(t)\, dt = F(t) - F(t_0) \qquad \text{(NB.5)}$$

(Section NB.3)

galilean velocity transformation equation: the equation $\vec{v}' = \vec{v} - \vec{\beta}$, which specifies how to calculate an object's velocity $\vec{v}'$ in a reference frame S' that is moving with velocity $\vec{\beta}$ relative to a frame S, if we know the object's velocity $\vec{v}$ in s.

gravitational field vector $\vec{g}$: the quantity that links an object's mass to its weight at a given place in a gravitational field, characterizing the strength and direction of the gravitational field at that point. This quantity also specifies the acceleration that all freely falling objects will have at that point, and thus it is sometimes called the **acceleration of gravity.** Near the surface of the earth, g has the magnitude $9.8\,\text{m/s}^2 = 22\,(\text{mi/h})/\text{s}$. (Sections N4.4, N10.1)

harmonic oscillator equation: $d^2x/dt^2 = -\omega^2 x$, where $\omega^2 = k_s/m$. This differential equation is the equation of motion for the object in a harmonic oscillator: its solution can be written $x(t) = A\cos(\omega t + \theta)$, where A and θ are constants determined by initial conditions. (Section N11.2)

hertz (Hz): the SI unit of frequency, defined to be one cycle per second, or s^{-1}.

Hooke's law: a force law stating that the force exerted by a spring increases linearly with the absolute value of the distance that the spring is stretched or compressed. For one-dimensional motion where $x = 0$ is the equilibrium position, this law reads $F_x = -k_s x$, where x is the object's position in an appropriate coordinate system. (Section N11.1)

hyperbola: a **conic section** with **eccentricity** $\varepsilon > 1$. A hyperbola is an open curve that goes to infinity at angles $\pm\theta_\infty$, where $\cos\theta_\infty = -1/\varepsilon$. (Section N13.1)

ideal pulley: a massless, frictionless pulley that changes the direction of a string without causing a difference in tension between the string on one side of the pulley and the string on the other. (Section N7.4)

ideal spring: a massless spring that exerts a force exactly proportional to the distance that it is extended or compressed from its relaxed length. (Section N11.1)

ideal string: a massless, inextensible, and perfectly flexible string. The tension forces exerted by each end of such a string are always equal. Not only is this model a reasonably accurate description of most real strings (or other connecting cords), but also its use greatly simplifies problems involving strings. (Section N7.3)

indefinite integral: the indefinite integral $\int f(t)\, dt$ is defined to be any (convenient) antiderivative of $f(t)$. (Sections N4.3, NB.4)

inertial reference frame: a reference frame in which isolated objects are observed to move with constant velocity (i.e., zero acceleration), in accordance with Newton's first law. Inertial frames move at constant velocities with respect to each other. (Section N9.3)

initial conditions: the value of an object's position and velocity at a given instant of time (usually $t = 0$). These values are needed to determine the object's position and velocity at other times from its acceleration. (Section N4.2)

initial phase θ: the constant θ in the general oscillation solution $x(t) = A\cos(\omega t + \theta)$ that expresses how far along the oscillation is in its cycle at $t = 0$. (Section N11.2)

instantaneous acceleration (of an object): see **acceleration.** (Section N2.3)

instantaneous velocity (of an object): see **velocity.** (Section N2.2)

inverse rule: if $f(t) = 1/g(t)$, then

$$\frac{df}{dt} = -\frac{1}{g^2}\frac{dg}{dt} \qquad \text{(NA.12)}$$

(Section NA.2)

Kepler's three laws of planetary motion: empirical laws devised by Johannes Kepler in the early decades of the 1600s to describe the observed motion of the planets. These laws state that

1. The planets move in ellipses, with the sun at one focus.
2. The line from the sun to the planet sweeps out equal areas in equal times.
3. The period of the orbit is proportional to the 3/2 power of its semimajor axis.

(Section N12.1)

kinematic chain: the chain of relationships linking an object's position to its velocity to its acceleration:

$$\vec{r}(t) \; \boxed{\begin{array}{c}\text{time}\\\text{derivative}\end{array}}\!\!\to \; \vec{v}(t) \; \boxed{\begin{array}{c}\text{time}\\\text{derivative}\end{array}}\!\!\to \; \vec{a}(t) \qquad \text{(N3.1)}$$

(Section N3.1)

kinematics: the mathematical study of motion. (Section N3.1)

kinetic friction force $\vec{F}_{KF}$: a friction force that opposes an already existing sliding motion along the interface between two solid objects. (Section N1.5)

lift force $\vec{F}_L$: a force arising from the contact interaction between a fluid and an object (such as a wing) moving relative to the fluid that acts *perpendicular* to the object's motion. (Section N1.5)

linearly-constrained motion: the motion of an object that is constrained to move along a line. (Section N6.6)

long-range interactions: interactions that act between particles or objects with macroscopic separations. Gravitational and electromagnetic interactions are the only long-range interactions currently known. (Section N1.5)

magnetic force $\vec{F}_m$: a force arising from an interaction between a magnet (or object carrying an electric current) and something else (a subcategory of an *electromagnetic* force). (Section N1.5)

motion diagram: a diagram that depicts an object's motion and (at least the direction of) its acceleration at various instants during its motion. A complete motion diagram of a moving object contains the following:

1. A single image of the object (to show its orientation).
2. Dots representing positions of the object's center of mass at equally spaced instants of time.
3. Arrows from dot to dot (depicting average velocities).
4. Acceleration arrows attached to the dots that correspond to differences between adjacent velocity arrows.

(Section N2.4)

net force $\vec{F}_{net}$: the vector sum of all external forces acting on a particle or extended object. (Section N1.4)

net-force diagram: a diagram used as a supplement to a free-body diagram that displays the vector sum of all the forces acting on a body (i.e., the net force on that object) as a double-line arrow. (Section N3.2)

newtonian synthesis: the unification of celestial and terrestrial physics accomplished by Newton's mechanics. (Section N1.1)

Newton's first law: *the center of mass of an isolated object moves with a constant velocity.* (From our modern perspective, we would say that this law is a consequence of the more fundamental law of conservation of momentum.) (Section N1.2)

Newton's law of universal gravitation: *the gravitational interaction between two particles or spherical bodies whose masses are m_1 and m_2 exerts a force on either body that is directed toward the other and has a magnitude of*

$$F_g = \frac{Gm_1m_2}{r^2}$$

where G is the universal gravitational constant and r is the distance between the objects' centers. (Section N12.4)

Newton's second law: *the net external force on an object causes its center of mass to accelerate in inverse proportion to the object's total mass m:*

$$\vec{F}_{net,ext} = \frac{d\vec{p}_{tot}}{dt} = m\vec{a}_{CM}$$

This law (stated here in its most general form) is a consequence of the definition of force as the rate at which a given interaction transfers momentum, the idea that the net momentum an object gains in time dt is the vector sum of the momentum transfers it gains from each interaction, the idea that internal interactions do not affect an object's total momentum, and the definition of the center of mass. (Section N1.4)

Newton's third law: *an interaction between two objects A and B exerts forces on each object that have equal magnitudes but opposite directions: $\vec{F}_A = -\vec{F}_B$.* This law is a consequence of the idea that an interaction transfers momentum from one object to the other. (Section N1.3)

noninertial reference frame: a reference frame in which isolated objects are observed to move with nonzero acceleration, in violation of both Newton's first and second laws, making it impossible to use the latter to link the true forces acting on the object with its motion. Noninertial frames *accelerate* relative to inertial frames. (Section N9.3)

nonuniform circular motion: the motion of an object that travels along a circular path with a varying speed. (Section N8.3)

normal force $\vec{F}_N$: the component of the force arising from a contact interaction between two solid objects that acts perpendicular to the interface between the objects. (*Normal* means *perpendicular* here, not *typical* or *usual*.) (Section N1.5)

parabola: a **conic section** with $\varepsilon = 1$. A parabola is an open curve that is essentially an infinitely long **ellipse** or an infinitely skinny **hyperbola**. (Section N13.1)

period T (of an orbit): the time that it takes an object to complete exactly one orbit around the system's center of mass. (Section N12.4)

period T (of an oscillation): the time required to complete one cycle of an oscillation. (Section N11.2)

phase rate ω: the rate at which the argument (sometimes called the *phase*) of the cosine in $x(t) = A\cos(\omega t + \theta)$ increases with time. One cycle of the oscillation corresponds to a phase change of 2π. The SI units of ω are rad/s $= s^{-1}$. (Section N11.2)

pressure force $\vec{F}_P$: a force arising when something solid presses on a fluid that is somehow confined. (Section N1.5)

primary: the more massive object in an isolated, two-object interacting system in which the mass of one object is much larger than that of the other. (Section N12.2)

product rule: if $f(t) = g(t)h(t)$, then

$$\frac{df}{dt} = \frac{dg}{dt}h + g\frac{dh}{dt} \qquad (NA.11)$$

(Section NA.2)

projectile: a falling object whose trajectory is "sufficiently short" and "sufficiently near" to the earth's surface that $\vec{g}$ is essentially constant. If air drag is also negligible, then we say that the object is **freely falling** and its motion is **simple projectile motion.** (Section N10.1)

projectile motion: the trajectory of a **projectile.** (Section N10.1)

restoring force: the force acting on an oscillating object that seeks to return it to its **equilibrium position.** If the object is moving along the x axis and $x = 0$ at the object's equilibrium position, then a restoring force will have x component $F_x < 0$ if $x > 0$ and $F_x > 0$ when $x < 0$. (Section N11.3)

reverse kinematic chain: the chain of relationships linking an object's acceleration, velocity, and position:

$$\vec{a}(t) \;-\!\!\boxed{\begin{array}{c}\text{time}\\\text{antiderivative}\end{array}}\!\!\!\to\; \vec{v}(t) \;-\!\!\boxed{\begin{array}{c}\text{time}\\\text{antiderivative}\end{array}}\!\!\!\to\; \vec{r}(t)$$
$$(N4.2)$$

(Section N4.1)

satellite: the lighter object in an isolated, two-object interacting system in which the mass of one object is much larger than that of the other. (Section N12.2)

second-law partners: a pair of opposing forces that **Newton's second law** tells us are equal in magnitude. Such forces must act on the *same* object and generally arise from different interactions acting on that object. (Contrast with **third-law partners.**) (Section N3.4)

semimajor axis a (of an **ellipse**): one-half of the widest distance measured across the ellipse. (Section N13.1)

semiminor axis (of an **ellipse**): one-half of the narrowest distance measured across the ellipse. (Section N13.1)

simple harmonic oscillator: any object moving in one dimension that experiences a net force given by the expression $F_x = -k_s x$, where x is the object's position in an appropriate coordinate system. An object that is connected to a fixed point by an ideal spring and that slides along a horizontal line on a frictionless surface is an example of a simple harmonic oscillator. (Section N11.2)

simple pendulum: a massive object (called a pendulum **bob**) connected to a fixed point by a massless, inextensible string. A simple pendulum behaves as a harmonic oscillator if the angle of swing is very small. (Section N11.6)

simple projectile motion: The motion of a **projectile** when air drag is also negligible. (Section N10.1)

slope [of a function $f(t)$] at a point: by definition the slope of a line that is tangent to the graph of the function at that point, that is, a line that touches the graph only at that point. The slope of this line has the value df/dt. (Section NA.3)

slope method: a method of drawing a graph of a function from a graph of its derivative that involves drawing a connected set of short line segments on the former graph at successive instants of time, the *slope* of each of which reflects the corresponding *value* on the derivative graph. (Section N4.2)

spring constant: the constant k_s in the harmonic oscillator force law. This constant has units of newtons per meter (N/m) and expresses the strength of the effective spring in a simple harmonic oscillator. (Section N11.1)

spring force $\vec{F}_{Sp}$: the force exerted by a spring on another object (this may be a compression force if the spring is compressed or a tension force if the spring is stretched). (Section N1.5)

static friction force $\vec{F}_{SF}$: a friction force that *prevents* two solid objects from sliding relative to each other along their interface. (You can think of this as being a "sticking force.") (Section N1.5)

statics problem: a type of physics problem where we are given that an object is at rest and we infer from this the value of one or more of the forces acting on the object. (Section N5.2)

sum rule: if $f(t) = g(t) + h(t)$, then

$$\frac{df}{dt} = \frac{dg}{dt} + \frac{dh}{dt} \qquad (NA.10)$$

(Section NA.2)

table of integrals: a list of indefinite (sometimes also some definite) integrals of various functions. (Section NB.4)

tension force $\vec{F}_T$: a force arising from a contact interaction that seeks to keep the interacting objects from separating. (Section N1.5)

the tension on a string: the magnitude of the tension force exerted by *either* end of an ideal string. (Section N7.3)

terminal speed v_T: the speed at which the drag force on an object is equal in magnitude to its weight. At such a speed, the net force on an object falling vertically downward is zero, and therefore the object will no longer accelerate, but rather will remain at that speed. Under normal circumstances, $v_T = (2mg/C\rho A)^{1/2}$, where ρ is the density of air, A is the cross-sectional area that the object presents to the air going by, and C is a unitless coefficient (≈ 1) that depends on the object's shape and surface qualities. An object falling from rest reaches this speed after $t \approx 2v_T/g$. (Section N10.5)

third-law partners: a pair of forces that we know are opposite in direction and equal in magnitude because of **Newton's third law.** Such a pair of forces represent the two ends of an interaction between two objects: therefore they must (1) act on different objects and (2) reflect the same basic interaction. Contrast with **second law partner.** (Section N3.4)

thrust force $\vec{F}_{\text{Th}}$: a force arising from the contact interaction between a fluid and an object (such as a propeller) that pushes the fluid backward. (Section N1.5)

time derivative $d\vec{q}/dt$ (of a vector $\vec{q}$): the limiting value as Δt goes to zero of the change $\Delta \vec{q}$ in $\vec{q}$, divided by the time interval Δt during which the change takes place:

$$\frac{d\vec{q}}{dt} \equiv \lim_{\Delta t \to 0} \frac{\Delta \vec{q}}{\Delta t} \equiv \lim_{\Delta t \to 0} \frac{\vec{q}(t + \Delta t) - \vec{q}(t)}{\Delta t} \qquad \text{(N2.5)}$$

(Section N2.1)

trajectory diagram: a diagram that is very similar to a motion diagram except that its purpose is reversed: instead of allowing us to determine an object's acceleration at various times based on its known motion, a trajectory diagram allows us to construct an object's trajectory based on its initial conditions and its known acceleration. (Section N4.6)

uniform circular motion: the motion of an object following a circular path at a constant speed. (Section N2.6)

unit vector: a vector with magnitude equal to 1 (no units) that we can use to indicate a pure direction. (Section N8.2)

universal gravitational constant G: a constant appearing in **Newton's law of universal gravitation** whose magnitude is 6.67×10^{-11} N·m^2/kg^2. (Section N12.4)

velocity (of an object): the time derivative of the object's position vector:

$$\vec{v} \equiv \frac{d\vec{r}}{dt} \equiv \lim_{\Delta t \to 0} \frac{\Delta \vec{r}}{\Delta t} \equiv \lim_{\Delta t \to 0} \frac{\vec{r}(t + \Delta t) - \vec{r}(t)}{\Delta t} \qquad \text{(N1.8)}$$

This formally defines in a mathematically supportable way what we mean by velocity at a single instant of time. This quantity is sometimes called the **instantaneous velocity** to distinguish it from **average velocity.** (Section N2.2)

vertex: the closest point on the curve of a **conic section** to one of the curve's **foci.** (Section N13.1)

weight: the gravitational force $\vec{F}_g$ acting on an object. At a given location in space and time, this force is proportional to the object's mass: $\vec{F}_g = m\vec{g}$. (Section N10.1)

x-velocity v_x, **y-velocity** v_y, **z-velocity** v_z: the components of an object's **instantaneous velocity.** Note that $v_x = dx/dt$, $v_y = dy/dt$, $v_z = dz/dt$. (Section N2.2)

Credits

Chapter 1 Page 7: © Amoz Eckerson; p. 16: © Jacksonville Journal Courier/Steve Warmowski/The Image Works

Chapter 3 N3.3a: Kevin Moan/SuperStock; p. 54(left): © Corbis/R-F Website; p. 54(right): © Everett Collection; p. 55: © Tony Savino/The Image Works

Chapter 4 N4.4: © Joel Gordon

Chapter 5 Page 83: © Paul Almasy/CORBIS; p. 87: © Galen Rowell/CORBIS; p. 91(top): © Allsport; 91(bottom): © Allsport; p. 92: PhotoDisc/Vol. 7; p. 93(left): © PhotoDisc/Vol. 67; p. 93(right): © Tom Bol/Outside Images/PictureQuest

Chapter 6 Page 110: © Eric Sanford/Index Stock Imagery

Chapter 7 Page 126: © Michael S. Yamashita/CORBIS; p. 127: © George Hall/CORBIS

Chapter 8 N8.5: © Allsport; p. 144: © Vince Streano/Corbis; p. 146(top): Courtesy Ford Motor Company; p. 146(bottom): © Richard Cummins/CORBIS; p. 147: © James L. Amos/CORBIS

Chapter 9 Page 154: © Corbis/R-F Website; p. 162: NASA; p. 165(top left): © Paul Souders/Corbis; p. 165(bottom): © Jeff Greenberg/Photo Researchers; p. 165(top right): NASA; N9.6a: © Corbis/R-F Website

Chapter 10 Page 179: © Zefa Visual Media/Index Stock Imagery; p. 181: © S. Carmona/Corbis; p. 183: © Allsport

Chapter 11 Page 194: © 2000 Richard Megna Fundamental Photographs, NYC; p. 200(left): © Corbis/Vol. 26; p. 200(right): Photo courtesy of Lani Lokendahle, Executive Director of the International Trampoline Industry Association, Inc.; p. 201: © Peter Vandermark/Stock Boston; p. 202: NASA

Chapter 12 Page 213: NASA; p. 214: NASA/JPL/University of Arizona ; p. 217: NASA/STScl/AURA; p. 219, 220: NASA

Chapter 13 Page 237: NASA.

Index

Entries followed by *t* and *f* refer to tables and figures, respectively.